AF778025

HOW TO HOT ROD SMALL BLOCK MOPAR ENGINES

®MARCA REG.

Covers All Chrysler, Dodge & Plymouth LA Series Engines—1964 To Present
• 273 • 318 • 340 • 360 C.I.D.

BY LARRY SHEPARD

HPBooks
a division of
PRICE STERN SLOAN
Los Angeles

ACKNOWLEDGEMENTS

Writing a book of this size is not an easy task and certainly not one that is done alone. Many, many people helped to gather the information, run the tests, record data, install parts and race the hardware. Any engine book starts with engine builders and we have had some of the best; Bill Hancock, Dave Koffel, Keith Black, Dick Landy, Ted Spedar, Maurice Petty, Joey Arrington, Jim Cavallero and cylinder head wizard Butch Elkins. Hundreds of racers involved in Chrysler's race and development programs over the years have also contributed heavily. A brief list of them reads; Dennis Maurer, Ritter-Weber, Paul Rossi, Ted Flack, Howard Comstock, Greg Luneack and Bob Glidden.

The largest thanks and gratitude go to Chrysler Engineering and the "Chrysler Race Group." The key managers of these groups were Dick Maxwell, Brian Schram, Tom Hoover, Bob Cahill and John Wehrly.

Finally, a special thanks goes out to my wife, Linda, who literally put this whole book together. I couldn't have done it without her.

ABOUT THE AUTHOR

Larry Shepard began working for Chrysler as a suspension engineer in 1968. By 1970, he had moved over as staff engineer of Chrysler's drag racing program, where he would guide the winning efforts of drag racing competitors, including Bob Glidden's legendary 340 Plymouth Arrow. He is currently a staff engineer with *Mopar Performance*.

Cover photo by Jim Kelso
Interior photography by Larry Shepard unless otherwise noted.

Although the cooperation of Chrysler Corporation and Mopar Performance is gratefully acknowledged, this publication is a wholly independent production of Price Stern Sloan, Inc. Mopar is a registered trademark of Chrysler Corporation.

Library of Congress Cataloging-in-Publication Data
Shepard, Larry.
How to hot rod small-block mopar engines / Larry Shepard.
p. 176 cm.
Includes index.
ISBN 0-89546-479-7
1. Automobiles—Motors—Modification. 2. Chrysler automobile.
I. Title.
TL210.S46 1989
629.25'04—dc20 89-11189
CIP

Published by HPBooks
A division of Price Stern Sloan, Inc.
360 North La Cienega Boulevard, Los Angeles, California 90048

© 1989 Price Stern Sloan, Inc. Printed in U.S.A.

10 9 8 7 6 5 4 3 2 1

NOTICE: The information in this book is true and complete to the best of our knowledge. All recommendations on parts and procedures are made without any guarantees on the part of the authors or Price Stern Sloan. Parts and procedures may not be legal for sale or use in some states. Check local laws. Authors and publisher disclaim all liability incurred in connection with the use of this information.

All rights reserved. No part of this publication may be reproduced, stored in a retrieval system or transmittted, in any form or by any means, electronic, mechanical, recording or otherwise, without the prior written permission of the publisher.

CONTENTS

INTRODUCTION

Building an engine is a complex undertaking. It involves precision skills, a sizeable investment and, in all cases, a lot of hard work. Of all the high-performance V-8s, Chrysler's small-block A engine is one of the easier engines to build, making it popular among amateurs and professionals alike. The important difference between amateurs and pros is that the professional is paid for his attention to details; the amateur is rewarded in the form of achievement for his.

If the amateur adheres to good building practices, it is quite possible he will produce an engine meeting or exceeding the output of the professionally assembled one. Any engine builder's success depends on two factors: *information* and *time*. The information is contained in this book. It was written with the performance enthusiast in mind, and contains large amounts of information generated by Chrysler's internal research and development engineers.

This book was also written with the assumption that the reader is already familiar with general overhaul, disassembly and assembly practices, especially those peculiar to the Chrysler A engine. Complete standards and techniques for these engines in stock form can be found in HPBooks' *How To Rebuild Your Small-Block Mopar* and in the Chrysler factory service manual for your particular engine. For factory manuals write: Chrysler Service Publications, 20026 Progress Drive, Strongsville, Ohio 44136. This book begins where those two books end—and that is on how to build a high-performance Chrysler small-block—very high performance.

There are a number of complex principles involved when building high-performance engines. Therefore, I urge you to read this information carefully and thoroughly. Once that task is completed, and you've gathered all the information you can about your particular engine, the next step is to concentrate on the second factor in successful engine building—*take your time!* Don't hurry! If you're not sure about something, don't go any further. Check it again. Then do it again just to be sure. Record all measurements accurately. I cannot stress the importance of good record-keeping. Make sure to jot down the measurements before and after machining. But the main

Perhaps the most famous (and successful) small-block racer was Bob Glidden's 340 Plymouth Arrow. Glidden and the Arrow were so dominating that NHRA factored them 3 times in a single season to slow them down. That didn't prevent Glidden from winning the championship, however. (Photo by Les Lovett/NHRA)

thing is to take your time, especially if you're new at this game. You're not going to win any prizes for being a fast engine builder. If you do a quick job, you'll get to do it again real soon—and it will be more expensive the second time! Remember, time and information are the "secrets" behind successful engine building—that and following instructions to the letter.

HISTORY

The small-block Mopar engine has its roots in the original A engine that was introduced in 1956 as a 277-cubic inch displacement (c.i.d.) V-8. It was enlarged to 301 and 318 c.i.d. in 1957. This 318 was built through the 1966 model year, and is one of *two* 318 Chrysler engines built. The one built from '57 through '66 is generally called the *old-style* 318; the other is called the *new-style* 318. The early 318 had the same bore and stroke as the *new-style* 318. Although similar, many parts from the old 318 don't fit the new 318 and vice-versa.

The original A engine 318 and its companion 277 and 301 engines are commonly called *old-style* A engines. Although they had a lot of potential, they were never developed for racing. Consequently, high-performance parts for old-style A engines are hard to find. Because of this I do not cover the old-style A engine in this book. From now on, all references to the A engine or 318 will mean the new-style version—those built from 1964 (273) or '67 (318) to present.

Origin—As far as performance is concerned, the real story of the Mopar small-block begins with the introduction of the 273 LA engine in 1964. As an *LA* (Low A) engine, it had a *shorter* deck and was lighter than the original 1956 engine. It was manufactured using modern, lightweight casting techniques. It also had a new wedge-style (valves-in-row) cylinder head.

Unfortunately, the 1964 model was equipped only with a 2-barrel carburetor, which left a lot to be desired in terms of performance. But while it may not have been king of the street, it began developing a reputation for durability, reliability and fuel economy—a reputation that stands today.

Richard Petty dominated the NASCAR circuit at the helm of a small-block Mopar-powered Dodge Charger. Shown here is his '78 version. (Photo by Don Hunter)

"Dandy" Dick Landy campaigned Chryslers of various sizes with great success. He was part of Chrysler's Performance Clinic program that toured the country in the Sixties. Today, he owns DLI in Northridge, California doing what he has always done best—building the fastest big- and small-block Chrysler engines around. (Photo by Bob McClurg)

In 1965, the 273 LA engine got its first 4-barrel carburetor package. It had a mechanical cam, single-plane intake manifold, low-restriction exhaust system and high-compression pistons, making it relatively advanced for its time and the performance climate in 1965. Big-blocks were then king, and the Chrysler Hemi ruled the streets and strip. The 273 was just too small and kind of got lost in the shuffle. However, those who ran lightweight A Body cars sure had fun with high rpm and good handling!

The LA version of the 318 was introduced in 1967 with a 2-barrel carb and mild hydraulic cam. It continued the LA engine's reputation for reliability and fuel economy, but it certainly wasn't a hot rod package in stock form. That came later.

In 1968, the performance side of the LA engine story took a turn for the better. The 340 burst on the scene and immediately established itself as *the* small-block to beat. In drag racing it was heavily *factored*—meaning its minimum competition weight was increased—in an attempt to slow it down, but it continued winning its share of races.

The biggest jump in small-block Mopar popularity and performance came in 1970 with the 340 Duster. It was light and very quick. At the peak of the big-block wars, it was really the only stock small-block capable of challenging big-blocks at the drag strip. To top it off, 340 Dusters were inexpensive. They became very popular for their combination of high performance, high rpm and low cost. They were quick, fast, agile and inexpensive. As a result, they were called "the workingman's hot rod." They were a favorite of the younger performance car nuts who couldn't afford the more expensive, less reliable models. However, they still toiled in the shadow of their big-block brothers, the 440-6 and Hemi. That situation would soon change, however.

In mid-1970, the 340 6-barrel package was introduced. It was initially offered in the Challenger T/A and AAR Cuda. With a total output of 290 hp, it was the most powerful LA engine ever produced. The whole program was developed to homologate the car, and the Mopar small-block, so it could compete in the highly competitive SCCA Trans-Am series. It was produced for only one year (1970), and is therefore a fairly rare engine.

In 1971 the 340 4-barrel package received a larger carburetor. The new Carter Thermo-Quad carburetor (TQ) flowed more air than the earlier '68-'70 Carter AVS. Because of this, the '71 340 Dusters were very fast. In stock form, they were the fastest of the Mopar small-blocks and certainly one of the quickest production small-blocks anywhere in this era. But the press wasn't interested—the 440-6 and the Hemi still ruled the quarter-mile drag strip.

Therefore, it's not surprising that when the 360 version of the LA engine was introduced in 1971, few people in the enthusiast press took notice. After all, it was offered only with a 2-barrel, primarily in very large cars. Though it continued the LA engine's reputation for reliability, durability and economy, it wasn't quite ready to step into the performance arena with the 340.

It was at this time, in 1972, that the well-documented and famous energy crisis swept the U.S. With it went the musclecar era for good. The 440-6 and the Hemi were gone from production. Strict emission requirements and low-compression ratios lowered engine outputs. The cars got heavier and slower. The automotive press soon lost interest in production performance cars because everyone knew that they were slower than in previous years. For performance, the automotive press turned to racing and hot rodded engines. What they failed to notice was that the production Mopar small-block was the new uncrowned king. The 440-6 and Hemi engines were no longer produced and even though the '72-'73 340's lost some compression and had a smaller intake valve, all the rest of the good A engine performance parts were still there. They weren't as fast as the 71's but they certainly were the class of the field in '72 and '73.

Jim Durham's 340 Duster is a frequent competitor in NHRA's G/SA class. (Photo by Jim Kelso)

The next performance surprise was the high-performance 360 that was introduced in 1974. This was virtually the *only* high-performance option introduced during the 10 year span from 1972 to 1981. The 360 Dusters got all the good 340 parts but the extra 20 cubic-inches really added performance, especially in the area of acceleration. These were the quickest production vehicles in 1974 and 1975 but no one seemed to take much notice. As far as the press was concerned, production performance was history. Gone were the days of the factory hot rods. Without the press' support, sales declined and the 1976 360 Duster was the last of the breed of high-performance small-block musclecars.

In the years 1976 and 1977, performance was almost non-existent. Things didn't start to look better until 1978 when the 318 4-barrel package was introduced. If this had been put in the '70-'71 Duster, it would have been a real runner, but this was 1978 and the Duster was gone, the high axle ratio was gone and the cars had gotten heavier. This was also the year of the truck. Fast trucks were "in." The 360 4-barrel package (as it was in 1974) was installed in a special Dodge pickup called the Little Red Express Truck. It was the quickest vehicle in 1978. It was again competing against its old enemies—the big-blocks—but this time the small-block would win. The Little Red Truck was the last Mopar small-block performance package. Times were changing, and so were the vehicles. The 360 was dropped from production cars in 1980. The 318 4-barrel has continued, but the emphasis isn't on performance.

Racing History—Racing the small-block Mopar started off very slowly. It was traditionally over-shadowed by the Chrysler big-blocks. In 1966 Chrysler introduced their first Mopar small-block "race" package. It was called the D-Dart. It used a 275 hp 273 c.i.d., which was the only A engine over one horsepower per cubic-inch in production. It had a big Holley carb, big mechanical cam and headers.

Although the 340 did reasonably well in "stock" class drag racing, the next major "modified" effort was at the Indianapolis 500 in 1969 with the STP team. Then it was sports car racing in 1970 with the Trans-Am series. At the time this was a big, all-factory series with Chevy and Ford battling it out. The 340's ran at 305 cubic-inches because of the SCCA rules. Even at the small displacement, they were the fastest at the end of the straightaway at almost every race.

In the mid-70's, NASCAR racers started winning with Mopar small-blocks. The big-blocks were being phased-out and Richard Petty started winning with the 355 cubic-inch Mopar small-block. He had more power and more reliability than the other small-blocks. But again no one really noticed. He had won with big-blocks before so everyone assumed he still had one.

In 1974, Chrysler decided to push harder on its small-block performance image. The NASCAR version was winning but in drag racing the big-blocks were still king. Therefore Chrysler started racing the 360 4-barrel package in Super Stock and did very well. They quickly demonstrated that the 360 was a force to be reckoned with. In the Stock and Super Stock classes at the time, the faster you went, the more you were weight-penalized, and it was soon apparent that the 360 was fast. Chrysler engineers soon determined that the best class to demonstrate the small-block's full potential was the Modified category. In 1976 and 1977, armed with the W-2 head and 295 cubic-inch versions of the 340, Chrysler attacked the Modified category and went faster than much of the competition. But still no one noticed! The automotive press was concentrating on the Pro Stock class. So the obvious choice was to put the now-proven Mopar small-block into Pro Stock. That's when Bob Glidden and his now-famous 340 Arrow entered the picture. The two went almost undefeated. So dominant were Glidden and the Arrow, that NHRA factored the car three times in one year.

The Mopar small-block domination of Pro Stock in 1978 and King Richard Petty's win at Daytona in 1979 with his Magnum (which he did on sheer horsepower advantage) were the two high-water marks in racing for the LA engine. It had always had more horsepower and better reliability and in a good car and with a good driver/crew it was a winner. That is still true today!

ABOUT THIS BOOK

The previous few paragraphs chronicle a brief history of the A engine cars and their racing. However, this book is written on hot rodding the Mopar small-block engine, not the cars. With this in mind, I have organized the Chapters around the major components of the engine, starting with the cylinder block. I will work through the various parts of the short-block and head assemblies and then cover the accessories like ignition and exhaust that bolt onto the finished cylinder head/short-block assembly. I have designed the sequence so that it is similar to the same order when building the engine. This book is not intended to be a parts catalog. Therefore there are few part numbers in the basic text. However, there is a parts listing at the end of each chapter, as well as a large section of *Tables* at the end of the book. The Tables section includes a listing of Performance Packages, Carburetor Modifications, and the Casting Numbers and Specifications of all related engine parts. I did not write this book to serve as a textbook on automotive and engine theory. There are several college textbooks such as Taylor and Taylor that cover that topic. It does include some racing design information or theory which is supplied primarily for interest or reference. Therefore, much of it has been separated from the basic text and is listed as a sidebar. In the performance business, it is more important to know which part works best than it is to know why it works. Getting the "part that works best" should take top priority. I have tried to follow this philosophy in the following Chapters.

I frequently use the terms "off-road," "street/strip" and "dual-purpose." Basically they mean the same thing and are interchangeable. They describe the same type of vehicle or application for the engine. On the other hand a *street machine* is a post-

'55 and newer hot rod with a modified engine and/or drivetrain while a *street rod* is a pre-'55 car with a 60's and newer engine/drivetrain installed. These are very general descriptions and there tends to be some overlap, but they will serve our purposes. I also use the terms "stock," OEM (original equipment manufacturer), and "production" to mean the same thing. Stock is an original car produced on the assembly line by the original equipment manufacturer in its original condition. Production is another word for the assembly process that produces stock vehicles. Any of these terms means "no-modifications-allowed" or "as built."

One final item to consider *before* you start any engine build-up project relative to performance: Most engine builders recognize that they must balance the engine's reliability/durability against its output. The added part of the equation is cost. As a rough guideline, keep in mind that an engine that makes one hp per cubic-inch can be built (rebuilt) for approximately $3,000 to $4,000. On the other extremes are the oval track NASCAR single 4-barrel 1.75 hp per cubic-inch durability engine and the 2.0 hp per cubic-inch, 8-barrel tunnel-ram drag race engines that cost from $15,000 to as much as $30,000. Always keep cost in mind when considering a change in your performance level. Try to get the best performance for your dollar.

There is one other very important factor to consider before getting started. It is essential that you find and establish a good relationship with a professional, engine machine shop capable of performing the necessary, and critical machining steps that I'll talk about throughout the book. I do not recommend, unless you have access to and are thoroughly familiar with this type of equipment, that you attempt any machining yourself. A slight mistake could mean disaster. What is important is that you are able to communicate effectively with your machinist, to be able to talk to him on his terms and fully relay what it is you need to have done to your block, head, crankshaft, rods or whatever. That's why I discuss all machining steps thoroughly. Not because I want you to learn to do them yourself, but so that you'll be familiar enough with the processes to tell your machinist exactly what needs to be done.

Generally in any performance engine project the cost of the hard parts is only half the total cost. Machining and assembly make up the other half. Therefore, assembling it yourself by following your reference library and the rest of this book will represent a substantial savings to you, the engine builder. Good luck!

Early Dodge, Chryslers and Plymouths are gaining popularity among street machine enthusiasts and car collectors alike. (Photo by Jim Kelso)

1 CYLINDER BLOCK

IDENTIFICATION

Identifying the cylinder block of a high-performance or race engine is an important first step. You may know what the block is because it's brand-new or because you just disassembled a known, used engine. On the other hand, you may not be that lucky. A basket of parts can be hard to identify. Looking in the junkyards for special pieces or specific engines can be tricky. And then there can be special versions of a given block such as the 340 "X" block and "T/A" block. Even more subtle differences are the new thinwall castings versus the older thickwall designs, especially in the 360 blocks. In all these cases, the block's *casting number* is one reliable way of identifying the actual block that you have. The T/A identification is cast into the side of the 340 T/A blocks right after the casting number. The 340 Race block, currently available, has an "X" cast into it on the front of the engine by the #2 cylinder. Each of the older blocks—273, 318 etc. has its own casting number cast into the side. However, keep in mind the casting number tells you only what the block started out as, not how it's been modified or machined. It could be overbored, decked, have an undersized crank, oversized tappets, etc. As you build the engine, you will need to check each of these items.

A general feature of all A engine blocks is that they are made of a hi-nickel alloy, the same material used for Chrysler's special race blocks. Most other manufacturers use this material only for special racing blocks, but Chrysler not only used the "good stuff" in its race block, (such as the 340 X block), but also in all of its production 318's, 340's and 360's. However, I have been able to establish the use of this material only back to the early 70's and the 340. The early 273 and the pre-'68 318's may not have used this hi-nickel alloy.

BLOCK FAMILIES

The A engine block family is divided into two basic groups. The original one consists of the 273 and 318. These two engines are virtually the same except for the bore sizes. The other, newer group consists of the 340 and the 360. These last two differ from the first group mainly because of the left side motor mount. However, the 340 and 360 use the same mounts. This allows the 360 to be bolted in to replace a 340 and allows the 318 to be

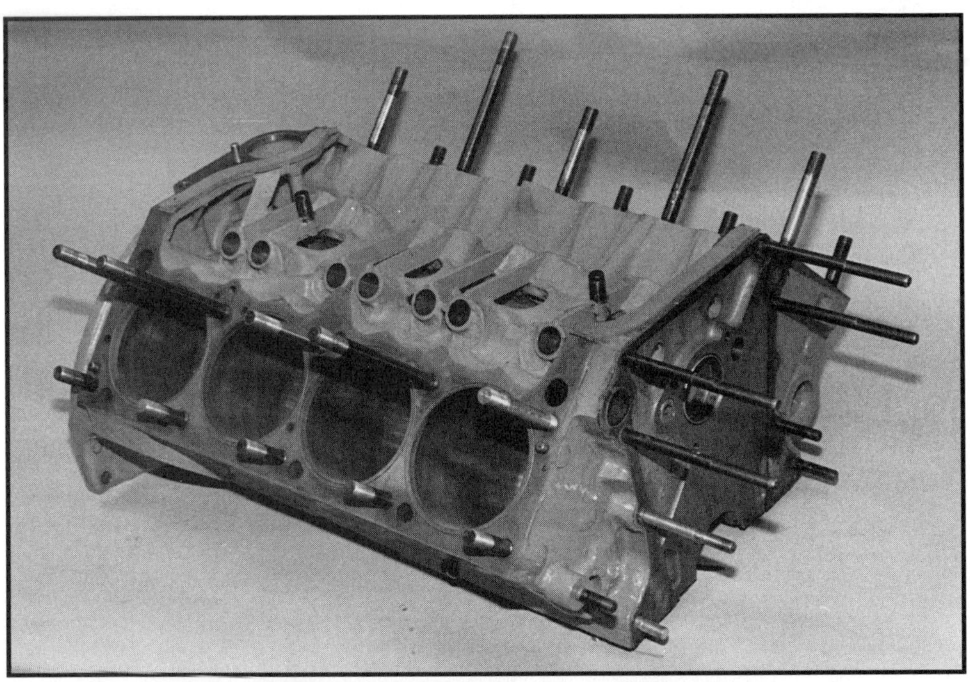

Dick Landy, famed Mopar racer turned Mopar engine builder, modified this 340 Race block from Mopar Performance for racing use. Landy opened up the cylinder bores from 4.040 to 4.10 inches, and added a Mopar Performance stud kit. For more information, contact DLI at 19743 Bahama St., Northridge, CA 91324. (Photo courtesy Hot Rod Magazine)

bolted in place of a 273. That is easy engine swapping. A swap consisting of a 340 in place of a 273, however, requires that one side mount be changed.

However, the A engine family can be divided into two groups somewhat differently. The 273, 318 and 340 share the same oil pan. The 360 has a special pan with a smaller diameter at the end seal. The 273, 318 and 340 also share the same 3.31 stroke and the same main bearing diameters. The 360 has the longer stroke at 3.58 and also larger main bearing diameters. I like the first method of grouping but judge for yourself. As long as you remember and note the differences, you won't have any problems.

The obvious visual differences between the 340 T/A and the Race 340 blocks and the stock A engine block is that the pan rail is filled in and the bulkheads are thicker. None use 4-bolt main caps in standard form. Only the specially machined 340 Race block comes with 4-bolt mains. The T/A and standard 340 Race blocks have the *provision* for 4-bolt mains but only two are machined. This means that the material is in the block bulkhead but the outside holes are not machined.

BASIC PREP

Once you have selected the block, it's time to start the detailed process of preparing it for racing or high-performance applications. The first step is to deburr it on all its outside and inside surfaces except for the top of the cylinder bore. Next, it needs to be cleaned thoroughly. You can't get it *too* clean and you can't clean it *too* many times. I would suggest that it be *hot-tanked*, which involves submerging the block in a hot acid solution for at least several hours. Most machine shops have this service available, or some shops use high-pressure, high-temperature steam systems. These work just as well. If you do hot-tank it, then the cam bearings will have to be replaced. This is a critical task best left up to the machine shop if you don't have the proper tools to do it. If you don't hot-tank it, assume they are okay. Check the cam bearing oil holes for proper alignment with the oil hole in the block. Also check the bearing shells' alignment front-to-rear. If it's a new block and it was stress-relieved in an oven, then you must replace *all* the freeze plugs and the main cap screws.

BASIC MACHINING

After the block has been thoroughly cleaned, measurements need to be taken and recorded to determine the extent of the first two machining operations. This is a procedure that you should handle yourself if at all possible, rather than let the machine shop determine it for you. That way, you'll know where you stand, and it will help you with parts selection.

Bore Measurement—The first step is to measure each bore with a *dial-bore gauge* to determine how straight and round they are and how much *taper* they have. These measurements determine the minimum amount of *overbore* required to put everything right. You may find that the block is already overbored. Once you have the information, you need to compare it to the piston sizes readily available and, determine the *cylinder bore size*. It's too expensive to generate custom pistons and rings in odd bore sizes like 0.047-in. over. Standard sizes like

A ENGINE BLOCK SPECS

For reference the A engine block specifications are as follows:
WEIGHT: Approximately 160 pounds.
BORE SPACING: 4.46 in.
MAIN BEARING BORE DIAMETER:
2.6925/2.6930 in. (273-318-340).
3.0025/3.0030 in. (360).

The casting number 3418496 identifies this block as a 360. The second part of the number "360" is not always there.

This casting number 4104230 identifies this block as a 318. Note the different ribs and freeze plug locations on this block compared to the 360.

0.020-in. over, 0.030-in. over and 0.060-in. over, are most common. When rebuilding a used engine, the amount of overbore is generally 0.030-in. or 0.060-in. over. These piston sizes are readily available for all A engines from most all aftermarket piston manufacturers. Rings for these oversizes are also easy to obtain. However, you should always keep in mind the block's capacity to be overbored. The 0.060-in. overbore is too large for some of the newer, thinwall-cast blocks.

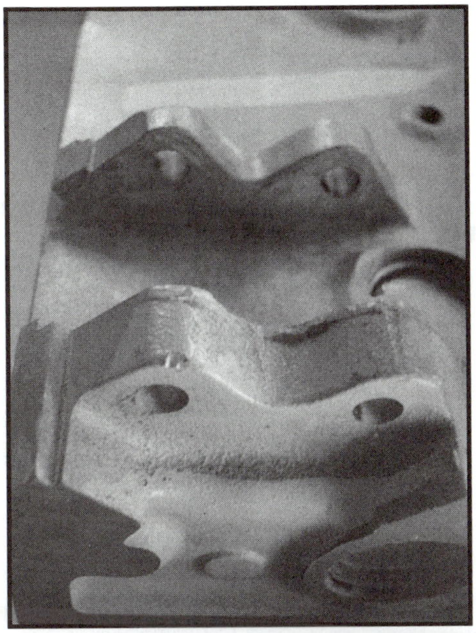

All the A engines use ears on the side of the block at the front to attach the motor mounts to the block itself. The 340-360 uses a different pattern of "ears" from the 273-318.

It's recommended that many popular small-blocks be overbored 0.010 in. to 0.030 in. maximum for high-output use or they will lose power. The older A engines (pre-1976) are stronger and can be overbored a total of 0.040 in. to 0.060 in. However, before boring out that much it is best to have the block *sonic-tested* to make sure it doesn't have *core-shift* (see sidebar p. 13). The newer blocks are thinwall cast and should only be bored 0.020-in. oversize unless sonic-tested and found not to have core-shift. If they don't, then they can withstand an overbore of 0.030 in. If you buy a Chrysler Race block from Mopar Performance, it can be overbored to a bore of 4.100 in., which is either 0.100-in. over or 0.190-in. over since it starts life as an unbored 318 with a standard bore size of 3.91 in. However, pistons are hard to find for the 4.100 in. bore size, rings are even harder. Therefore, I recommend either a 4.060-in. or 4.080-in. overbore because of parts availability.

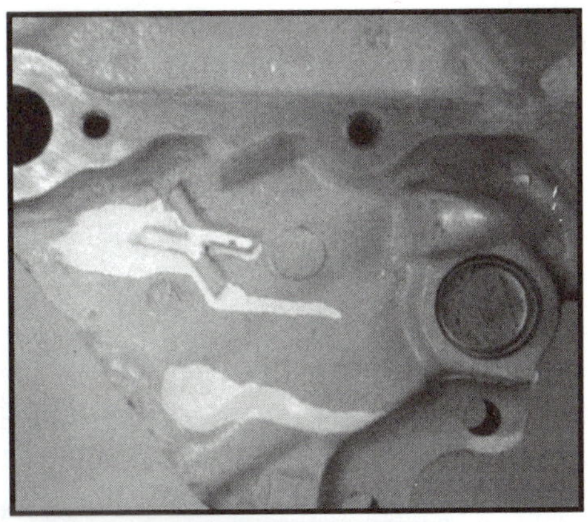

The Chrysler Race block for the A engines has a large "X" cast into the right bank at the front.

Align-Boring & Honing—Before boring the cylinders, you'll need to check the alignment of the main bearing bores, especially if you are using different, higher-quality main caps. Quality main caps are important on high-output, durable racing engines and maximum durability engines. The 340 Race block ("X") is the only one that has them initially. Otherwise, you should add them and then *align-bore* and *align-hone* the block.

The very best machine shops will have an *align-bore bar*. They are extremely accurate and must be made specifically for a particular engine's

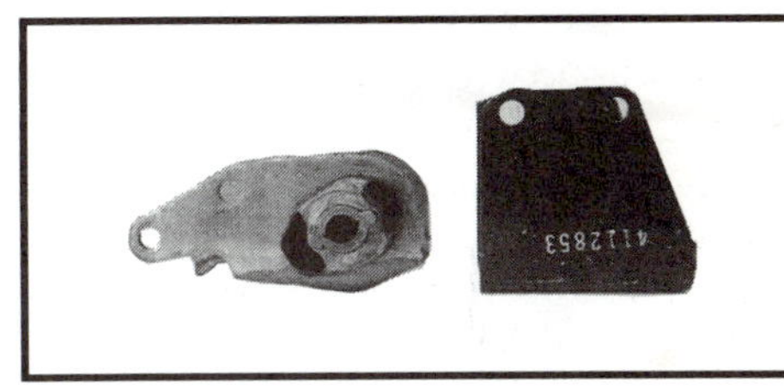

Motor mounts themselves vary with application. The new style passenger car mount is shown on the left while the current truck mount is shown on the right. Note the fail-safe style of the thru-bolt on the passenger car mount.

Cleaning the block is extremely important and should be done several times before final assembly. A large shop will have a special parts washer—high temperature and high pressure. If doing it yourself, take the block to a pressurized, do-it-yourself car wash. Spray block with plenty of grease cleaner and wash it down, then wash it again!

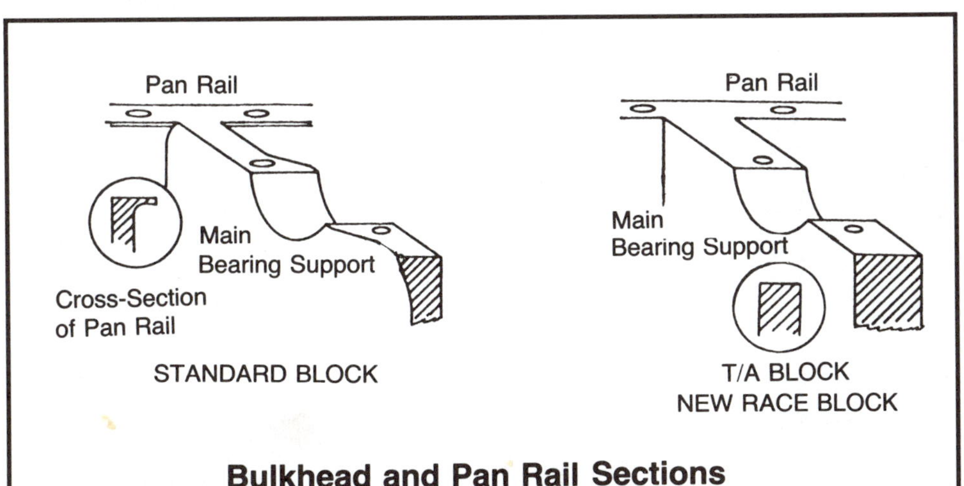

A engine standard blocks have pan rails and main bearing bulkheads undercut. This makes crank installation and block machining easier. Undercutting is typical of production small-blocks. The '70 Trans-Am block and the 340 Race blocks have solid (filled-in) pan rails and bulkheads. (Drawing courtesy Mopar Performance)

main journal size. For instance, the A engine family would need two bars; one for the 273-318-340 and one for the 360.

If your machine shop doesn't have an align-bore bar, you can use a long, steel straightedge. Most machine shops should have one. But using one will determine only if a block is bad, not if it is perfect. When using this method, check for a difference along the straight end of less than 0.003 in. If it's 0.003 in. or more, it should be align-bored.

The crankshaft can also be used to check for alignment. Assuming that the crank is straight (see next Chapter), install it into the block using old main bearings and a small amount of oil. Torque down the main caps to specification, then rotate the crankshaft in the block. It should rotate smoothly and easily, either by hand or with an inch-pound torque wrench. The wrench will record a specific turn-over torque. If you notice a tight spot in the 360-deg. turn, either by hand or on the torque wrench, then the block should be align-bored or *align-honed*. If just a few thousandths of an inch need to be removed, then align-honing will suffice. If more than that needs to be removed, then align-boring is necessary, followed by honing.

Align-boring is also necessary if the main bearing caps are replaced. It is a process similar to resizing the big end of a connecting rod. Material is removed so main bearing bores are realigned and made round to equal and exact size. This may move the crank up slightly in the block by a few thousandths-of-an-inch, but that shouldn't cause any problems. You shouldn't have to align-bore more than once. More than that and the timing chain fit will get sloppy.

Cylinder Boring—Boring the block is generally a two-step process. The first step is the boring itself. The second step is the honing. Any race or high-performance engine should always be bored and honed with *honing plates*, which every machine shop should have. A honing plate is a 1-in. thick plate of steel with the head bolt holes and cylinder bores (slightly oversized) machined in. It is bolted to the block with the exact same size head bolts and torqued down to the eventual specification. This is important, so the bolts will have to be shimmed. The machinist should already know this. The plate simulates the stresses and distortions normally caused by a torqued cylinder head, so the boring and honing will be more accurate. It's

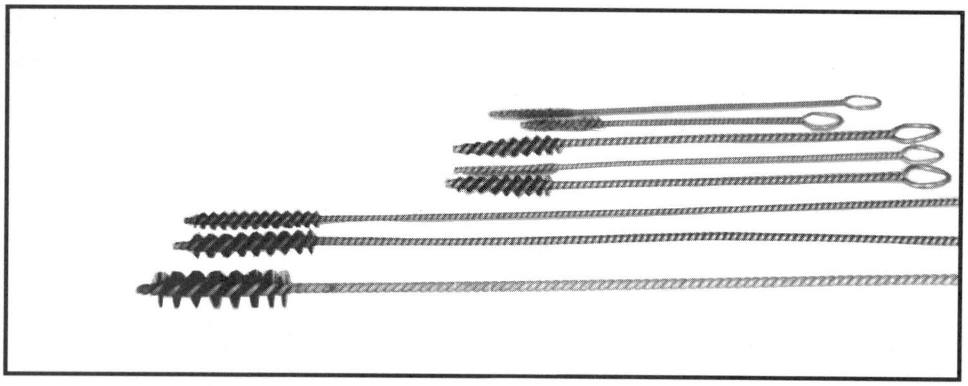

To clean the block properly, you will need a large assortment of bottle brushes with varying lengths and diameters.

A long-handled, large diameter brush is needed to clean both main oil galleys.

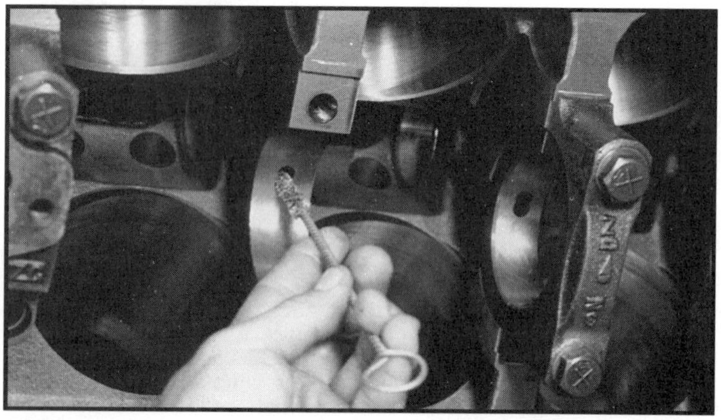

All oil passages should be checked for dirt and other objects even if the block is new. The vertical holes to the cam journals and the main oil feeds, which come down to the mains from the right oil galley, are doubly important.

If building a new block for racing, a small parts package of various plugs will be needed. If you stress-relieve a block, all the core plugs must be replaced.

also recommended that main caps be installed and properly torqued before the block is honed. Both the plate and main caps will help ensure that the cylinder walls are bored straight and round. Having the bores straight and round with no taper at wide open throttle is a critical factor to obtaining maximum performance from an engine. Therefore, discuss both of these steps with your machinist to make sure that he adds the plate and main caps and torques them properly. Give him the torque specs to be sure. Sometimes, they don't do it automatically, and it's always best to leave nothing to chance.

To determine if you need align-honing, place a machinist's straightedge in the saddles and measure any clearances with feeler gauges. It should be less than 0.003 in. Note that this straightedge is no common ruler. (Photo by Michael Lutfy)

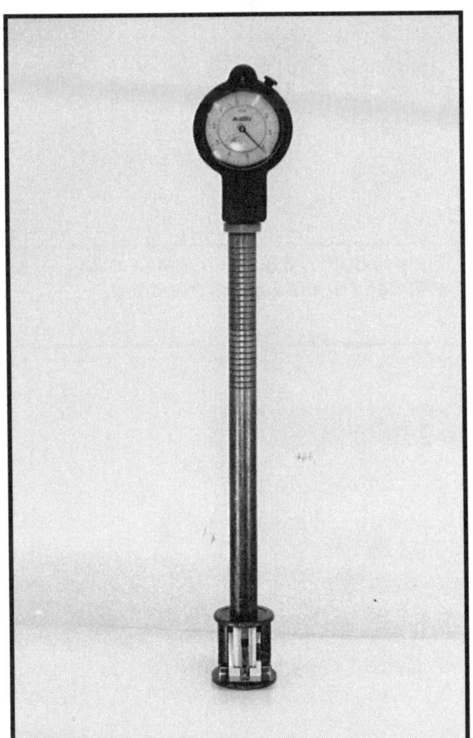

A dial-bore gauge is also used to accurately measure the cylinder bore's diameter. It needs to be used carefully. If you are honing the block yourself, then you must have one.

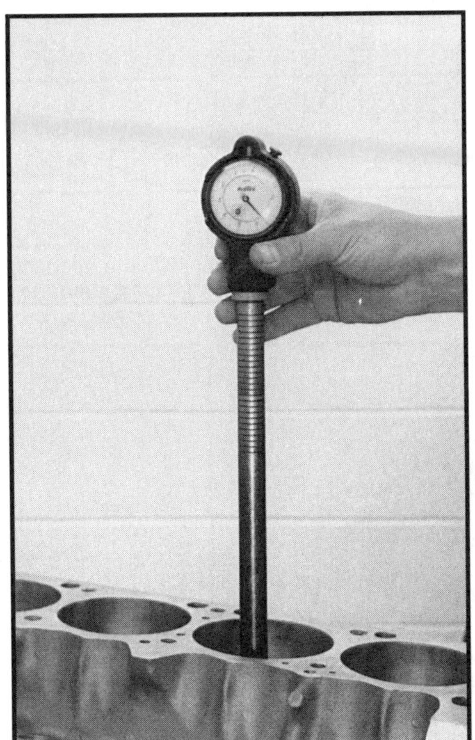

The dial-bore should be used to measure the cylinder bore at three heights (top, bottom and middle of piston travel) and in two different directions at each height.

Decking—The next machining operation done to the block is to *deck* or *mill* off the top or deck surface on each bank of cylinders. The question is how much to mill? The maximum would be around 0.060 in. to 0.075 in., but the less you have to cut, the better. The only way to tell is by measuring. Don't cut until you're sure! Remember you can't put it back on. The best way to measure is to install the crank, rods and pistons into the block. Since the block has already been bored to the new piston size, they should fit. Torque the rods and main caps down to specification, then rotate the crank to bring a piston up to *top-dead center* (TDC). Measure the distance from the flat of the piston to the top of the block,

A dial-bore gauge measures the bores of the torqued main caps in two directions on either side of the parting line. This is also part of the measurement to determine the amount for align-boring and honing.

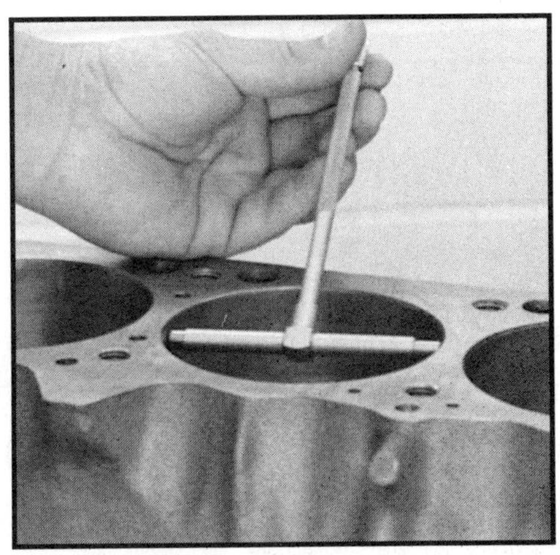

You can use a snap gauge and a micrometer to measure the cylinder bores but they are not as accurate as a dial-bore gauge so they should only be used as a guide.

parallel to the pin centerline. This measurement is called the *deck height*. For example: if the piston is 0.025 in. below the deck surface and you want it to be zero, then mill the block deck surface 0.025 in. Keep in mind that you also have to consider the *piston-to-head clearance* when milling the block (for more info, turn to page 40). Also, remember that whatever is milled off the deck must be added to the head milling number when determining how much to mill the intake manifold or intake side of the head for proper intake manifold fit.

Once you've determined the milling measurement, discuss the steps thoroughly with your machinist to make sure both of you are thinking along the same lines. However, don't turn the block over to him just yet.

Before sending the block out to be milled, remember to pull the cylinder head dowel pins out. Also, if milling 0.020 in. to 0.030 in. off, ask the machinist to make several cuts, not just one big one. It takes a little longer and may cost a little more but there will be less block distortion. Also check the decks to be sure that they are parallel to the crank's centerline.

Studs vs. Bolts—If the block was stress-relieved, the main cap bolts must be thrown away. New bolts are readily available, or you might want to consider *main cap studs*. If the engine is expected to make 2.0 hp per cubic-inch or more, I recommend studs. Otherwise the stock bolts from Chrysler are fine.

On very high-output engines in the 1.8 to 2.0 hp per cubic-inch category, I recommend *cylinder head studs*. The extra threads offer better clamping strength onto the block, which decreases the distortion of the cylinders. Studs generally promote longer gasket life, although they should not be considered a cure for gasket problems. They do make it hard to get the head off in some in-car installations.

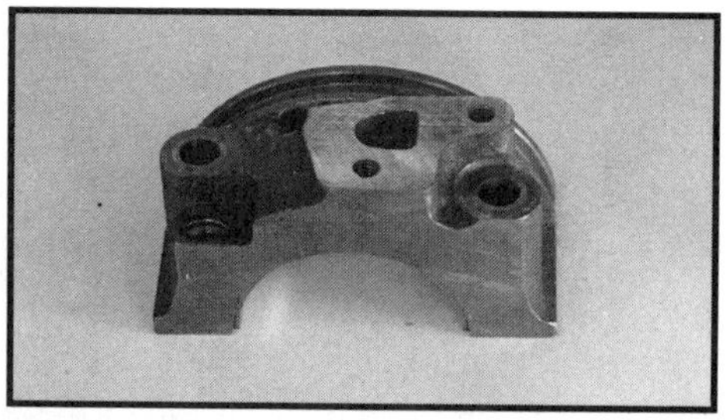

The unique #5 cap also holds the oil pump in place and holds the rear main seal. And it's all one piece.

Regardless of which head bolt or stud used, the top of each bolt hole should be *chamfered*. The chamfer (countersink) needs only to extend down 1 thread in depth. This applies to both head bolt and main cap bolt holes. Each bolt hole should have a bottoming tap run down to be sure that all the threads are there and that they are clean and fully cut.

You will note that after the block has been bored that, much, if not all of the chamfer at the top of the cylinder bore has disappeared. This is desired. *Do not put a chamfer back on the cylinder bore*. Usually, you'd have to bore 0.060-in. over or mill the block ex-

The five main bearing caps with the #5 to the right. They should not be switched or swapped or reversed and if they are replaced, then the block with caps should be align-honed.

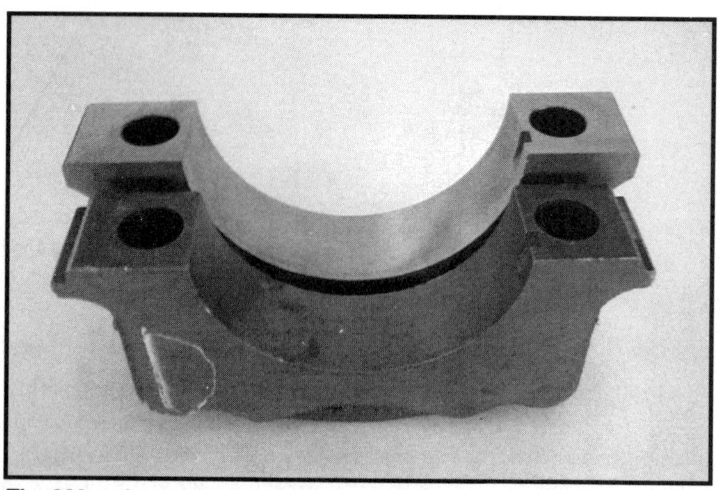

The 360 main cap (to the rear) has a larger diameter and the bolts are spaced farther apart than the 318/340 main cap (in front).

The 360 main cap screw (top) is shorter than the 340 screw (middle). The screw at the bottom is the special one that is used on the #2 and #4 mains to secure the windage tray.

cessively to remove all of the chamfer, but it's not worth going this far just to remove the chamfer. If it disappears, then break the sharp edge of the top of the bore with a piece of fine sandpaper. Without a chamfer, it will be more difficult to install the rings, so more care must be given to that operation.

Tappet Bores—The preparation of the *tappet bores* depends on the type of cam and tappets selected. First, check the tappet clearance itself. For hydraulic and mechanical tappets use 0.001 in. to 0.0015 in. clearance as a guideline.

If a hydraulic cam is used the tappets must have oil. This means that the tappet bores can't be *bushed*, the *tubes* can't be added to the 2-4-6-8 side galley and the oil can't be blocked to the left galley at the front of the left side of the block. Roller tappets usually require more clearance (0.002 in.-plus) but the tappet bores should be bushed

While the block can be hand-honed, almost every shop today uses a Sunnen CK-10. Basic set-up is shown. The CK-10 delivers a better job for less effort and cost, and is very accurate, however keep in mind it's still only as good as the operator.

Ready to hone. Note the thick, steel honing plate bolted to the top of the block to simulate cylinder head stress. The honing plate is an absolute must if you want to make a lot of power at high rpm. Main caps should also be installed and torqued to spec prior to honing.

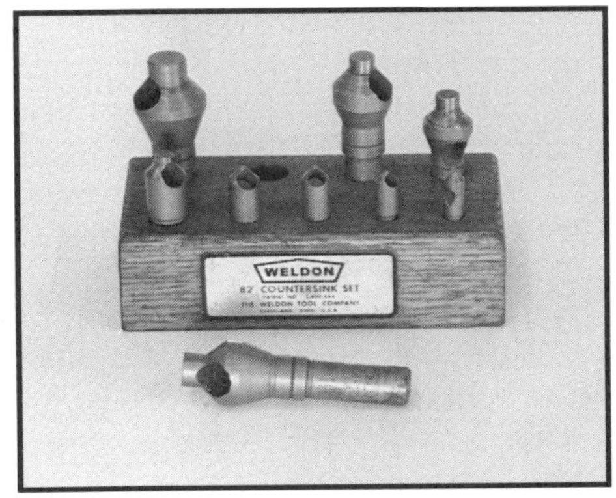

All bolt holes should be countersunk slightly. The special countersinks shown here make job easy and controllable.

Countersinking head bolt holes requires a countersink of proper size and a two-handed electric drill for better control.

Resulting cylinder wall cross-hatch pattern from honing operation performed by DLI. Actual micro-finish is a function of piston ring style and design. Consult ring manufacturer. (Photo by Michael Lutfy)

Deburring rough edges near machined surfaces requires a high-speed grinder.

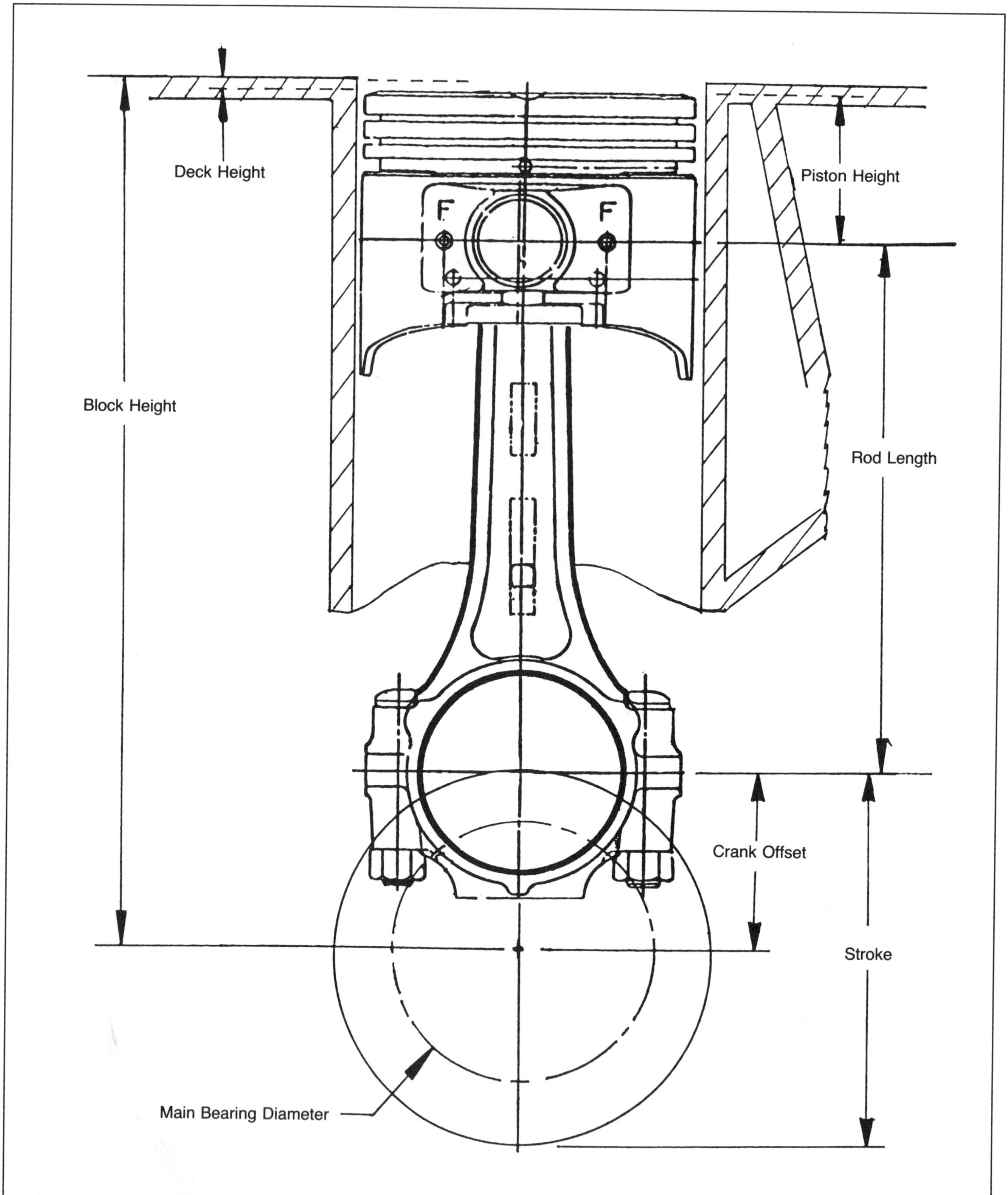

General short-block terminology. Block height and main bearing diameter are rarely used in discussions because they are so hard to change. The stroke is twice the crank offset. Rod length is the rod's center-to-center distance. Piston height is the distance from the center of the piston pin to the top of the piston. Deck height is the distance from the top of the piston to the top of the block. Deck height can be positive or negative.

or tubes need to be added to the right side. Roller tappets use *guide bars*, which need to be checked for clearance to the block during normal operation. If *mushroom tappets* are used, or if there's any possibility that they might be used in the future, then the bottom of each tappet bore should be *spot-faced*. This requires a back spot-facing tool and a bushing like an upside-down mushroom lifter with a hole in the center. The bottom of the tappet bore only needs to be flat and square with the centerline of the tappet bore.

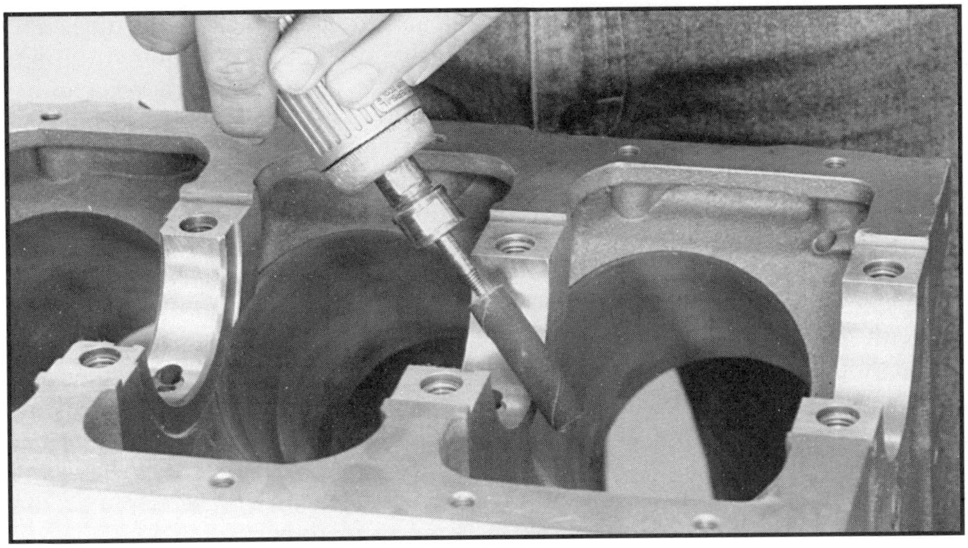

Radiusing main saddles requires a high-speed grinder with a sanding roll, (called a tootsie roll). Just break the edge, don't remove any material. The shape of the tootsie roll allows for better control.

Special areas, like the top of each cylinder bore, should be done by hand. Don't create a big chamfer, just lightly break the edge. A small, short file works best.

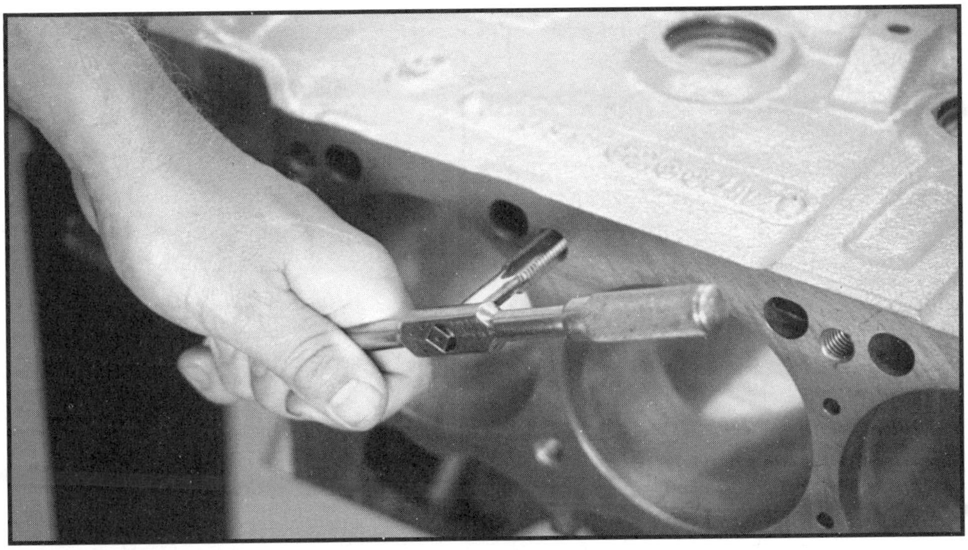

All bolt holes in the block should be cleaned with the proper size tap, preferably a bottoming tap. Don't forget to clean the block again *after* deburring and tapping.

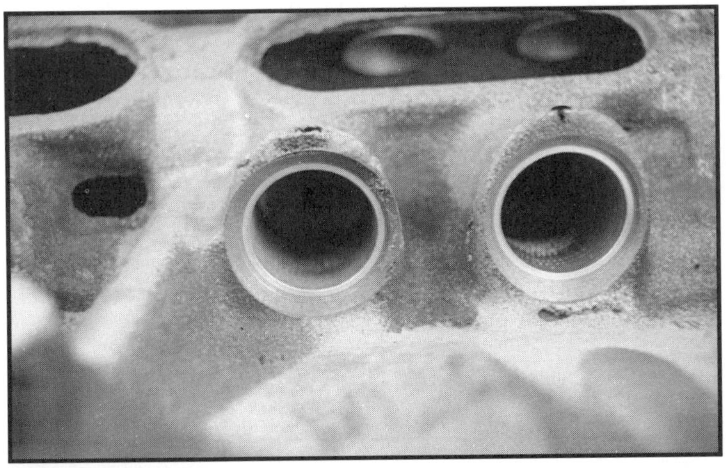

These are bushed tappet bores. Obviously hydraulic tappets cannot be used. Bushing must be done by a precision machine shop and can be expensive.

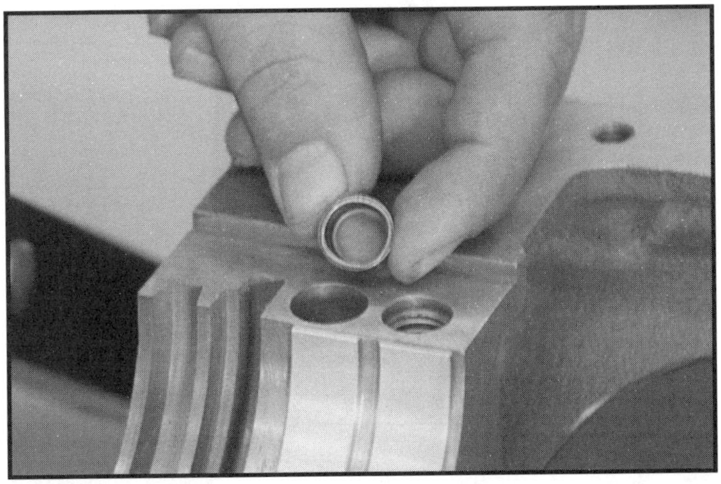

The oil passage plug must be driven into place on a new block and checked to see that it's in position on an old or used block.

Most high-output race engines need to block the oil to the tappets. They don't use hydraulic cams, and there are many reasons to block the oil. One main reason is that the engine will be low on oil pressure once it warms up if the tappet clearance is too high. To block the oil, bush or insert tubes to the right side oil galley.

Bushings are expensive and more difficult than tubes to install. A machine shop needs to index each tappet hole, checking to be sure that it's perpendicular to the centerline of the camshaft. Then the machinist bores the tappet hole oversize. After that, the bushings are pressed into the oversize holes.

Installing Tubes—Tubes are relatively inexpensive and much easier to install. Once you have the tubes, the drill and ream that's required (all are available from Chrysler), you can install the tubes yourself. Using the ream, enlarge the right side galley halfway from either end. Remove the oil galley plugs. After reaming, fit the tubes from either end; then re-install the plugs. Next, pound the round-nosed, tube-peening tool through each tappet bore on the right side to dent the tube in properly. Then, with the long drill provided in the kit, drill all oil galleys from the mains, into the right galley tube. The last step is to block off the left side with a threaded plug at the front of the block.

Final Check—The last main step is to insert the cam that is going to be used into the block (don't forget to lube it well) and check how easily it rotates. Some new cam bearings can be a little oversized and some cam journals can be slightly larger than normal. This can make the cam tight and cause it to rotate with a great deal of effort. At this point, the easiest solution to this problem is to have the machinist hone the tight cam bearing(s) so the cam rotates freely.

At this point, the basic prep of the block is complete. It should be cleaned again in preparation for the next step. Note that I haven't added any oiling passages to the block. They are not generally needed, but I'll discuss them later in Chapter 9 on *Oiling Systems*.

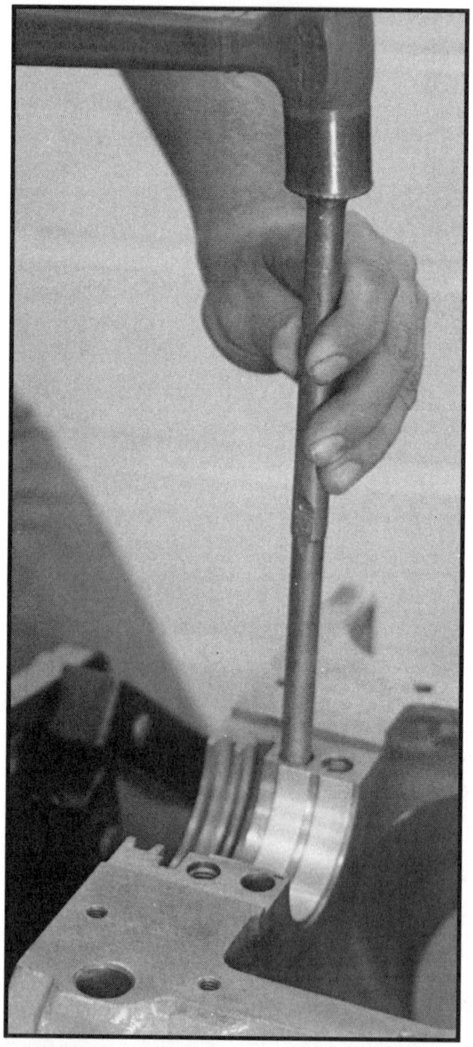

A long punch is required to drive the plug into position. Be sure to use the proper size punch so plug isn't distorted or damaged during installation.

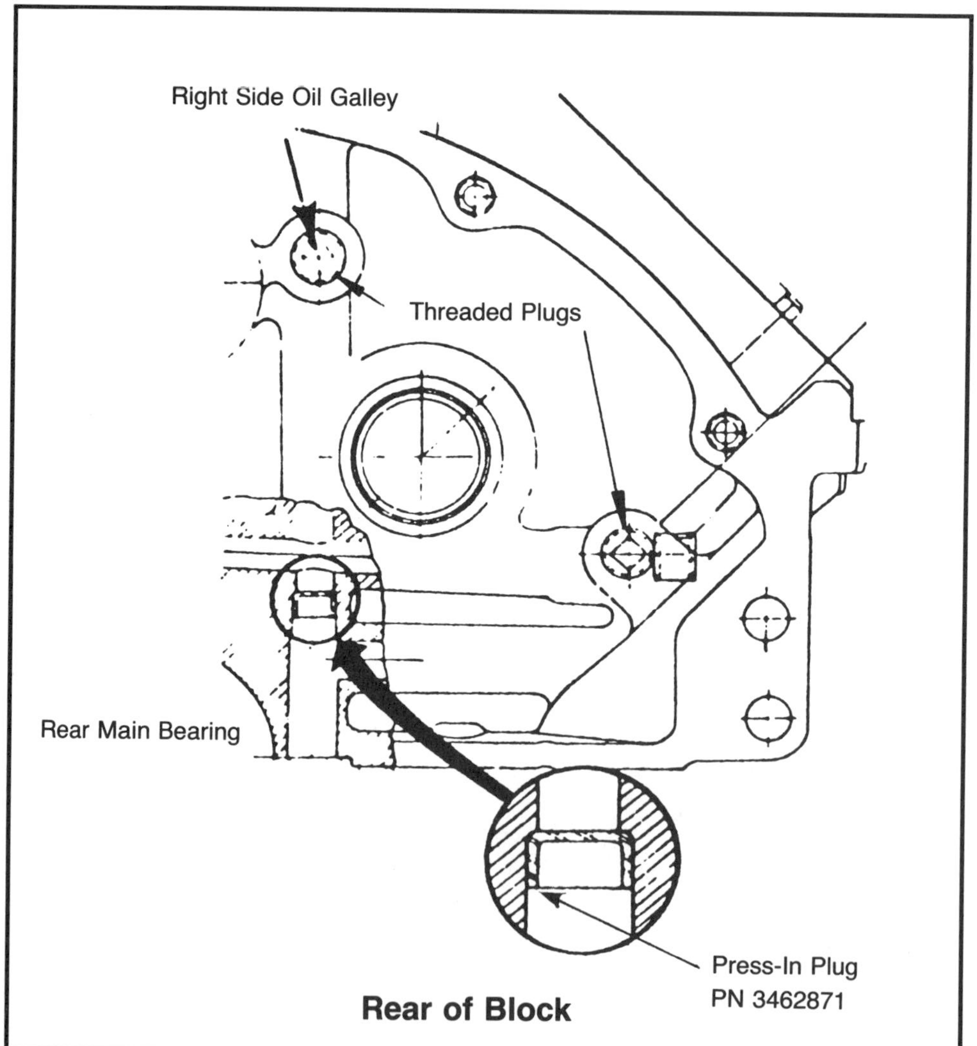

Special press-in plug must be in proper location. There is a machined step in the block that the plug seats against. It must be checked in any new block, or used block without a known history. It should be replaced if the block is stress-relieved. (Courtesy Mopar Performance)

SPECIAL CONSIDERATIONS

The basic prep of the block covers almost everything needed to build a race engine. However, there are a few exceptions that you may or may not want to consider. It depends on how far you want to go and what you're using your engine for. If nothing else, it's a good idea to be familiar with these steps.

Sleeving—This procedure is usually done only when the block is damaged. It may have broken a rod that knocked the bottom out of the bore or it has a badly scratched bore. It could have a pin-hole leak or rust. Whatever the reason, *sleeving* the one cylinder should fix it. If done properly, the re-

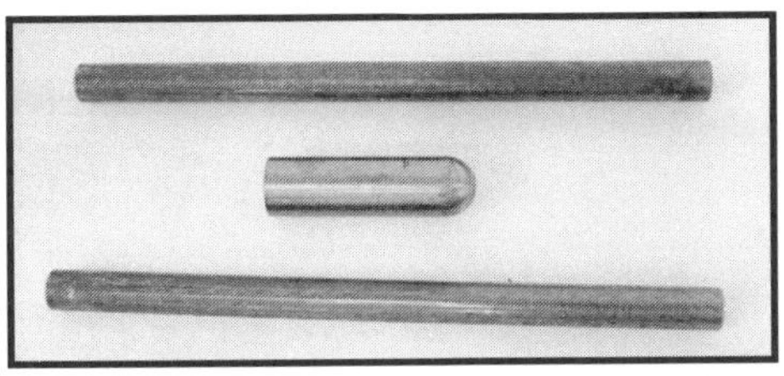

The peen and tube kit for blocking oil to the tappets is similar to bushing the tappet holes, but much easier to do and less expensive. Both tubes go in the right side galley—one from the front and one from the rear. The peen is then driven down through each right side tappet to "peen" the tube in so all tappets fit.

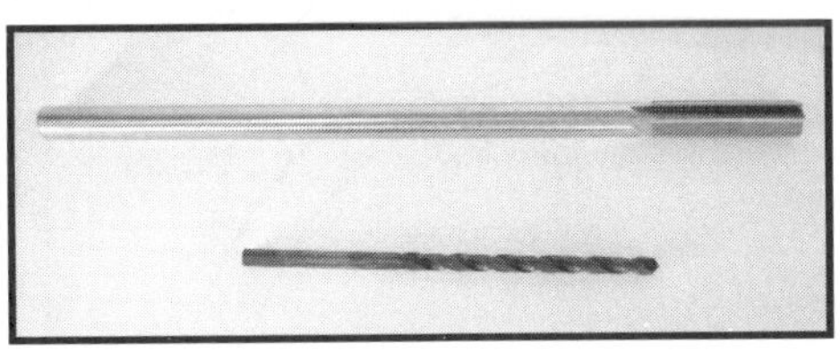

A drill and ream package is required with the peen and tube kit unless you have a set already. The ream is extra long to reach for either end of the block to the middle to enlarge the oil galley (main oil feed) for the tube installation. The drill is extra long to drill the oil passages up from the mains (all 5) into the tubes once they are in place. Otherwise the crank will not receive any oil.

Four-bolt mains can only be added to the 340 T/A blocks or the Race blocks. The four-bolt vertical cap is the most popular style.

STRESS RELIEVING

There are two types of blocks that you'll build your engine with—either a *seasoned* block (one that has been used), or a *green* block, which is brand-new. Seasoned blocks have already been relieved of the stress caused when the block is cast, or roughly machined. A brand-new engine will adjust itself with each cycle (power run or drag strip pass etc.) to relieve itself from these stresses. However, during this process the block "moves." These minute block movements cause the cylinder bore to shift and it will no longer be straight or round. This effect on the cylinder bore can cause a loss of power. If building a competition engine for racing only, consider stress-relieving the block if it is new or green. The only brand-new A engine block casting that is stress-relieved by the factory is the 340 Race block. Chrysler stress-relieved very small batches of production 340 blocks in the early 70's, but they were only used in racing or sold through their Direct Connection (now called Mopar Performance) program. The problem is they can't be identified from any other 340 block. As a general rule, the block will stress-relieve itself after a period of time.

However, I do not recommend stress-relieving a *used* block. It's not worth it. In many cases, a used block is already stress-relieved! Generally fifty to hundred drag cycles, 5 to 10 short-track races or 5,000 to 10,000 street miles will stress-relieve a block. But keep in mind that thermal cycles (hot-cold-hot, etc.) are more important than mileage. Race cars don't require as much time to relieve a block as a street engine because they operate at higher outputs and higher loads which makes the forces greater. Drag cycles are the best because of the high loading combined with the thermal cycles. Many racers today are duplicating these cycles and load on dynos. Engines that are stress-relieved by any of these "in-use" methods are called "seasoned" blocks. They make very good blocks from which to build a high-output race engine.

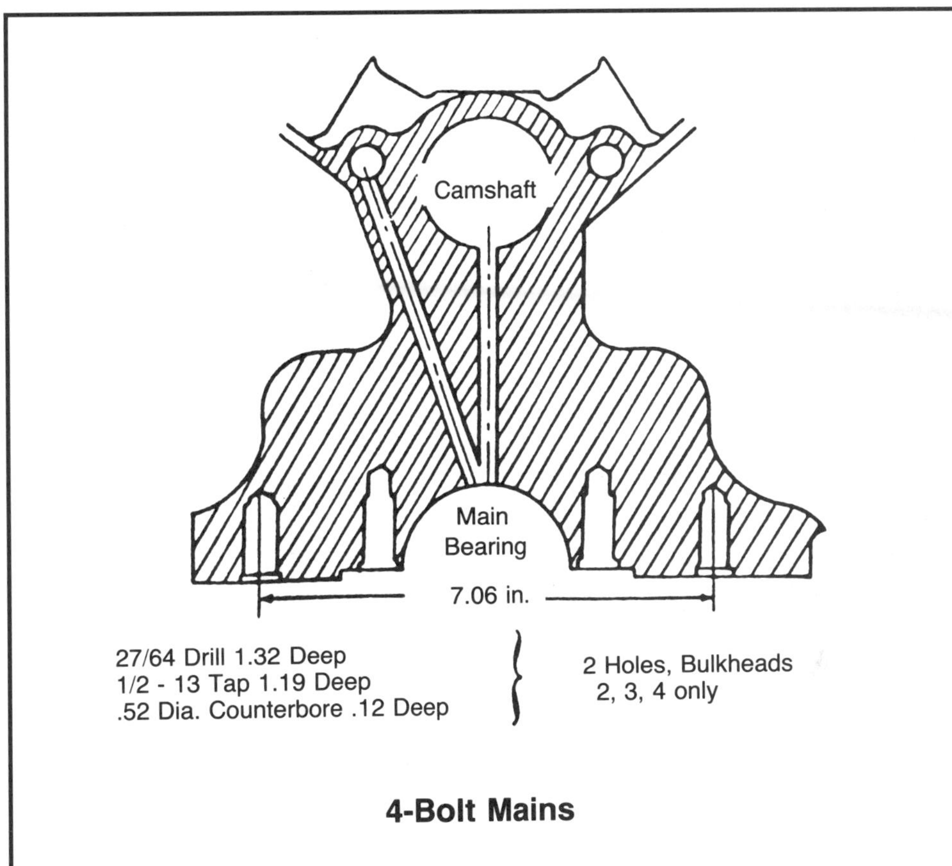

Four-bolt main installation. This can be done only to a Mopar Race block. There are two different methods of adding the second set of bolts. The one shown is the most popular and the preferred method. The second method angles the outer bolts. (Courtesy Mopar Performance)

built engine should be as good as it was previously.

However, sleeving becomes tricky if you want to install eight sleeves at the same time. This is usually done when the engine is worn-out or an odd bore size is desired. Once the engine gets to 0.060-in. over or more, and the piston clearance is on the high side, the engine is effectively worn out. Sleeving back to a stock or 0.030-in. overbore is a way of saving it.

For example, if you want to build a 305 cubic-inch engine with a 3.31 stroke from a 318 or 340 block, you need a 3.83-in. bore. None of the engines in the production-based A series use this bore size. Therefore, the only way to get the 340 or 318 engine to this size is to sleeve all 8 cylinders. While most engine shops can sleeve one cylinder without any noticeable power loss, only a few professional shops can sleeve all eight and have it still perform as well. What seems to happen is that the machining operation induces stress back into the block like the original casting process did. It then needs to "relieve" itself again. This often requires two rebuilds to get back up to power. There are shops that get it perfect the first time out. There's no substitute for quality machine work and an experienced professional shop to perform this process.

Four-Bolt Mains—For most racing applications, 2-bolt main cap blocks are adequate. The use of 4-bolt mains is generally overrated and not necessary in most cases. As a general guideline, consider 4-bolt mains only if building an engine in the 2.0 hp per cubic-inch range, that will run for long periods of time (such as in endurance racing) over 8,000 rpm. Four-bolt mains should be used only on the 340 T/A and the 340 Race "X" blocks. Don't try to add them to a production

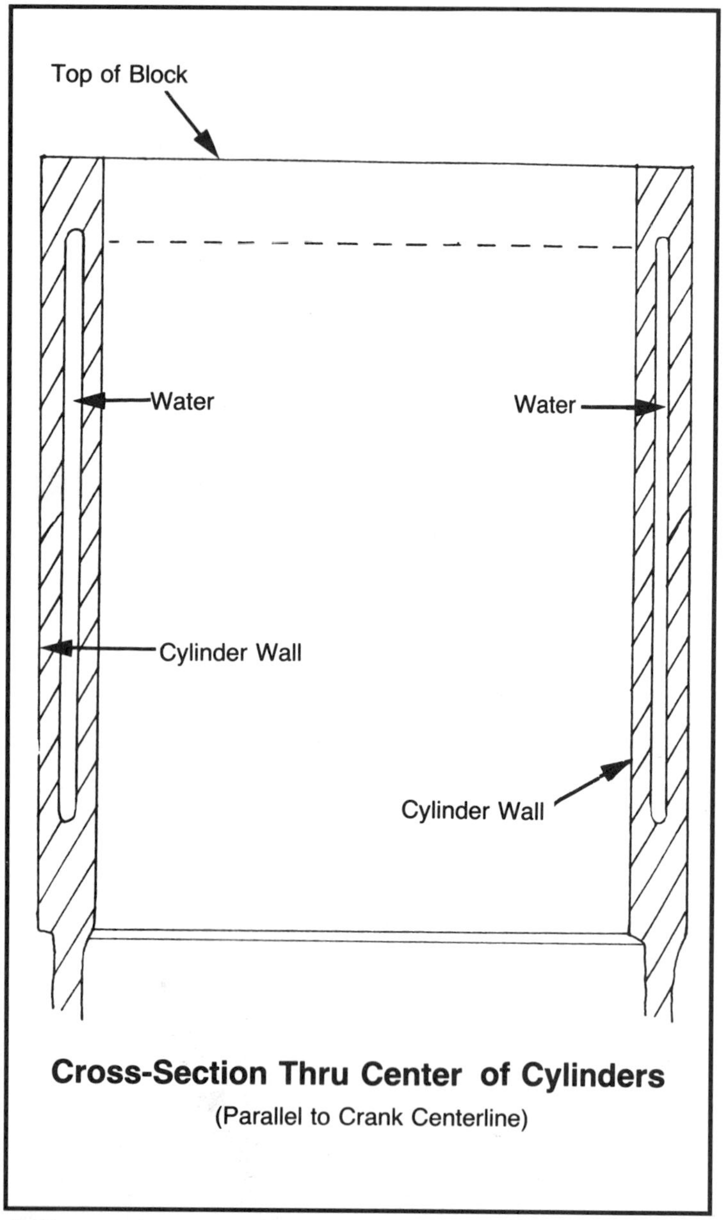

Unlike many large small-blocks the A engine has water between cylinders (i.e. the bores are NOT siamesed). This gives better cooling 360 deg. around cylinder bore which boosts power output and durability.

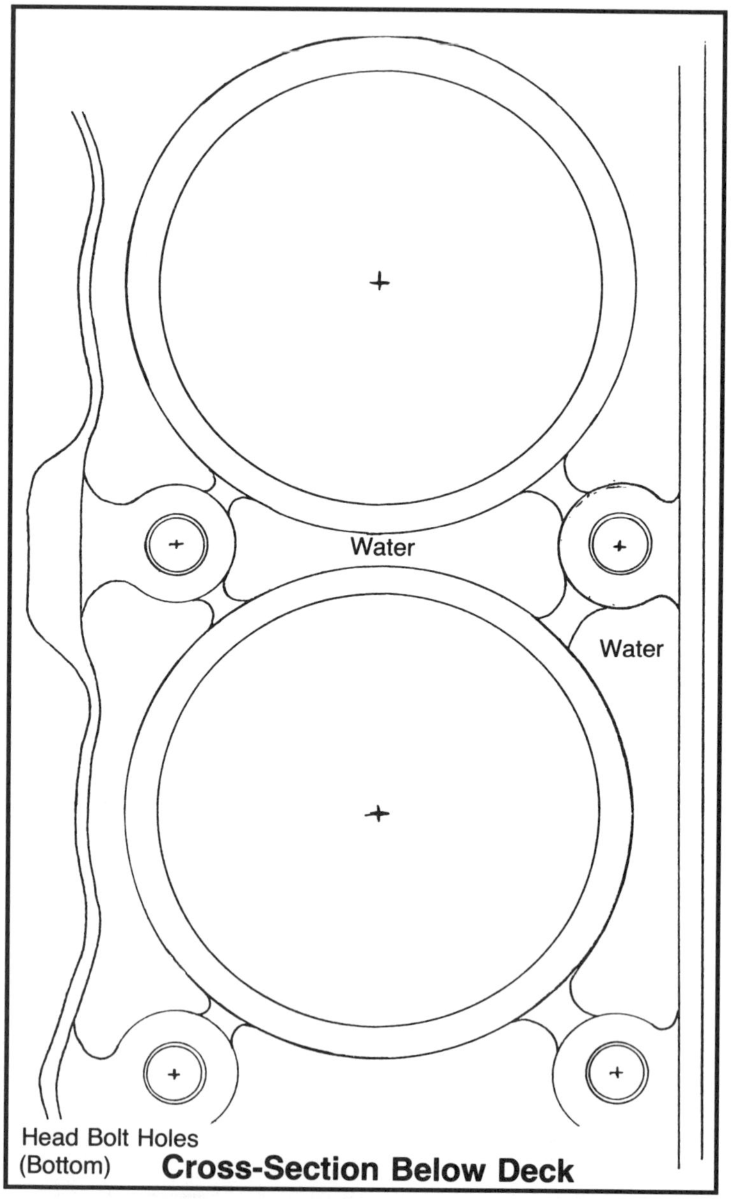

Having cooling all-around the cylinder bore is one of the secrets to making both power and durability from the engine. Another hidden secret are the posts that tie the head bolt holes to the cylinders. This adds stiffness and rigidity with minimal affect on cooling.

block. There are kits to do this but the whole procedure is expensive and the block really isn't any stronger when done. If you need 4-bolt mains then you also need thicker bores, thicker deck, thicker bulkheads—in other words, you need a Chrysler Race "X" block, which comes with 4-bolt mains anyway.

O-Ringing—The head gasket may not be considered as *part* of the actual block, but it is critically related to it. I'll discuss head gaskets in depth in Chapter 5 on *Cylinder Heads*. However, very high-output racing A engines using very high-compression ratios (over 12.0-to-1) can be difficult to seal. One of the solutions is to *O-Ring* the block. To O-Ring a block, a very small trough (0.032-in. wide and 0.017-in. deep) is machined around the cylinder bore at 4.325 in. minimum diameter. A small, stainless-steel wire (0.032-in. dia.) is added that is slightly taller (slightly less than 0.016 in.) than the trough is deep, so it sticks above the deck surface. With the correct head gasket, these O-Rings work very well at high-compression ratios. Actually, it is more common to O-Ring the cylinder head. If you do that, then don't O-Ring both.

SONIC TESTING

It's always been very difficult to determine in advance which engine is going to be the "good" one. The differences may be small, but it may be the difference between winning and losing or setting "low time" at the meet. You could build four engines just alike, run them in the same car and the one that went the fastest would be the one you raced. You could call this the trial-and-error approach. Quite often, you won't have time to do this. *Sonic testing* is one method to determine which engine out of a group is going to be slightly better than the other. Sound waves are used to determine the thickness of the cylinder bores, round ones or ones that have the thick side in the major thrust direction. If you're building an engine to run in the 1.8 to 2.0 hp/cubic-inch area, then the block should be sonic-tested. However, even if you only have one block and are only building one engine, sonic testing can still be very useful. It will tell you in advance what to expect from the engine in general terms. It can also indicate whether to overbore or not and how much overbore would be safe. Blocks that have a lot of "core-shift" (bore thicker on one side than the other), should not be overbored as much as a block that has round bores.

CHRYSLER PARTS LIST

BLOCKS

318 Bare Block	P4179728
360 Bare Block	P4201315
340 Race Block	P3870873
340 Race Block with 4-Bolt Caps on #2, 3, 4	P4120637

NOTE: Only Race blocks have provisions for 4-bolt mains.

PLUGS

Cylinder Block Hardware Plugs (freeze plugs, dowels, oil galley plugs, cam bearings, main bearing bolts)	P4007099
Tappet Bore Drill & Ream Package	P4120602
Tappet Bore Tube & Peen Tool Package	P4120603

NOTE: Race engine only

MAIN CAPS

4-Bolt Main Caps	P4120206
2-bolt style bolts	P4120096
4-bolt style bolts	P4120207

Two dowel pins at the back of the block align it with the bellhousing (trans. case on an automatic). If they don't align properly, clutch/torque converters, etc. fail prematurely along with other poor performance characteristics and failures. To align properly, offset dowels are required.

RIGHT—Another style of four-bolt cap—this one by Diamond Racing Engines—has angled outer bolts. Note the studs for main cap screws. Studs can be added to any A engine block.

2 CRANKSHAFT

TYPES OF CRANKSHAFTS

All A engines feature five main-bearing, 90-deg. crankshafts. The crankshaft generally comes with the engine. For example a 340 will use a 340 crank. There are two basic production cranks—3.31 and 3.58 stroke. The 3.58 is only used in the 360.

Basically,there are three types of crankshafts—*billet*, *forged* and *cast*. Only the cast and forged are used in production A engines.

Billet Cranks—Billet cranks are specially made aftermarket pieces that are mostly used for racing. However, billet cranks are also used by most automotive manufacturers to explore new stroke considerations and new engine output levels.

The billet crank is whittled out of a large, round, bar or *billet* of solid steel. A crankshaft lathe is used to do the machining. Virtually any stroke can be machined into a billet crank and for any engine. Billet cranks are *very* strong and very expensive. They also do not generally come from most manufacturers completely finished and are quite heavy. Supplemental machining by a machine shop is often necessary to lighten them.

Forged Cranks—The forged crank is probably the most common type of crank used in racing. It was the only type used in the A engine family up to the early 70's. The forged crank is very strong and adapts to racing well. These cranks are often chosen by many racers, especially amateurs, because they are strong, less expensive than the billet and readily available. The only drawback is the limited selection of strokes currently available. Only the 3.31-in. and the 3.46-in. strokes are easy to get. There was once a 2.96-in. stroke, forged crank that was sold by Chrysler, but it was not used in production and it is not easy to find today. Forged cranks were phased out of production A engines by the end of 1974 and replaced by cast cranks. However, they are readily available from Chrysler's Mopar Performance division in the 3.31 and 3.46 stroke lengths.

Cast Cranks—The A-engine cast crank was introduced with the '71 360 engine. The 360 has never used anything except a cast crank in production. By 1975, all A engines used a cast crank. The automatic transmission 340 engines got one in '72; however the manual transmission cars still used a forged crank until the end of '73. Cast cranks made their first appearance in the 318. By the end of 1974 all production forged-crank A engines were gone. This hasn't been a problem for the A engine, because these production cast cranks have proven to be very strong, and are suitable for most racing applications. They can withstand the stress of Super Stock and bracket racing.

The cast crank is the least expensive crank, and is also the lightest. It is effective in dampening out torsional vibrations common in cranks used at high speed or high loads, but this is hard to observe. The biggest drawback is that 340 and 360 cast crank engines need to be externally balanced (see sidebar p. 18).

Bolt Patterns—The most identifiable feature of the crank is how many bolts hold the flywheel or flexplate on. The most common number is six. The pattern, diameter and size of the 6-bolt crank flanges on the A engine are all the same. The next possibility is an 8-bolt pattern. These come only in the Chrysler race A engine cranks (and 426 Hemis). Eight-bolt cranks obviously require a different flywheel or flexplate than the production ones.

This is the Mopar Performance 8-bolt crank flange, forged crank that has been modified by Dick Landy and DLI. A full .300-in. of material was shaved from the counterweights to reduce windage. It was destroked to 3.23 in. to produce 341 cubic-inches when mated to a 4.10-in. bore. (Photo courtesy Hot Rod Magazine)

The 8-bolt crank is used to gain more *clamping load*, which is needed to run higher rpm or high outputs safely. The 8-bolt cranks are recommended for drag racing engines running over 8000 rpm and 7500 rpm-plus oval track engines. They should also be used on any engine where output reaches or exceeds 1.8 hp per cubic-inch. Obviously, a billet crank can be made with either a 6- or an 8-bolt crank flange, but because most billet cranks are used for racing, then an 8-bolt is recommended. Most other small-block engines only use a 6-bolt crank flange. Only the 413-style H.D. Chrysler truck engines and 426 Hemis use the 8-bolt crank in production, but those are B engines.

Many 4- and 6-cylinder engines use dowel pins in the crank flange, in addition to the bolts. However, the A engine V-8 has never shown any need for more than 8 flywheel bolts.

While inspecting the crank flange area, be sure to check the pilot hole in the center of the flange. If you are using a flywheel/manual transmission, then a pilot hole and pilot bushing must be installed. If the hole is machined into the crank it's an easy enough job to put the pilot bushing in at a later time. Over the last fifteen years, the use of manual transmissions has decreased. This means that the pilot bushing holes may not be machined or may only be partially machined. This is the time to check and fix as required. If you are using an automatic transmission, then you have no need for the pilot hole or the pilot bushing.

BASIC CRANKSHAFT PREPARATION

In spite of the crankshaft's apparent simplicity, it has more clearances and measurements to check than any other single part, and can require a great deal of machining. Therefore, it is best to enlist the aid of an excellent machine shop, preferably one that specializes in crankshaft machining. After all of the machining is done, the crank needs to be checked before it is installed in the block. Much of this checking is often overlooked. This can be disastrous or cause a loss of performance.

VISUAL INSPECTION

The first step after a thorough cleaning is to examine the crank visually. All the oiling passages in the crank from the main journals to the rod journals should be checked to make sure that they go all the way through. This can be done with a long-handled, small-diameter bottle brush. These passages should be cleaned thoroughly, both before machining and prior to installation. The oil hole on the journal's bearing surface should be radiused lightly with 400- or 600-grit sandpaper.

CRANKSHAFT SPECS

The general specifications for A engine crankshafts are as follows:

CRANKSPIN DIA.—2.125 in.
MAIN DIA. 273/318/340—2.50 in.
MAIN DIA. 360—2.81 in.

The common strokes are 2.96 in., 3.31 in., 3.46 in. and 3.58 in. The 3.58 in. only comes in a cast crank. The 2.96 in. was the stroke of the 1970 305 Trans-Am engine based on the 340.

These are billet logs of standard aircraft-quality steel, and the ultimate 4340, 300M steel, the finest material available for making crankshafts. Crank grinders, such as Henry Velasco, start with this blank and begin the long process of machining it down. (Photo courtesy Super Street Machine Magazine)

Crank Journal Radius—The next feature of a crankshaft that can be visually inspected is the *crank journal radius*. All production crankshafts have an undercut radius on the rod and main journals. This weakens the crankshaft somewhat. For high-output racing use, I recommend that you use a *full-radius crank*, which increases crankshaft strength without adding weight. Full-radius racing cranks are available from Chrysler and can be included as a feature on a billet crank, or a crank machinist can perform the step. For A engine race cranks, the 8-bolt crank flange and the full-radius journals occur together. The full-radius cranks require the rod and main bearing shells to be radiused on the outside next to the crank cheeks where the full-radius is located. This operation can be done by hand using a bearing knife, but it's best to let a machine shop do it because it's easy to go "too far" with the radius and inadvertently "nick" the bearing surface.

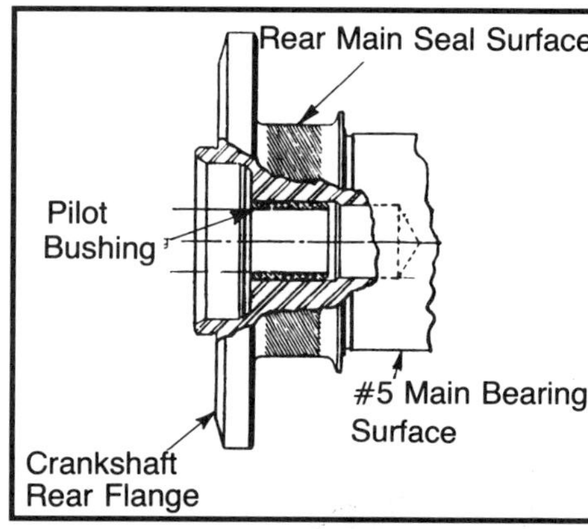

Cross-section of the rear crank flange. The pilot bushing is only installed on manual transmissions. The thin lines machined into the rear main seal surface were very important with the old-style rope-seal but less-critical with today's neoprene seal.

This is a cast crank for a 360, typical of all A engine cranks. Although this one has 6 flywheel bolts, race 8-bolt designs are also available.

BASIC CRANK MACHINING

The following machining steps are listed to give you an understanding as to why and how they are done, because unless you are an experienced machinist with access to the sophisticated equipment required to perform these procedures, you'll be putting the fate of your crank into the hands of a good machine shop. Depending on the application, some or all of these steps may apply. If the crank machinist is experienced, he'll be able to recommend which steps are necessary.

Crank Grinding—If the crank is used, then the machinist will most likely grind both the rod and main journals undersized. Usually, 0.010 in. is taken off each, hence the name a "10/10 crank." If this is done, then oversized bearings (0.010 in.) are needed to compensate.

Cross-Drilling—Cross-drilling is the process of drilling an extra oil path to the rod journal from the other side of the main journal. Some people think this process is an unnecessary step, especially for the A engine. This is because the oiling system of the A engine has proven sufficient to lubricate the crank and rod bearings properly for drag and NASCAR racing engines. I recommend that you consider cross-drilling as an option, not as a necessary step. For street and bracket racing go with standard crank drilling.

Grooving—Grooving the crankshaft means a radial groove is machined around the crank journal. This is a definite no-no! Grooving weakens the crank and causes premature failure. Use grooved main bearings to get more oil around the crank journal, but don't groove the crank itself! You'll be adding stress to one of the most highly stressed areas in the crank, which is a quick way to break expensive parts.

Knife-Edging & Radiusing—If lightweight pistons and rods are used, then the counterweights on the crank will be too heavy to balance the assembly properly. Therefore, weight will have to be removed from the crank. For years, the standard practice was to drill a hole in the counterweight to solve the problem. However, as the

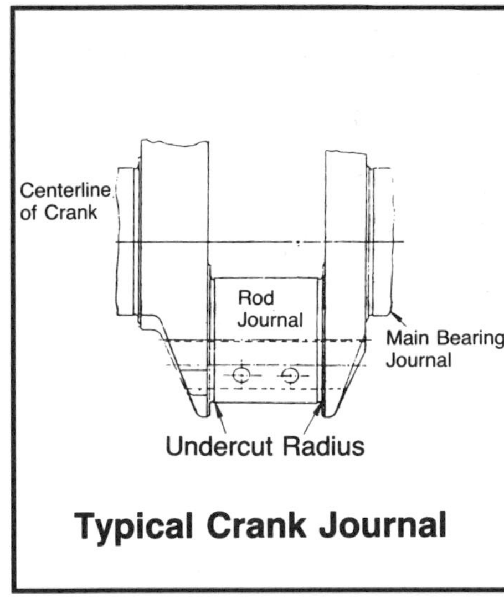

All production crankshafts use an undercut radius on the rod and main journals. Only the race cranks (Chrysler/Mopar Performance) have full-radius journals. Obviously with full-radius cranks, edge radius clearance has to be given to the bearings next to these full-radius corners.

The crank should be measured carefully with a micrometer. Each rod and main journal should be checked.

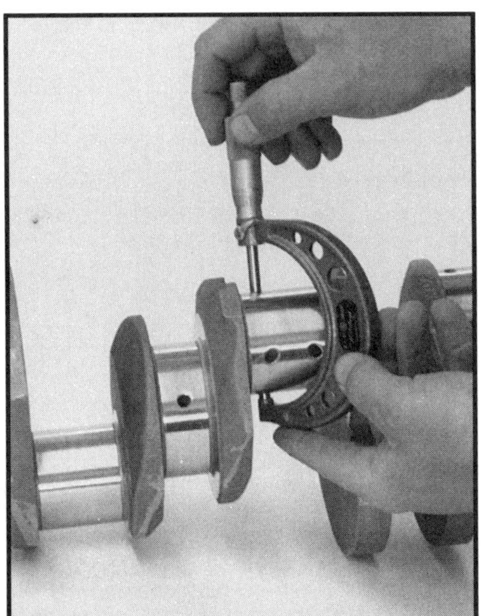

A 2-3 in. micrometer is used for the crank measurements. Each should be read carefully and written down on an engine build-up sheet.

Extreme care must be used in radiusing the crank's oil holes. One slip and you could "nick" the crank journal. Masking the journal with "racer-tape" can help. The small radius helps distribute oil flow and pressure around the journal. However, a large radius decreases cross-sectional area that's available to carry load and can weaken the crank. Note small radius on journal to the left.

pistons got lighter, the holes got bigger. For removing a small amount of weight it was a good solution. For large amounts of weight *radiusing* or *knife-edging* may be better solutions, although these procedures are somewhat expensive. Radiusing is the machining of the counterweights to decrease their radius—how far out from the centerline of the crank they extend. Knife-edging removes material from the front and rear area of the counterweight in a tapering fashion to create a "knife-edge." The knife-edge does not change the initial radius of the counterweight. Knife-edging is the more popular procedure and also the more expensive of the two. It should be done only by a professional crank shop that specializes in this procedure.

Balancing—One of the most often overlooked crank machining operations is *balancing*. Production engines are balanced at the factory. However, because the crank has been machined and the pistons and rods are new (and usually lighter), then the factory balance is obviously off and the whole bottom end assembly will have to be rebalanced.

For any racing, high-output applications, the engine should be balanced as an entire assembly. This means fully prepped crank, rods, bearings, pistons, pins and vibration damper should be balanced at the same time. Once balanced internally, no external balance should be added or used on the flywheel or torque converter with some exceptions (see sidebar on *External Balancing*.). The machinist will balance all eight pistons to the lightest piston/pin assembly, and the lightest rod assembly. Since he can only remove metal, the heaviest pieces are machined to weigh the same as the lightest. Once balanced, the weight of one piston/pin/rod assembly is used as

EXTERNAL BALANCING

External balancing seems to be a mystery to some racers. It is used because of a cast crank or a heavy piston and rod assembly or a combination of both. External balancing doesn't necessarily make an engine faster or slower. By the same token, neither will balancing an engine internally. However, I still recommend that a race engine be internally balanced because it is more convenient.

With a manual transmission, a race engine is likely to use an aluminum flywheel. Aluminum flywheels are very difficult to use with external balance because the drilled holes used to create the external balance assumes steel or cast iron is being removed. Because the aluminum is so much lighter, these drilled holes would have to be enormous! With an automatic transmission, the Chrysler weight package can be added to any torque converter. The problem here is a race car rarely uses one converter forever. Racers are always looking to go faster and converters are commonly changed to gain this result. With external balancing, you must add weight. It's not really expensive but it can be somewhat of a pain! External balancing also makes it harder to borrow converters from other racers to test or to use in emergencies because they would have to use the same external balance as yours, which isn't likely to happen.

With this basic knowledge of balancing, it's easier to explain why the 318 cast crank doesn't use external balance while the 340 and 360 cast cranks do. The easiest comparison is between the 318 and 340, which basically share the same crank. The size of the counterweights were designed for the production piston and rod assembly's total weight in forged steel. Cast iron is lighter than forged steel. Therefore the counterweight of the cast-iron crank is not as heavy as the 340 piston and rod assembly. Malloy metal could be added but this is not cost-effective for factory production. On the other hand, external balancing is. Comparing this to the cast crank 318, one would assume that this engine would also need external balance since it has the same counterweight as the 340. But the 318 does not because the 318 rod is lighter than the 340 rod and the 318 piston is lighter because of its smaller bore. The result is that the lighter 318 piston and rod assembly balances off the lighter counterweight in the cast-iron crank, hence internal balancing.

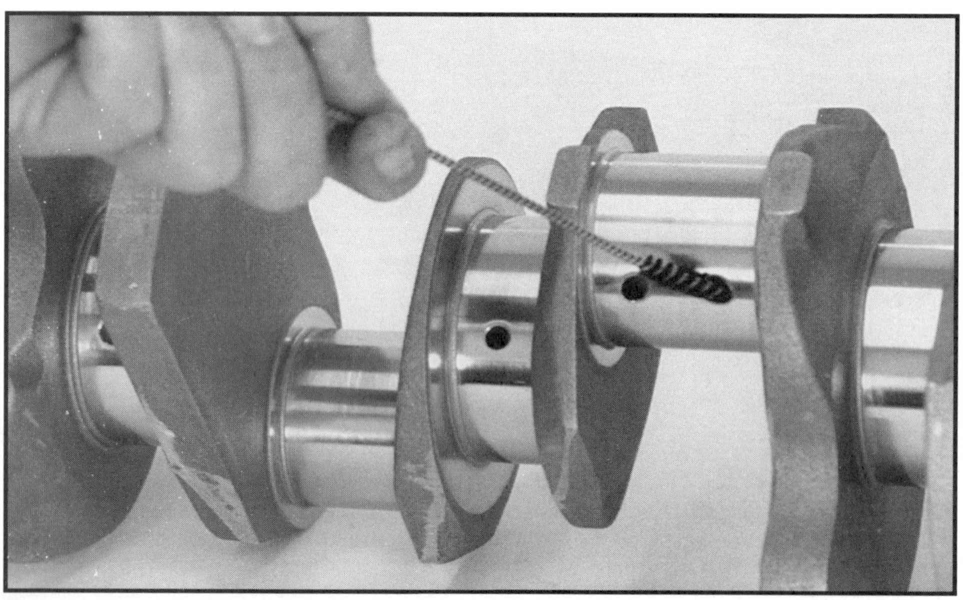

After crank has been deburred and radiused, it should be cleaned. Pass bottle brush through each oiling hole, all the way through to the next journal.

Cross-drilling involves drilling a hole through the main journals so bearings will have 360 deg. of oiling. The drilling needs to be precise, and should only be done by an expert. Consider it as an option. Fully grooved main bearings are a more cost-effective solution. (Photo courtesy Super Street Machine Magazine)

a *bob weight*. This is used as the rotating weight to balance against the crankshaft's counterweight. If lightweight pistons and/or rods are used, then weight is removed from the counterweight on the crank (see *Knife-edging & Radiusing*). A billet crank is typically lightened in the counterweight area also. A cast crank, (except most 318's) will have to have weight *added* to balance internally. This is done by adding *Malloy metal* to the counterweights.

For racing engines, it's best to balance parts, especially if swapping pieces. Note the two large holes drilled in the bottom of the counterweight. They were part of the production balance done at the factory.

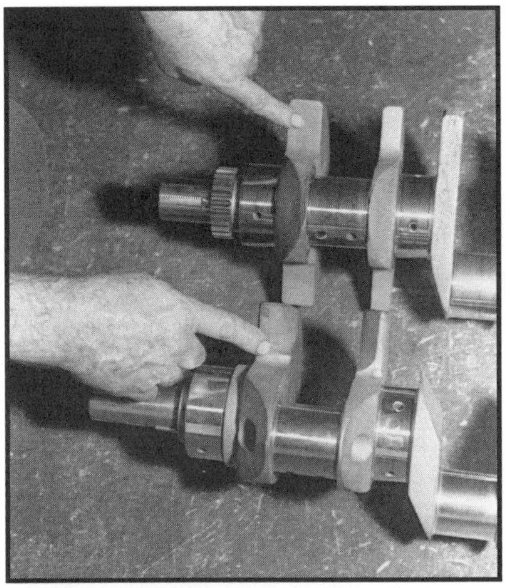

Crank on top is not machined, while the bottom one has been knife-edged. Knife-edging reduces windage as well as weight. (Photo courtesy Super Street Machine Magazine)

This 360 cast crank was externally balanced at the factory. Externally balanced torque converters or flywheels are not desired on race engines or dyno engines. To internally balance the 360 (or 340 cast crank to a lesser extent) you have to add Malloy metal. It looks like steel but is heavier than lead. Holes are drilled parallel to crank centerline and the Malloy slugs pressed-in. Once sized and located, they should be spot-welded in place.

Here's a comparison between the ultimate racing crank (left) with center weights removed, knife-edged, drill-lightened and cross-drilled and an unmachined one. (Photo courtesy Super Street Machine Magazine)

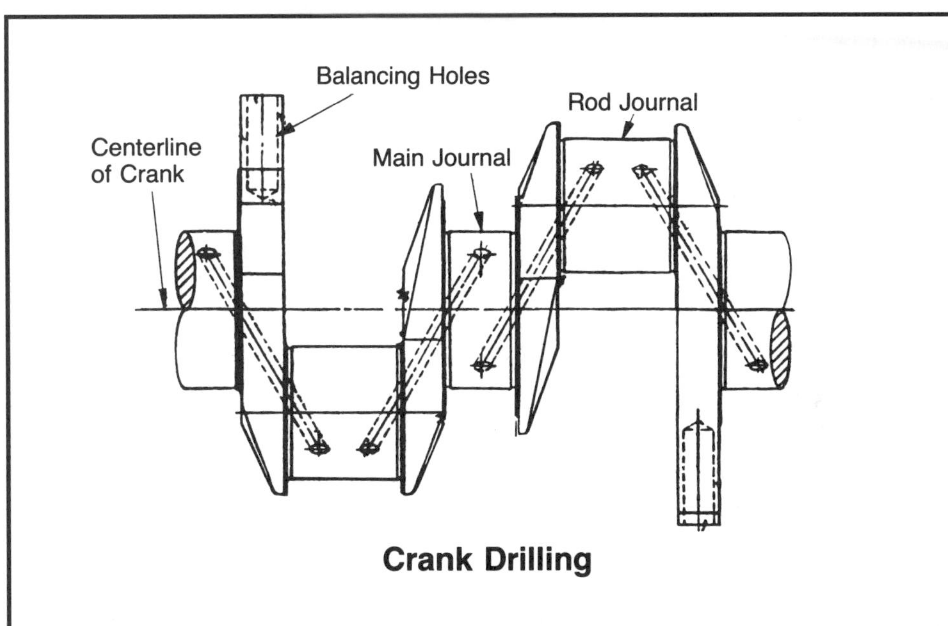

Crank is drilled to distribute oil pressure. Oil and oil pressure are fed to each main journal directly. The rods are oiled from the mains by passages drilled in the crank from the center of the main to the center of the rod bearing journal.

Polishing—The last operation performed on the crank is polishing. It should be done by hand using 400-grit sandpaper first, followed by 600-grit sandpaper. Be careful not to remove any metal. Polishing should remove less than 0.0005-in. It's more important to polish a cast crank than it is a forged one. It's also more important to polish a new one than a used one, assuming that the old one was okay when removed.

SPECIAL CONSIDERATIONS

There are many aspects of a crankshaft that don't fit into the topics I've covered. However, they are just as important and must be considered when selecting and preparing a crankshaft.

VIBRATION DAMPER

The vibration damper presses onto the front of the crankshaft. It also serves as the "zero" for the timing mark. There are two different 6-bolt patterns used to bolt the front pulley on the damper. The new ones are symmetrical and equally spaced, while the older ones have one hole offset. The cast cranks also use an offset-weighted damper except for the 318. They are all the same overall diameter although some are "thicker" than others. This means that the same timing tape can be used to degree all the dampers, which is the easiest method to do so. There are also specially degreed dampers.

With any production-style damper, you should mark the rim of the outer balancer relative to the inner hub. This makes it possible to check whether or not the damper has slipped and thrown off the timing mark. Check the damper against TDC of the piston, then mark it for reference.

As of July 1986, NHRA no longer allows cast-iron dampers in Super Stock or Super Gas. All must be made of steel or aluminum. In my opinion, the Fluidampr® is the best unit. Aluminum dampers have been known to decrease engine durability. However, just be sure to use a damper, whichever is legal for your class of racing. The engine's chances for survival are reduced if a damper is not used. Generally, oval track engines should use larger

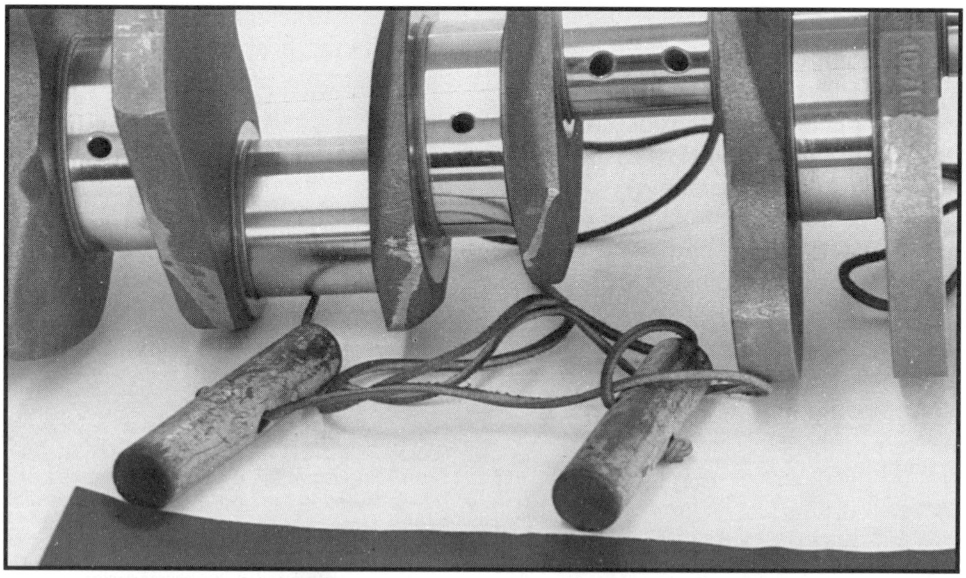

A strip of 600 sandpaper, a long rawhide thong and two wooden handles can be used to polish the crank by hand. There are polishing machines that are used by some shops, but this procedure saves some money and the results are the same. Be careful not to remove any metal.

One full turn plus another of the rawhide around the journal holds the strip of sandpaper tight. Grasp the handles to pull the thong back and forth, spinning the sandpaper around the journal.

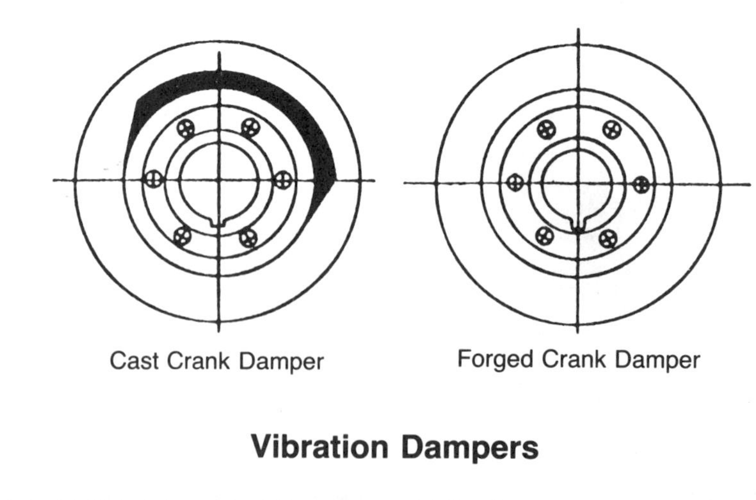

Vibration dampers are pressed onto the front of the crank. The cast crank 340 and 360 engines use a special offset damper in that a large flange is added to part of the damper. The forged damper is symmetrical. They should not be switched. (Courtesy Mopar Performance)

dampers than drag race engines, because those engines often run 500 miles at wide-open throttle (WOT). Again, I recommend the Fluidampr®.

LONG STROKES

In V-8 engines, absolute stroke length is not as important as the *bore-to-stroke ratio*. Modern V-8s are typically over-square. This means that the bore is larger than the stroke. The definition of where a short stroke ends and the long stroke begins tends to vary, but the dividing line is in the area of a 0.80 to 0.85 bore-to-stroke ratio.

The long stroke engines make more torque. The 360 engine with a 0.89 ratio is considered to be a long stroke A engine. These high-torque engines make the best street engines and are also great for bracket racing. When built right, they can go very fast, even when mildly tuned. Large-bore and short-stroke engines make the most power per cubic-inch, but this makes it more of a race engine than a street engine. The 340-based, 305 cubic-inch Trans-Am engine had a bore-to-stroke ratio of 0.73. This is one of the lowest ratio (most over-square) engines based on a production model that still has parts available from the OEM.

REAR MAIN SEAL

There are two rear main seals that have been used in the A engines since the beginning: rope and rubber. The rope seal is the old one: cheap and easy to install. The rubber or *neoprene* seal is currently popular and is the better seal to use for racing. It has less turn-over torque than the older rope seal. The rubber seal must be installed carefully so that it doesn't go in backwards. The scraper that rides on the crank seal journal must be pointing inward so that it scrapes the oil off the journal and dumps it back into the engine. The rubber seal is also more likely to spin in its housing than the rope seal. This can be hindered by putting a small dab of RTV behind the rubber seal as it's inserted into the housing.

BEARINGS

The proper bearing to use in an A engine is determined by what the engine is going to be used for. *Babbit bearings* are used in drag engines and dyno engines while *tri-metal bearings* are used in oval track racing engines. Aluminum bearings can be used in combination with both and make a good bearing for high-performance street engines.

The oiling system greatly influences the choice of bearings. In many high-mileage, maximum-durability applications, oil leaks out of the end of the crankshaft. It's not usually a problem in race engines because of their limited operational time, but it can be troublesome for high-performance street engines. There is a special #5 rear main bearing available from Chrysler that helps solve this problem.

The A engine has larger bearings than many other engines, especially other small-blocks. This increases durability, especially in NASCAR racing. Obviously, larger bearings last much longer. However, you'll lose this advantage if you grind the crank journals undersize, but you'll be no worse off than other small-blocks.

A high-output, high-rpm drag racing A engine does not need as much bearing area as it has for most applications. Therefore, the crank can be ground undersize. This can extend the life of most A engine cranks greatly if done with care. From a performance standpoint, crank grinding can be an added cost that only returns a limited gain. For a professional racing class, it should be considered, but for an

A vibration damper should always be used on the engine. The dampers on A engines press onto the front of the crank and are held in place by a large bolt. While all the same diameter, dampers can be tricky. Note that the damper on the left has some metal missing from the outer ring. This is for a cast crank engine. The damper on the right is for a forged crank engine.

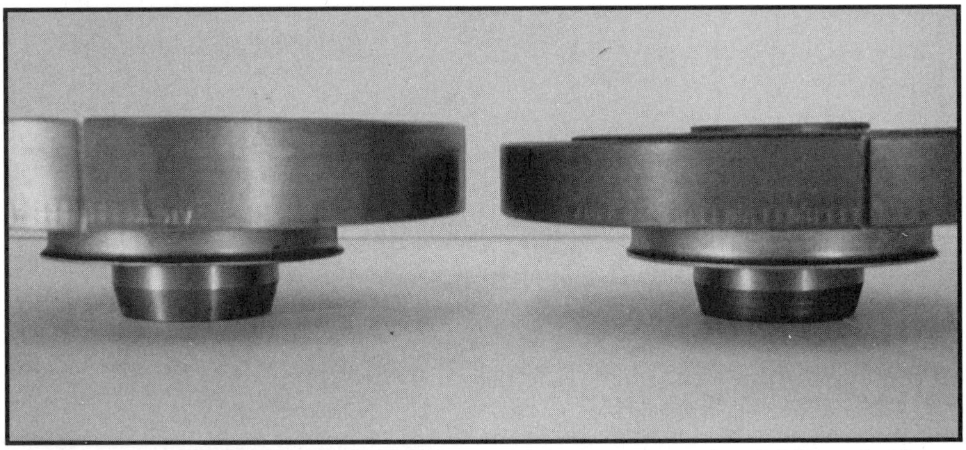

Thickness is another feature of dampers to consider. Note that damper on left is much thicker and could cause pulley alignment problems.

amateur racer, it's not worth it. The reason is that every time you regrind the crank, you get less use out of it, because regrinding weakens the crank. Regrinding is expensive, and eventually the cost of numerous regrinds (plus total engine rebuilds) is not justified.

OFFSET GRINDING

Offset grinding is a special procedure that increases the crank throw without welding and without using billet cranks. Because the A engine rod journals are larger than those of a small-block Chevy, they can be reground to accept the Chevy rods, which are common and relatively inexpensive. But the main advantage to offset grinding is that by removing the majority of the material from one side (the side closest to the crank's centerline), the center of the rod journal is moved away from the crank's centerline, which *increases* its stroke.

Another version of a cast crank damper. Note the added weight to the lower left.

NHRA now requires SEMA-approved dampers in most of their faster catagories-even Super Gas. The "Fluidampr" is SEMA-approved and carried by Mopar Performance. It does its job over a wider range of frequencies than standard rubber and steel ring dampers.

Generally, the amount of offset equals the amount of stroke gained. The result is called a *stroker crank*, which increases the displacement (and torque) of a small-block A engine. However, remember that this is a very critical, very special, crank-grinding procedure that should only be performed by a highly specialized crankshaft machine shop.

There are many other considerations concerning crankshaft preparation that involve checking numerous clearances, such as end play and main bearing clearance. I've listed those procedures in Chapter 7, *Blueprinting*. It is at that time, during the "preassembly," that I'll cover those measurements.

CHRYSLER PARTS LIST

CRANKSHAFTS

Forged 340 Crank (3.306 in.)	P4120504
Forged 355 Crank (3.454 in.)	P4120312
(Both are full radius, 8-bolt flange, race cranks).	

MAIN BEARINGS (#3 Main)

F-77 Tri-Metal—Std.	P3690686
F-77 0.001 U/S	P3690999
Babbit—Std.	P3690656

MAIN BEARING (#1, 2, 4, 5)

F-77 Tri-Metal—Std.	P3690685
F-77 0.001 U/S	P4007016

MAIN BEARING SETS

318/340 Babbit—0.010 U/S	P4286939
'71-'73 360 Alum.—Std.	P4286944
'71-'73 360 Alum.—0.001 U/S	P4286945
'71-'73 360 Alum.—0.010 U/S	P4286947
'74-'80 360 Alum.—Std.	P4286948
'74-'80 360 Alum.—0.001 U/S	P4286949
'74-'80 360 Alum.—0.010 U/S	P4286951
'74-'80 360 Tri-Metal—Std.	P4286956
'74-'80 360 Tri-Metal—0.001 U/S	P4286957
'74-'80 360 Tri-Metal—0.010 U/S	P4286959

ANTI-LEAK REAR MAIN BEARING

273/318/340 Babbit	P4349042
360 Tri-Metal	P4349043
Plastigage Assortment	P4286919

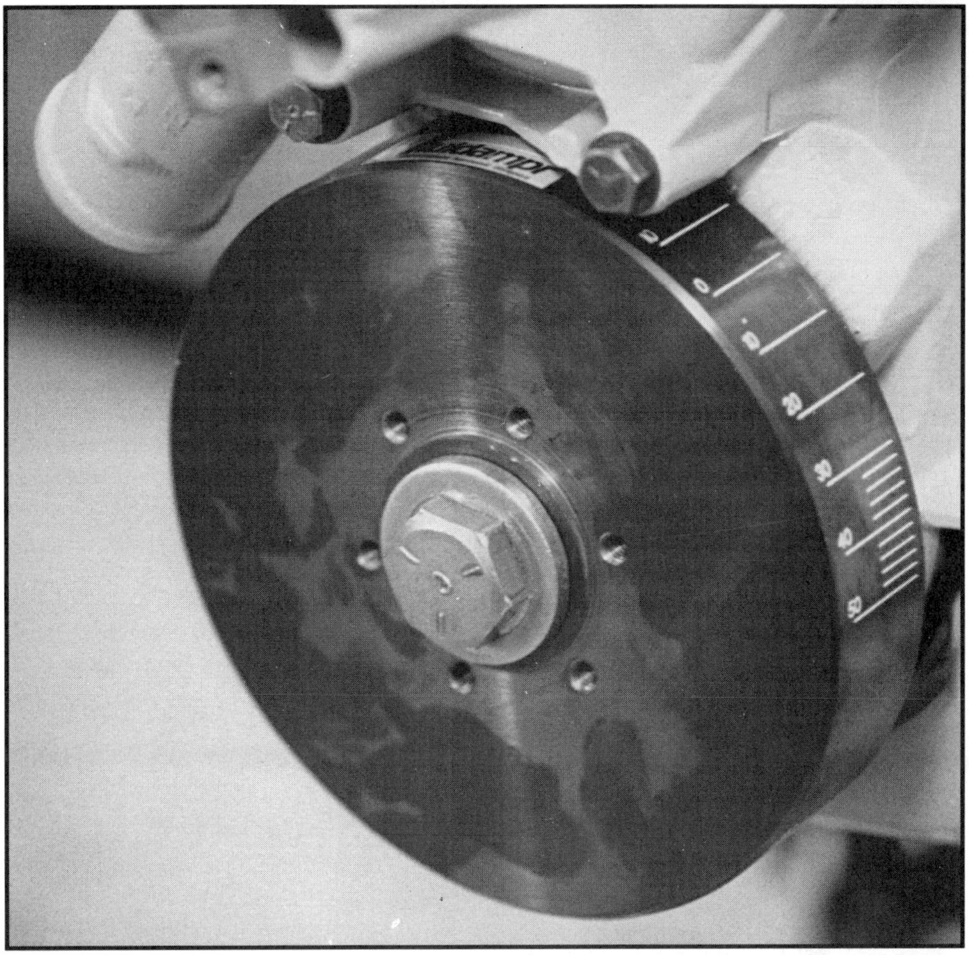

Any time dampers are swapped or replaced, recheck TDC as indicated by the damper and timing tab versus the true TDC on the #1 cylinder.

Henry Velasco, known as "The Crank Master" in drag racing circles, is performing a precision offset grind on a racing crankshaft. (Photo courtesy Super Street Machine Magazine)

3 CONNECTING RODS

Though this chapter follows that of the crankshaft, the selection and preparation of rods and pistons (covered in the next chapter) should occur at the same time as the crank. In fact, a machinist will most likely request that the block, crank, rods and pistons be delivered together, because the machining and prep of all these components directly affect one another.

ROD SELECTION

Unlike the block or crankshaft, there are many different types of connecting rods available for the A engine family. Not only are there several different styles available from Chrysler, but there are also quite a few aftermarket manufacturers. Choosing the best rod for a race engine can be easy or difficult, depending on how much rod theory you get into. Rod selection directly affects the crankshaft, block, piston, pin and rings and indirectly, the cylinder head and intake manifold. Other performance characteristics that need to be considered are intake volume, runner length, cam centerline, duration and carburetion. Then there's the application, which is the most important consideration of all.

TYPES OF RODS

Stock Rods—The stock rod is a fairly simple device. Its function is to transmit the downward forces acting on the piston to the crankshaft. All the A engine's rods are the same length: 6.123 in. center to center. The rod's big ends are very similar and all use the same rod bearing. Also, all A engine production rods are forged.

There is a difference between the forgings of the 273/318 and the 340. The 340 forging is bigger and stronger but also a little heavier. It was also used in the 4-barrel version of the 360 cubic-inch engine. Most other 360's use the same forging as the 318.

The 273, 318 and 340 utilize a full-floating pin design while the 360 and the newer ('74 and up) 318's use a pressed pin. Floating pins have a bronze bushing in the small end of the rod. A pressed pin has an interference fit between the pin and the small-end of the rod: steel-on-steel. Selection of pins depends on the rpm range. For more on piston/pin/rod combinations, turn to page 38 under *Piston Pins*.

Racing Rods—Production rod forgings are not right for all applications. Most race engines reach output levels and rpm plateaus that require something special. Custom rods are expensive and usually used only in racing. There are several styles available; so the question is, which one to use? The general guidelines are: use stock forged steel rods up to 8000 rpm; custom steel rods from 7500 to 8800 rpm; and aluminum rods for any higher rpm.

For racing purposes, consider Chrysler's P3690641 special rod. It utilizes the same basic forging as the 340 but it's made out of 8640 material. This material is much stronger than standard rod material, yet it doesn't make the rod any heavier. It also features premium 3/8-in. bolts and a pressed pin. It is the rod recommended for most racing applications.

For high-output NASCAR endurance racing, the Chrysler P3614514 is a good, heavy-duty forging that is Magnafluxed, shot-peened and polished. It features high-strength 7/16-in. SPS rod bolts and nuts, and is bushed for a floating 0.984-in. diameter pin. This rod was used in the 355 GN engines that won so frequently on the NASCAR circuits in the 70's. It's very strong and very durable; however

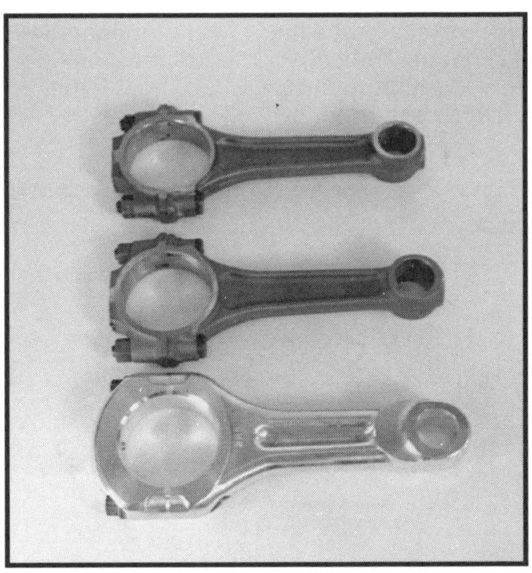

All three rods share the same center-to-center distance of 6.123 in. The top rod is a production one, in the middle is a high-performance version made of special 8640 steel while the bottom one is aluminum.

it is also heavier than a stock rod. By today's standards, it's too heavy for "Pro" racing.

Like pistons, lightweight rods increase performance, although not quite as much. However, lightweight rods are popular in all areas of high-performance except endurance racing. The reason is obvious: lighter weight reduces strength, which eventually reduces durability.

Generally, it is best to purchase lightweight custom rods, because a stock rod can't be lightened to the weight of a custom rod without weakening it. Removing weight from a stock rod is restricted to removing the balance pads from the top and bottom of the rod and radiusing both ends. However, be careful not to decrease the minimum cross-section in those areas.

The standard production rod has a cross-section in the mid-beam area shaped like an **I**. The Carrillo's cross-section in this area is in the shape of an **H**. This makes them look totally different. For oval track use, the Carrillo rods need two 3/32-in. holes drilled into the pin boss at 45 to 50 deg. off of vertical, in the center trough between the two sides of the **H.** Because of the **H** construction and the fact that the holes weren't drilled into the sides of the **H,** the strength of the rod won't be affected. The holes should then be chamfered. To save the trouble of drilling these holes, contact Carrillo or the custom rod manufacturer and have them drill these holes for you.

All race rods have a bronze bushing in the small end to allow for full-floating pins. In used race rods, check the bushing for size. They sometimes wear out and need to be replaced.

Aluminum Rods—Aluminum rods are lighter and have more *column strength* because of their larger, cross-sectioned area. They also have more bulk to damp out certain vibrations. Column strength is needed for blown engines (supercharged or turbocharged). However, aluminum rods generally have *low fatigue resistance*. Fatigue doesn't mean they're not strong. It means they get "tired" after

ROD RATIOS

Basically, rod ratio is the *ratio of the rod's length to the crank's stroke.* There's a lot of disagreement about what the rod ratio should be and how it affects things. Actually, rod ratio theory didn't come into play until the mid-70s. It first came up in drag racing and soon worked its way into other types of racing.

The 273, 318 and 340 all share the same 1.85 rod ratio. The 360 has a unique 1.71 rod ratio. As a very general guideline, use the following ratios:

SHORT GN TRACK—1.60 to 1.70
LONG GN TRACK—1.70 to 1.80
DRAG RACING (unlimited) 1.80 to 1.85
DRAG RACING (limited) 1.65 to 1.75
ROAD RACING—1.65 to 1.75

All these different numbers indicate one very important fact about rod ratios. There is no *ideal* rod ratio. Basically, the lower the rod ratio, the more torque the engine will make. The higher the rod ratio, the more power the engine will make at high rpm. That's why a NASCAR race engine, which runs constantly at high rpm for long periods of time, uses a higher rod ratio than a road racing car, which uses a variety of rpm coming off corners and on straightaways.

Back in the real world, changing the rod ratio isn't as easy as just deciding what you want it to be. In a given engine with a given stroke, shortening the rod to lower the rod ratio requires the pin to be moved down in the piston. This will make the piston heavy. In this same engine, lengthening the rod to increase the rod ratio requires the pin to be moved up in the piston. This limits the oil ring and oil ring grooves.

Do your calculating before buying parts. If the physical parts of the engine limit rod ratio selection, change the the cam's lobe centers 2-4 deg. to help offset the limitation. Cams are the easiest thing to adjust, much easier than rod ratios. If you need a lower rod ratio, making the cam's lobe centers smaller (closer together) by 2-4 deg. will help obtain the desired performance.

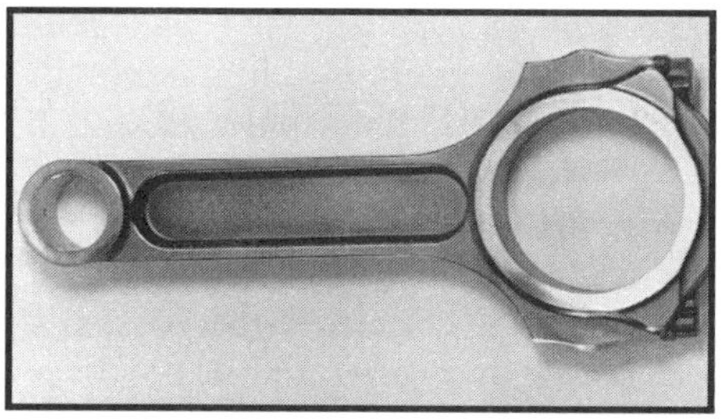

The race, steel connecting rods from Mopar Performance and other special aftermarket manufacturers use an I beam construction more like a production rod. Contrast it to the H beam used in a Carrillo.

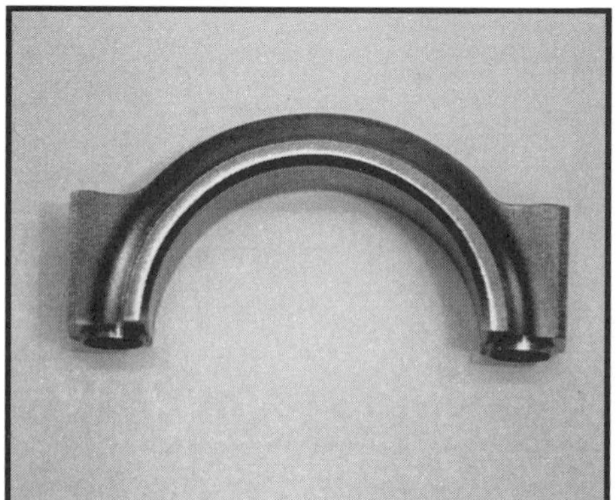

This is a special steel cap used with Carrillo rods. Note the cap pilots on either side that extend above the parting line to pilot the cap onto the rod.

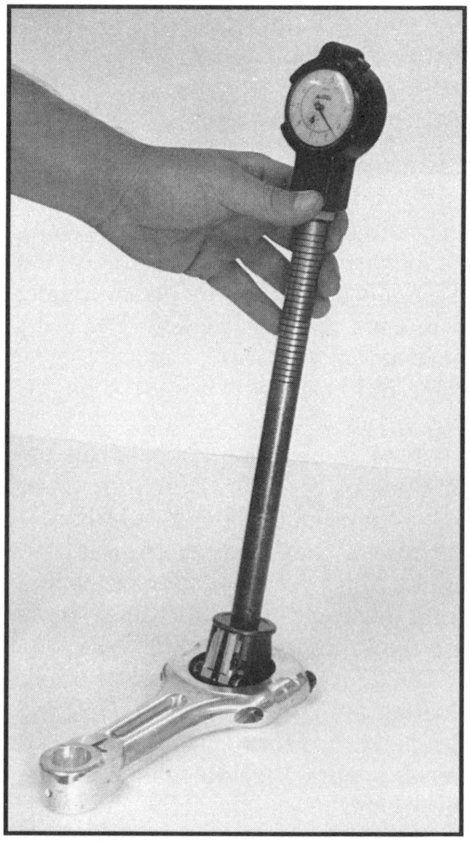

Inside diameters and roundness need to be measured prior to reconditioning. You can use either a snap gage, or a dial-bore gauge like the one shown here.

A rod vise is used for holding and torquing bolts when the rod isn't in the engine.

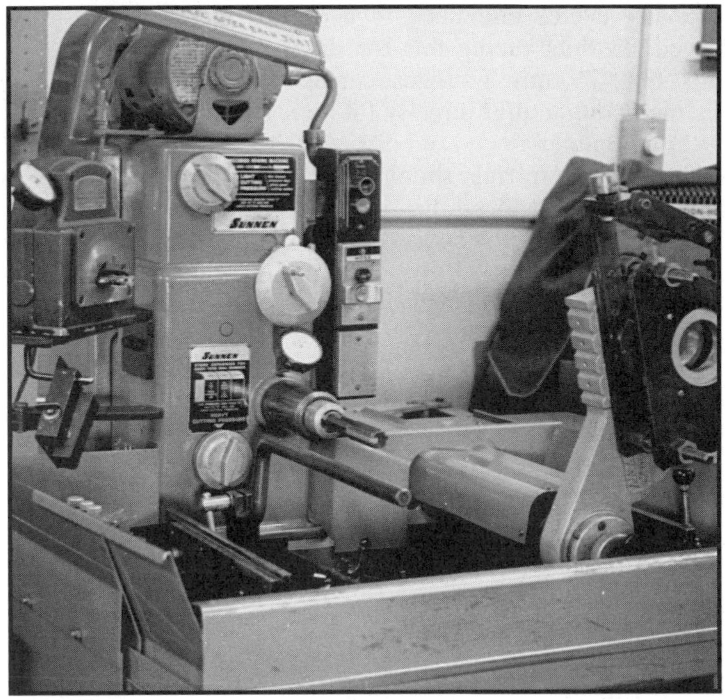

Rods are reconditioned and sized on a special Sunnen machine.

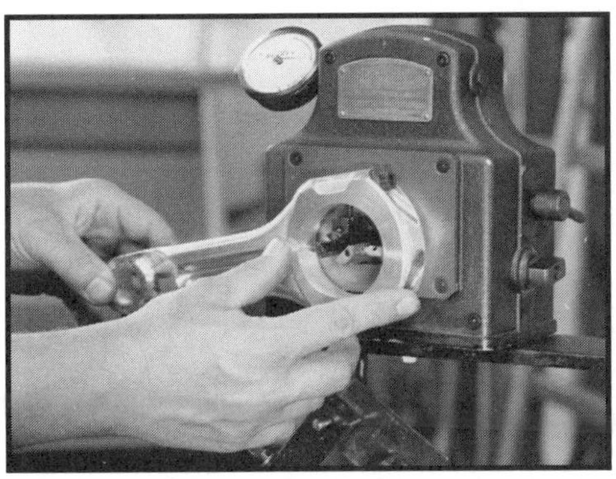

First measure the big end of rod for size and roundness. Note that this particular rod is aluminum and should only be resized by the original manufacturer. However the measurements and operations are the same.

Once the size of the rod's big end is determined and the amount of material to be removed is calculated, the rod is installed on the machine's cutting bar, similar to a hone. Note the machine's torque support. Resizing the big end is not typically done on a new rod, only checked unless something special is desired. Consult the manufacturer before modifying the big end of an aluminum rod.

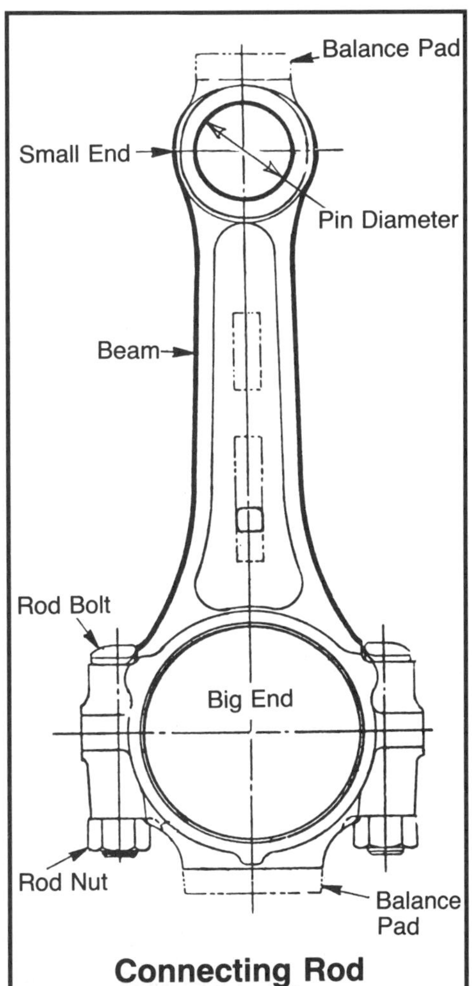

Typical connecting rod and its features. Balance pads may be partially ground off.

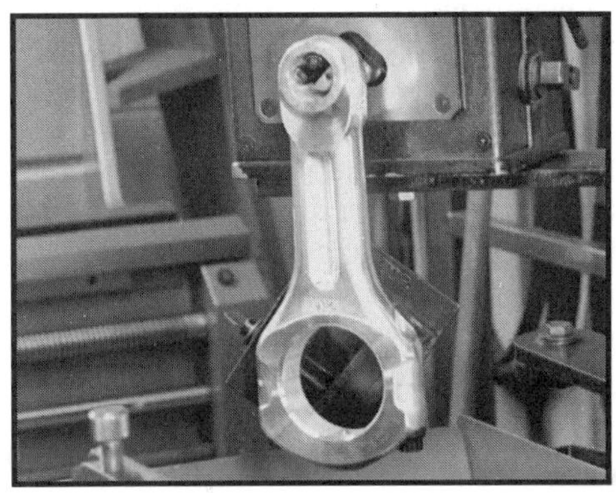

Check the small end of the rod next. Be sure to measure in several directions.

If the small end requires machining (honing) it is put on the machining bar as shown. Note the torque support. This is an aluminum rod and made for a full-floating pin. The pin's size and desired clearance determine the rod small-end inside diameter.

so many cycles and need to be replaced. In drag racing this translates into about 75 runs. I suggest contacting the manufacturer directly for their latest guidelines.

The aluminum rods should use the following clearances: pin clearance 0.0006-0.001 in.; rod bearing vertical clearance 0.002-0.003 in.; and rod side clearance 0.018-0.020 in.

ROD PREP

Once you've decided on which type of rod to use, the next step is to prep them. Once again, this will require a machinist, but there are steps you'll have to perform as well. The preparation of each type of rod is somewhat different.

Stock Rods—Standard, stock rods are prepped on a Sunnen rod reconditioner. The machinist will first measure each rod for: straightness; parallel bores; center-to-center distance; and the size of both the large and small end of each rod. All rods should be Magnafluxed for cracks. If any fail the Magnaflux test or the straightness test, they should be replaced. Once a full set has passed inspection, they can be deburred and polished. Be sure to check the rules for the class that you plan to run because rod polishing and deburring is not allowed in all racing classes.

The next step is to have them *shot-peened*, a process where the rods are bombarded with tiny pellets of metal shot to strengthen the outer rod surface. This is usually done with the original rod bolts removed and the machined surfaces masked off. Once these pieces are received back from shot-peening, the next step is to *resize* the big end. The big end's internal diameter (ID) should be between 2.2500 in. to 2.2505 in. For high rpm use, it's recommended that the big end be sized two-tenths undersize or 2.2498 in. Now check the the small end for clearance. If a floating pin is used, the clearance in the rod to the pin should be 0.001 in. If a pressed pin is going to be used, the clearance should

To check the squareness of the rod's big end to its small end, another special tool is required. It doesn't machine the rod. It only allows you to measure it for squareness.

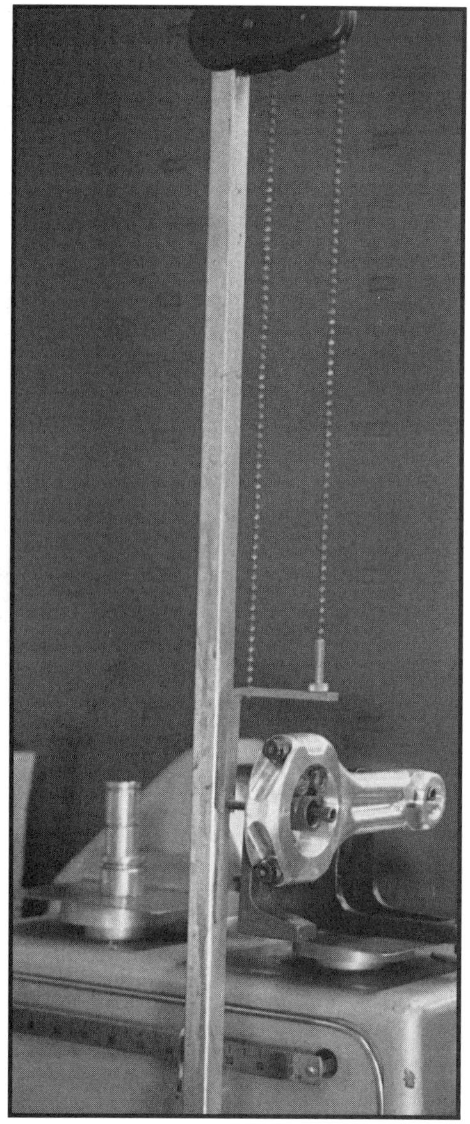

To balance the engine and balance the rods one to another, each end of the rod needs to be weighed by the machine shop. Note the tall fixture to hold the rod square and parallel, to get the truest reading. The gram scale balance is weighing the small end.

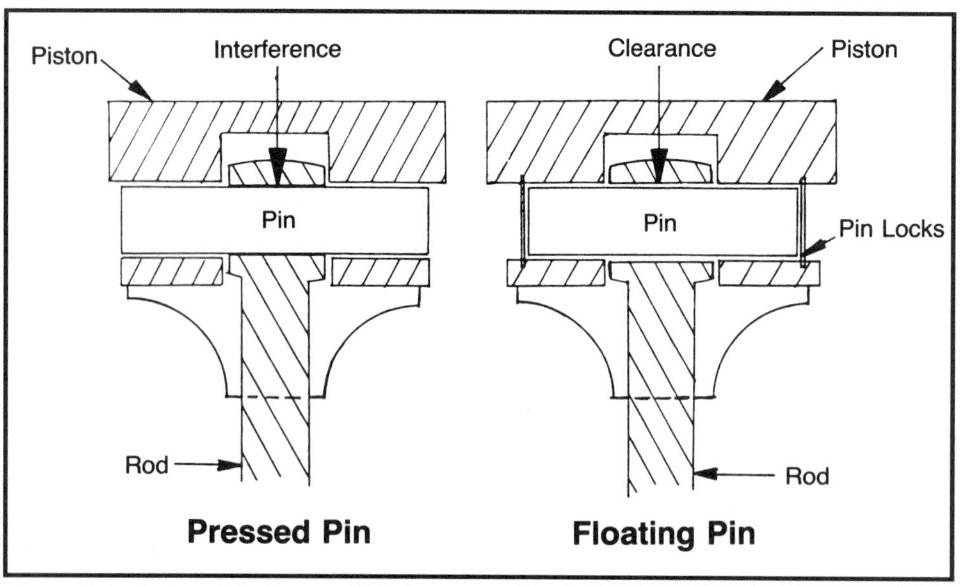

Pressed-pin has an interference fit between small end of rod and the pin. Floating pin has clearance between the pin and small end of the rod. This "float" requires pinlocks be added to the piston to locate the pin. Although not shown, the floating pin also needs a bushing in the small end of the rod. Floating pins are usually used in racing, if you have a choice.

To weigh the big end, the rod is reversed. Note complexity of holding fixture used to eliminate errors and false readings.

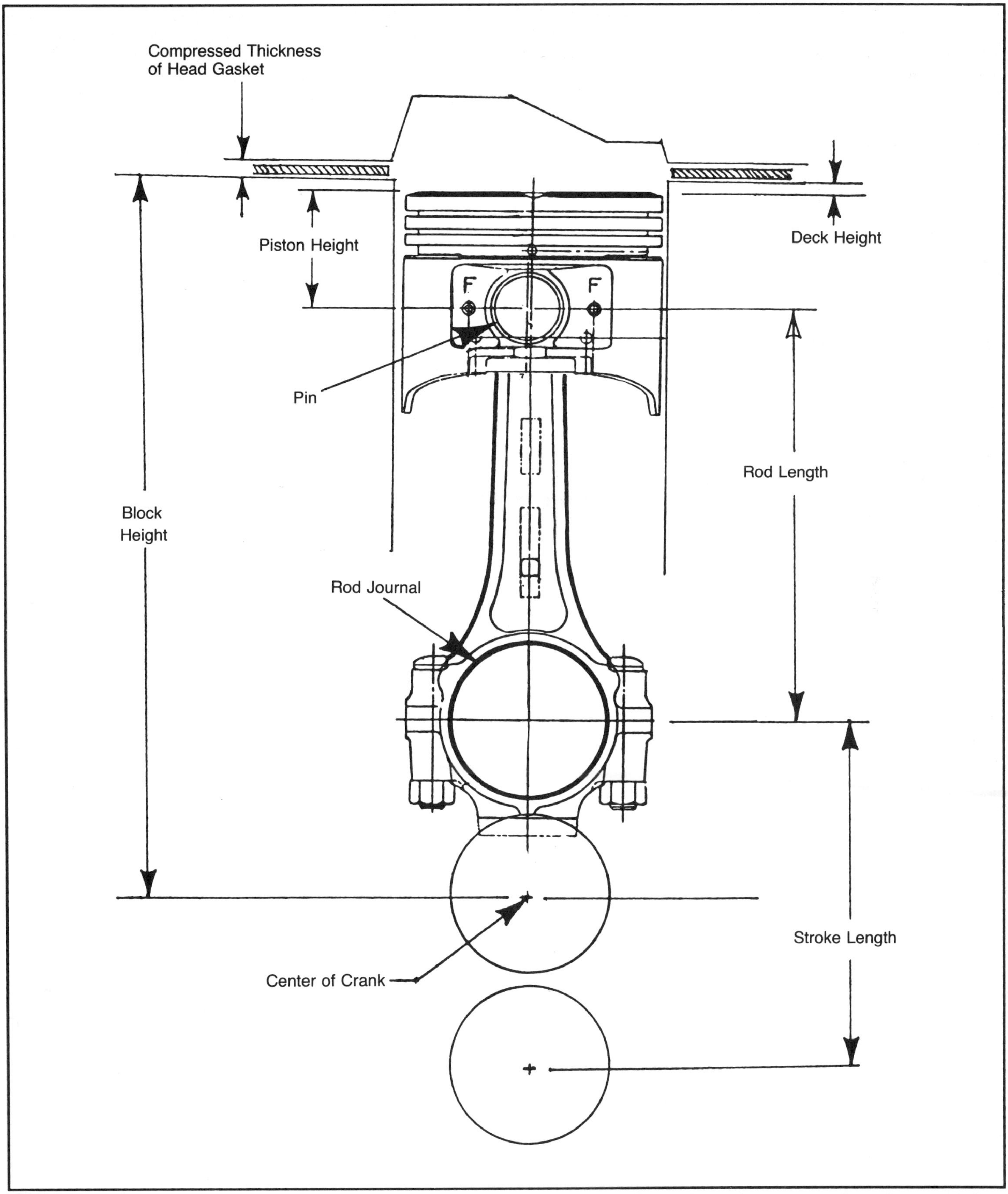

On a stock or production-based engine, the items shown are not changed. In many racing engines using a non-stock crank, rod and piston combination, most of these items may change. The change is determined by the type of racing, length of track, car weight, etc. The most easily changed areas are gasket thickness and deck height but even these simple changes will affect valve-to-piston clearance and piston-to-head clearance.

be 0.001 in. *interference* fit. The rod is now ready for assembly.

Use new high-strength rod bolts and nuts such as those from Mopar Performance or from aftermarket suppliers such as ARP or SPS. Stock bolts are the weak link in a stock rod. For racing and high-performance use, they should be replaced when the engine is built. When the rod bolts are replaced, be doubly sure that the head of the bolt is seated squarely against its machined-flat surface on the rod. Also note how the nut takes torque. It should torque-up to spec quickly and stay there. If not, the bolt head may be seating itself against the rod, or the bolt or nut may be "too soft." Don't use anti-seize compounds on the bolt threads.

Some production-based rods may have a small hole in the big end of the rod toward the top and to the outside of the journal. A little research may show that the original rod bearings had a matching small hole through the bearing shell. This hole is for oiling the cylinder wall at low rpm. High-performance bearings won't have this hole because you don't want any extra oil on the cylinder walls at high speed. If you have the hole in your rod, it won't hurt anything. If you don't have one, don't add one.

Racing Rods—Floating pins used in high-output racing engines, need to have more oil. The best way to do this is to drill a 3/32-in. oil hole on the top of the rod. After it is drilled, it should be chamfered. Some specialists recommend two holes—one on either side, toward the bottom of the pin bore at say 5 and 7 o'clock positions. Remember, these "side holes" are not recommended for stock rods, because they weaken it. The top of the pin bore is a lightly stressed area and therefore the small hole doesn't affect the rod strength. It will break somewhere else first if over-stressed.

Balancing—The rods in a performance engine should be balanced. The first step in this process is to weigh the rods. The 273/318 rods weigh approximately 726 grams while the 340 rod weighs 758 grams. Once the eight rods are weighed, the lightest rod is used as the standard and the other seven are lightened down to this weight by grinding off metal.

Side Clearance—With the rods together, they can now be checked in matched sets on the crankshaft for rod side clearance. The rod side clearance should be between 0.009-0.017 in. with steel rods. The clearance can be as high as 0.020 in., but these engines run low oil pressure and tend to use more oil because they blow so much onto the cylinder walls. It's very difficult for the oil ring to control this much oil, especially at high rpm. If your side clearance is above 0.020 in., the crank or rods should be replaced or repaired before further engine building is done. This doesn't apply to aluminum rods, however.

Rod Bearings—The rod bearings for all A engines are basically the same. There are different sizes and materials but the diameters are all the same. The best rod bearing clearance for racing or performance applications

To drill the pin oiling hole in the top of the rod, a Bridgeport is typically used. A drill-press can also be used with extreme care. Note that this particular cutter drills the hole and chamfers the hole in one step.

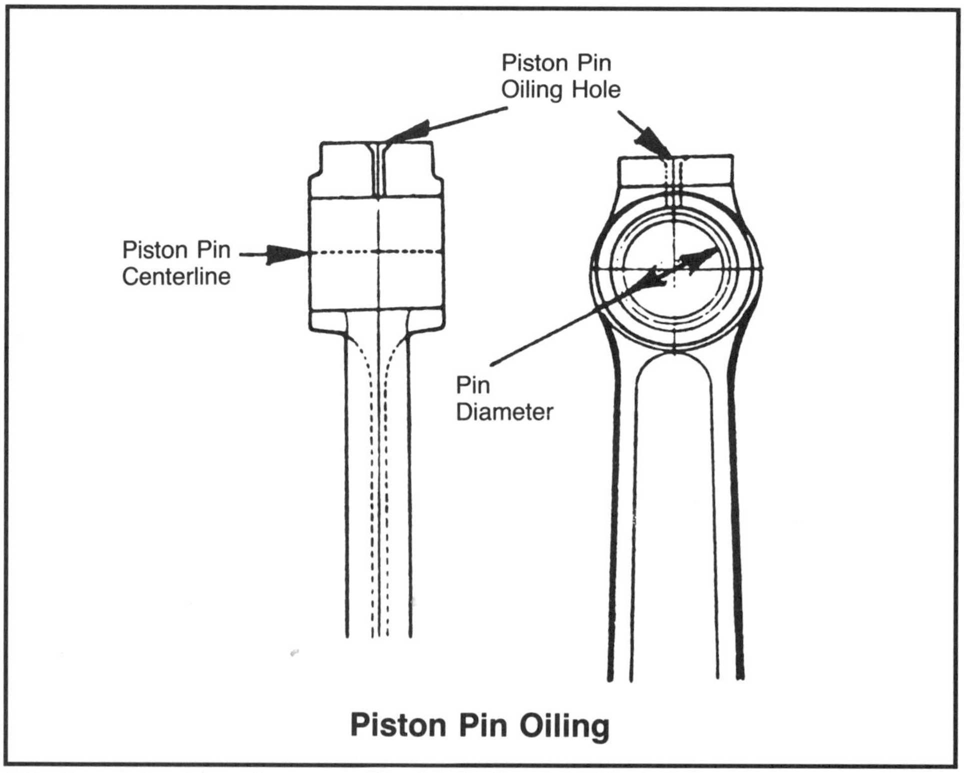

A floating pin has to receive some oil for lubrication. A 3/32-in. hole in the top of the rod is the best way to add oiling without weakening the rod. (Courtesy Mopar Performance)

is between 0.002-0.003 in. There are three types of material used in rod bearings—*babbit*, *aluminum* and *tri-metal*. Tri-metal bearings are hard. Babbit bearings are soft. Babbit is recommended for drag racing while tri-metal is recommended for oval track endurance racing. The aluminum bearing is a compromise and might be used in any racing engine. It's also common to use aluminum main bearings and babbit or tri-metal rod bearings to suit the application. Before deciding, check to see what's available. All three materials are not always offered in the size that you may need.

Bearing preparation is pretty straightforward once you have checked the actual clearance. If a full-radius crank is used, then the outside edges of the bearing shells need to be chamfered. The sides of the rods on the same crank throw do not need to be done on the side facing the other rod.

Keeping the rod side clearance under 0.017-in. (0.020-in. for aluminum rods) helps keep oil off the cylinder walls, which helps oil rings seal and promotes performance and durability. High side clearance, in conjunction with higher racing clearances on the journal, can cause very low oil pressure. If this problem is detected after the engine is assembled, the only solution is to rebuild it.

CHRYSLER PARTS LIST

CONNECTING RODS	
H.P. Special Forging	P4349617
Alum. Rod	P4349220
BEARINGS (F-77 Tri-Metal)	
Std.	P3690683
0.001 U/S	P4007018
0.002 U/S	P4349023
0.010 U/S	P4349024
BEARINGS (Babbit)	
Std.	P4007026
0.001 U/S	P4007027
ROD BOLTS	
3/8 in.	P4120097 Set

In racing it is very common to pin the bottom shell in the rod bearing. Most aluminum rods come with the provision in the cap. Racing machine shops have a simple fixture for modifying the shells to accept the "pin" in the rod's cap. Generally this process is unnecessary in properly sized rods but it doesn't seem to harm anything.

By design, the bearing shell is held in the round bore by the amount of "crush" designed into the bearing. Typical crush is around 0.001 in. Simply put, this is the amount that the bearing shell sticks up above the parting line of the rod or cap. Racing bearings have slightly more crush than standard bearings—up to 0.001 in. to 0.003 in. total crush.

There's one oddball point in bearings. If the crank has been ground offset, (stroked or destroked), then new bearings will be required. However, these new bearings won't be A engine bearings anymore. They'll now have to be the size that was selected to undergrind the crank. Be sure to remember this when ordering new rod bearings. With offset grinding, the mains are left as is. So the mains still take A engine bearings but the rods don't. This can be a costly mistake.

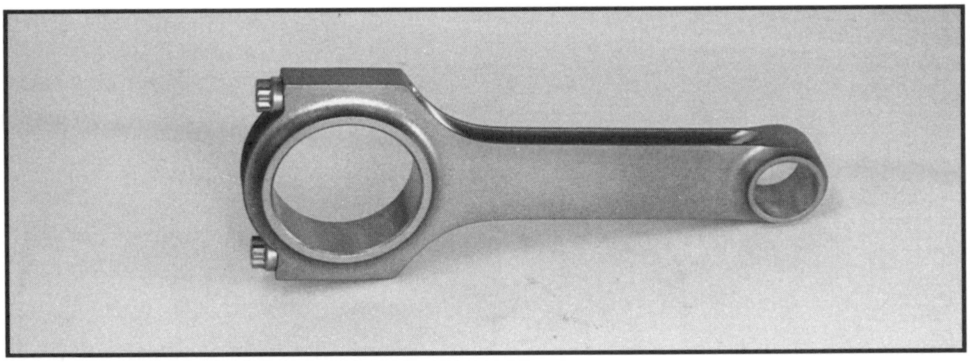

Standard rods are considered to have an "I" cross-section. The Carrillo steel rod shown above has an "H" cross-section.

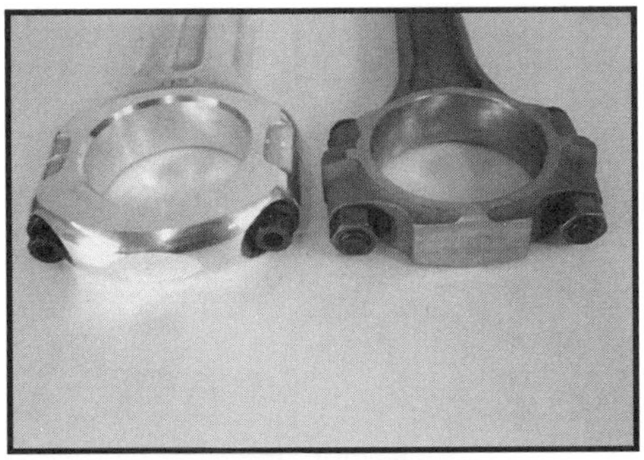

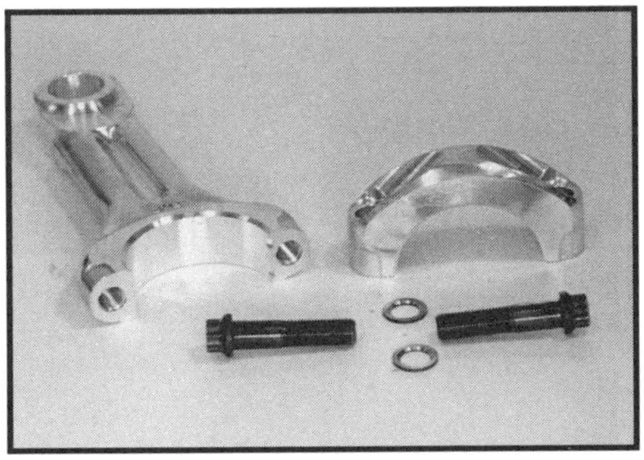

Standard rods use a rod bolt that is pressed down from the rod forging and a nut that attaches on the bottom to hold the cap on. The aluminum rod on the left uses a special bolt that threads directly into the rod itself. There is no nut.

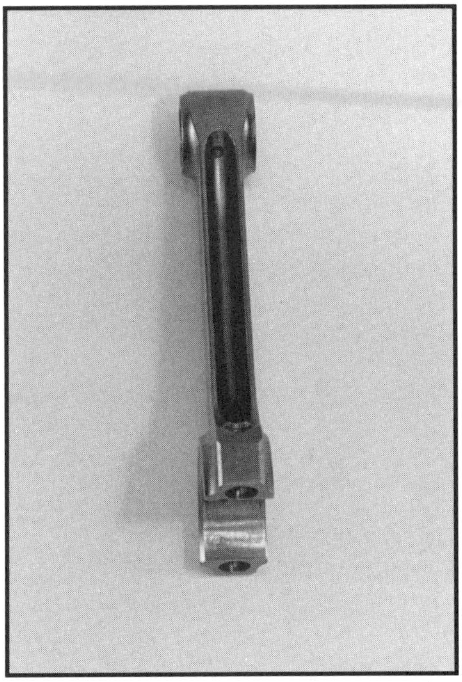

The Carrillo steel rod featuring its H-beam design. Note the pin oiling hole in the side. This should only be done to a rod of this type.

4 PISTONS & RINGS

PISTONS

Next to camshafts, pistons are the most popular internal aftermarket performance item of an engine. There are many different manufacturers and each one offers something unique, but you must consider many different factors before selecting a piston design. Two of the most important factors are not directly related to the piston itself. They are *compression ratio* and *valve-to-piston clearance*. The valve-to-piston clearance is more closely related to the camshaft and its phasing. The compression ratio should be selected based in part upon what gas is going to be used, and whether or not the engine is going to be turbocharged or supercharged. These two decisions should be made early in the engine building program, *before* the pistons are purchased. The piston relates to many parts in the engine and can't be chosen without considering how these other parts affect the piston. First, let's discuss the piston's elements.

PISTON THEORY

The basic science relative to a high-performance, racing piston hasn't changed much since the early days of D. Boertman's 383 and Ed Hamburger's 340. The key to piston theory can be summed up in one word: *lightweight*. A piston must be strong, yet as light as possible. To satisfy this design requirement, a piston is generally manufactured from aluminum or an alloy of aluminum. This is true for all modern-day (50's and newer) performance engines. Since all pistons use aluminum as the material, the type of construction, *cast* or *forged* is where piston design differs most.

Forged vs. Cast—All A engine production pistons are cast. Racing pistons are typically forged. Forging the aluminum makes it stronger for the same area or thickness, yet it can be lighter. There are cast and forged pistons available for just about any bore size, and the choice of which one to use is based on many factors. Cast pistons are used in production engines. Forged pistons are generally used in racing. However, cast pistons can also be raced in many cases. Cast pistons cost less and are more readily available. Forged pistons are stronger, more expensive and typically have a better performance "ring line-up." Generally forged pistons are available for higher compression ratios and have more design flexibility relative to pin location, skirt design and overall shape.

Piston Weight—The stock piston weights for the following stock engines are as follows:

- 318 c.i.d.: 592 grams
- 340 c.i.d.: 719 grams
- 360 c.i.d.: 584 grams

Observe that the 340 piston is the heaviest. This would seem to make sense because at 4.04 in. it has the largest bore diameter. But the 360 bores at 4.00 in. are almost as large as the 340, and yet its pistons are the lightest of the three. Why? Because the 360 has a longer stroke, but uses the same rod length and same block height. The longer stroke means that the pistons must be shorter, which makes them lighter. Keep in mind that these weights are only production numbers. The weights for all-out race pistons are much lighter.

Replacement pistons are designed by the aftermarket to replace the stock, production piston. They are *not* racing pistons. Generally they weigh the same as the stock piston that they are

Three typical A engine pistons. At the top left is a domed, high-compression piston. At the bottom middle is a race flat-top and at the upper right is a production flat-top with a dish, in this case from a 360 c.i.d engine.

replacing, unless they are oversize. Then they will be slightly heavier.

Racing, however, is where weight really becomes a factor. It has been proven that the lighter the piston the faster the engine or car will go. This is especially true in drag racing, short-track racing and in engines with automatic transmissions. Obviously, pistons can only be so light before they fail, so there are limits.

Piston weight is usually discussed in metric units of grams rather than ounces. The weight is that of the piston only, without the pin or rings. A good, pro-style drag racing piston should weigh in around 400 grams. An oval track piston should weigh 520 grams because it needs more strength for durability and can't be as light. Both these racing pistons are *domed* at a 4 in. bore. Flat pistons weigh a little less than domed pistons while large bore pistons tend to weigh more. Conversely, smaller bore pistons, such as the 3.91 in. used in a 318, are lighter. Much of the lightening process is done by removing small pieces of metal, so it is always good to weigh all the pistons before modifying and always balance "down" to the lightest one.

One of the common ways to lighten a piston is to make the top of the piston thinner, but there is a limit. With a forged piston in a drag engine, the minimum thickness across the top of the piston is 0.250 in. For the spot in the valve notch pocket, it can be as low as 0.150 in., but not if the intake and exhaust notches run together to form a trough. It is better stay in the 0.175 in. or higher range for that type of setup. (Keep in mind these minimums leave no room for error. For safety, it's recommended that an extra 0.050 in. be left.) Don't allow the valve notch to "break-into" the ring land. With a cast piston, you can't be this thin! These minimums should be increased by an additional 0.050-0.100 in.

Piston Skirt—The bottom part of the piston below the pin is called the *skirt*. It holds the piston upright in the bore and guides it up and down the cylinder walls. There are two basic designs, each with its own variations. The *rigid skirt* is very stiff and extends 360 deg. around the cylinder bore. The *slipper skirt* (sometimes called a TRW skirt) is currently the most popular. It has two *tails* (180 deg. apart) that stick down below the basic piston to form the skirt. Chrysler's testing has indicated that the slipper skirt popularized by TRW and Venolia will make more power and offer more durability than other current designs. Most racing aftermarket manufacturers offer this type of skirt or a variation there of. The rigid-skirt pistons require more piston-to-wall clearance to keep from scuffing. This extra clearance causes the piston to tilt more in the bore, which makes it hard for the rings to seal.

The piston skirt transfers loads onto the cylinder bore walls as it moves up and down. The stiff skirts can push on the walls so hard that the cylinder walls may split or crack. There is no evidence to suggest this is a problem with the Venolia/TRW skirt designs. This problem only occurs in very high-output engines and/or engines with very large bores.

As mentioned earlier, the piston skirt design has a lot to do with "how it runs." Each design has a clearance (piston-to-wall) that it likes to run best at. You should be able to get this number from the manufacturer. If you run the piston too tight, it will scuff the cylinder wall. This can smear some of the aluminum from the piston onto the cylinder wall or move the aluminum around on the skirt as though it had been hit by a rough file or machined by a dull tool. In any case, engines with this problem will tend to run hot, be down on power and have higher blow-by percentages. However, blow-by

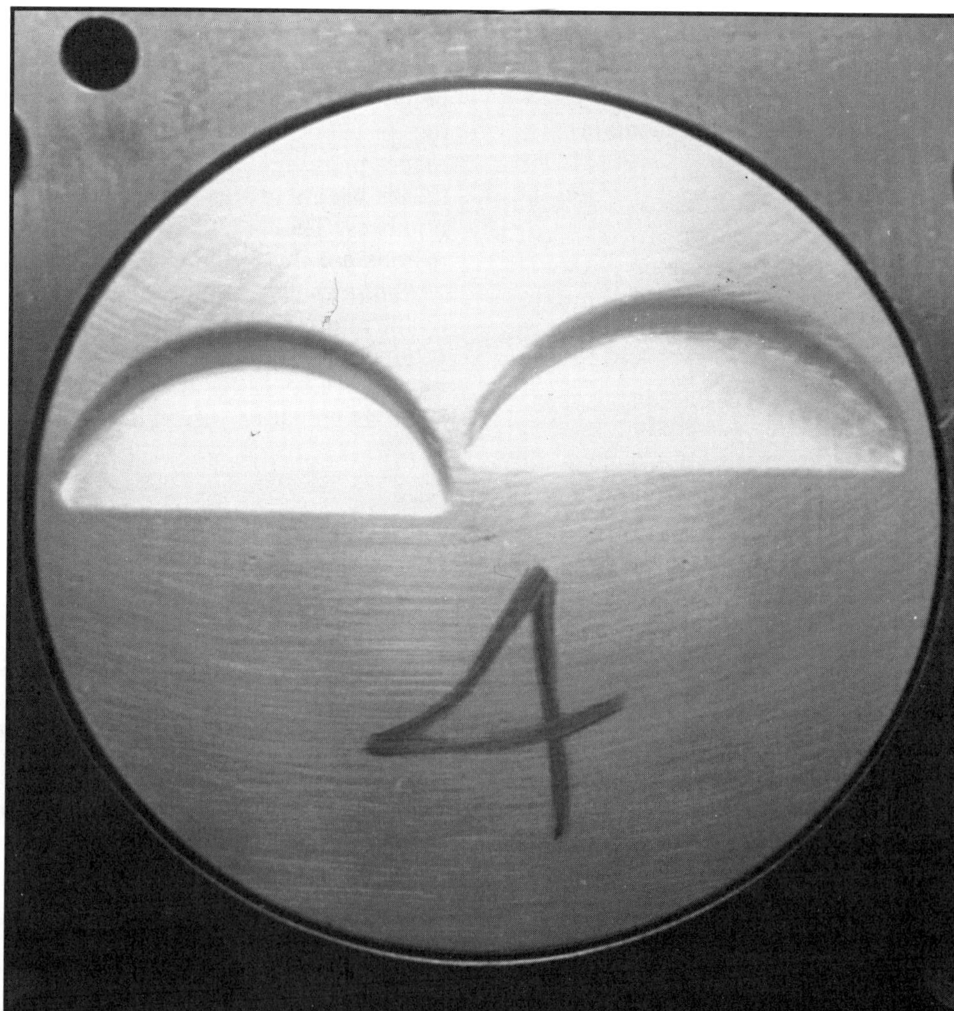

The top of a racing flat-top, forged piston. Notice that there is no chamfer on the outside of the piston or on the top of the block although both are radiused slightly. The smaller diameter of the exhaust valve moves its notch closer to the piston centerline.

can't be used to determine if the engine is scuffed. In mild cases, the blow-by doesn't increase enough to give a strong indication. Unfortunately, the solution is always the same—rebuild the engine and replace the scuffed piston(s).

DOMED PISTONS

Domed pistons are used to increase the compression ratio of an engine. They are much more effective in smaller engines than in larger engines. In fact, it's possible to obtain an 11.5-to-1 or 12.0-to-1 compression ratio in both the 340 and 360 with notched flat pistons, but almost impossible to do this with a 273 or 318. Domed pistons are necessary to achieve compression ratios over 12.0-to-1, regardless of small-block engine size.

Most lightweight domed pistons feature a "hollow dome" design. This means that the underside surface of the piston top is recessed to match the dome shape. While this makes the piston lighter, it also means that the dome can't be milled off to create a flat piston. If you want a lightweight, flat piston, you will have to buy it as such. Machining too much off the dome will weaken it to the point of collapse.

Domes increase the engine's compression ratio: the bigger the dome the higher the compression ratio. The domes can be tall and thin or short and fat. They obviously are not the same and don't work the same. The *height* of the dome above the flat of the piston is one consideration, but not the main one. The leading factor is *volume*. The volume of the dome is discussed in cc's, similar to the cylinder head volume. However, unlike the head volume which can be measured directly, the dome volume must be measured indirectly.

Measuring Dome Volume—The method for measuring piston dome volume is usually called the "1/2-in. down fill" volume. On a 426 Hemi engine, a "1-in. down fill" volume is used because the Hemi dome heights typically run 0.750 in. to 0.900 in. The A engine domed pistons aren't nearly this high. They generally run in the 0.250 in. to 0.400 in. area so the 1/2-in. down fill method works well. Keep in mind that the 1-in. down method would work on an A engine, it's just requires more "cc" fluid and time.

To understand the 1/2-in. down fill method, picture a flat disk, 0.500-in. thick with the exact same diameter of the cylinder. Assuming zero deck height, if a flat piston is lowered in the cylinder bore 0.50 in., then our 0.500-in. thick flat disk is the result (volume above the piston top to the block's deck in this position). The volume of this flat disk is easily calculated based on its diameter and thickness, both of which are known quite accurately in our engine. So if you actually measured this volume in the engine and the piston is perfectly flat, the measurement and the calculator should give the exact same answer. This would mean that the dome volume is zero, or that the piston is flat. This sounds pretty simple but there are a couple of tricks! First, let's put valve notches in the flat piston. No other changes. This would be typical of a Super Stock engine. These notches usually are several cc's in volume, so the actual cc measurement is larger by the same amount than the calculation for a perfect flat disk. Mathmatically this would mean that the piston was "dished" or has a negative dome.

Now if you have an actual domed piston with a known dome volume of 10 cc's, let's see what happens. Because the dome rises up into the imaginary disk and won't let as much "cc" fluid in, the measurement of the 0.500-in. down fill volume is 10 cc's smaller than the calculation. Mathematically subtracting the measured volume from the calculated volume will yield the dome volume. Now the trick comes in with a real engine. A "real" engine with domed pistons will also have valve notches. If the valve notches are 5 cc's each (10 cc's total) and are machined into the 10 cc domed piston,

Both pistons are flat-top designs but the one on the right is a cast, production piston while the one on the left is a forged racing design.

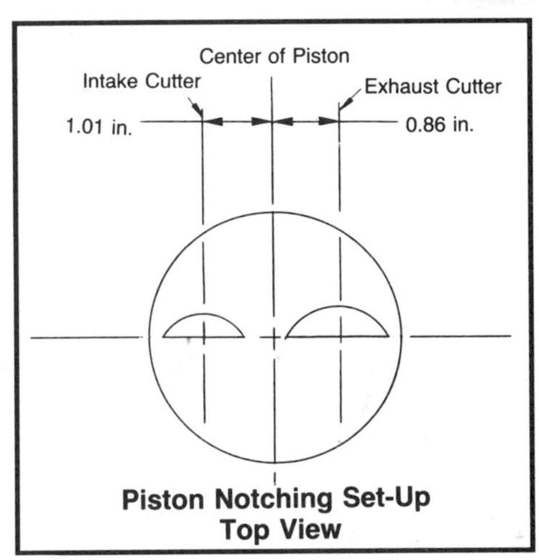

Noting the location of the intake and exhaust valve notch cutters in the top view of the piston. Realize that only half the individual pistons will look like this (#1, #5, #4, #8,) while the other four will be mirror images since the location of the intake and exhaust valves are reversed.

the measurement and the calculation are the same. This means that the notches have offset the dome, and the compression ratio would be the same as with a flat piston with no notches.

Once you know how large the dome is, or how much larger the dome is than the piston notches that have been cut into the piston top, you can now look at the shape of the dome itself.

Dome Shape—Domes are forged with a squarish, sharp-edged shape. Once the valve notches are cut, the whole top of the piston has many sharp edges. The ideal shape is round and smooth, so all sharp edges should be radiused. This obviously makes the dome slightly smaller, which lowers the overall compression ratio. In the larger small-blocks, such as the 340/360 cubic-inch versions that can be built with compression ratios over 13.5-to-1, this trade-off is a very good one. It's also a plus for oval track racing engines, which run an average compression ratio of 12.5-to-1 unless they are limited to 9.0-to-1, which is becoming more popular. In these engines, the loss of compression isn't as important, but the improved combustion chamber shape will increase the engine's output.

Reversing—For drag racing purposes, stock pistons are usually reversed on the rod. This means the piston notch or arrow points to the rear of the engine instead of the front. The stock piston is machined with the pin offset and installed in such a manner as to load the skirt against the cylinder wall. This approach makes the engine quiet but slightly increases the friction. For racing, reversing the piston switches the offset which cuts down on the load the piston imparts to the cylinder wall. This will make the engine noisier, especially when it's cold, but it will decrease the amount of friction and thus increase the amount of power available. Forged racing pistons are usually machined on-center, or at least much closer than stock, so this reversing trick doesn't work on forged, racing pistons. Additionally, keep in mind that intake notches must be below intake valves and a dome won't fit into the combustion chamber reversed!

PISTON SELECTION

Pistons are not selected on their merits alone. The choice is also based on other aspects of the engine. The camshaft and fuel octane mentioned earlier are just two of the many considerations. However, the most important physical characteristic of the piston is the *bore size*. This is where the trail to proper piston selection begins.

Typical domed piston with valve notches and spark plug relief. Note the chamfer around the outside of the piston top to provide added piston-to-head clearance. This chamfer is also found in production pistons although not quite as large.

This is a basic piston dome shape. Valve notches extend dome angle below piston top.

Another style of racing piston with a uniquely shaped dome. Notice that there are only two ring grooves.

Bore Size—In the A engine family, there are only four *stock* bore sizes to consider—3.63 in., 3.91 in., 4.00 in. and 4.04 in. However, each of these four could be bored 0.020 in., 0.030 in., 0.040 in. and 0.060 in. oversized. Most aftermarket pistons are 0.030 in. and 0.060 in., while the other two sizes are available from Chrysler or can be custom-made. Chrysler is out of many of these oversize pistons for older-style A engines, (pre-1974) so check to see that they are available before you start machining. Also check to be sure that piston rings are available for the bore size that you select. In many cases, special rings are harder to find than the pistons.

Most of the time the condition and machining of the block will determine which overbore to use. If the block is being rebuilt, then it will have to be overbored and new oversize pistons will have to be installed. If it's in good shape and it's the first rebuild, then it can be overbored 0.020 in. and it should "clean up," meaning the bores will end up straight and round with no taper. In poorer shape it may take 0.030-in. over before the block cleans up. On the second or third rebuild it will most likely need to go out to 0.060-in. over.

If a new block is used, it may be able to clean up at the stock bore, especially with the slightly increased clearance used in racing applications. In racing you can usually get two or three rebuilds at a given bore size before the piston-to-wall clearance gets so high that it's time to bore out the block to the next size. If you start with a new block and build-up the engine at the stock bore size, then rebuild it later at 0.030 in. and later still at 0.060-in. over, you will obviously get more use out of that engine. Many racers build an engine at 0.060-in. over the first time. I don't recommend this, because the minimal performance gained isn't worth the decrease in the number of runs, except when it comes to engines built for Super Stock and Stock Eliminator NHRA class racing. NHRA rules allow Super Stock engines to run 0.060-in. over and Stock engines 0.035-in. over. Highly competitive class racing requires the engine be pushed to the limit. The loss of total runs or the decrease in the total number of rebuilds physically allowed by the block is a necessary evil. It comes with the category.

Compression Ratio—The next consideration for proper piston selection is the piston's basic *compression ratio*. To know the engine's actual compression ratio you need much more information than just the piston you're going to use. You'll find that I also discuss compression ratio in the *Cylinder Head* and the *Blueprinting* chapters, in terms of how it relates to those two areas. But the piston also has an important effect on compression ratio. Generally, there are three basic compression ratios for pistons. There's 8-to-1, 10-to-1 and 11.5-to-1. Remember, this is not the final compression ratio of the engine. The first two are production ratios and use basically a flat-top piston. The last one, 11.5, is only used in high performance or racing and generally means a domed piston of some sort. The key to selecting one of these ratios is the type of gasoline that will be used. Generally, pump gas will dictate an 8-to-1 compression ratio. The 10-to-1 engines will need some form of racing gas or racing gas blend with unleaded premium. If a 10-to-1 piston is used, but the actual engine blueprints to 9.5-to-1, then other mixtures using octane boosters are possible. The 11.5 compression ratio requires straight racing gas. It's typically purchased in 55-gal. drums or at the race track. It's the most expensive and hardest to find gas.

Once the gas has been determined, the pistons can be selected. There are many aftermarket manufacturers of pistons. If you have the pistons custom-made, you can also select the piston skirt design, pin height, ring style and, to a certain extent, weight.

Forced Induction—The exceptions to the general piston selection guidelines are the turbocharged and supercharged "street" engines. The supercharged engine can use stock or replacement cast pistons for moderate boost pressures up to 7 psi. For 10 psi and up, forged pistons are recom-

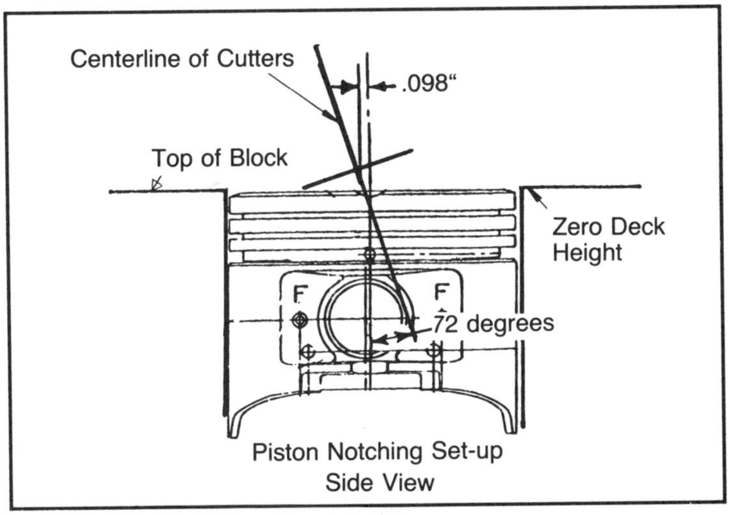

Most pistons used for race engines will have to be notched for valve clearance. Cutter sizes are not the same for the intake and exhaust valves but the offset and angle remain the same.

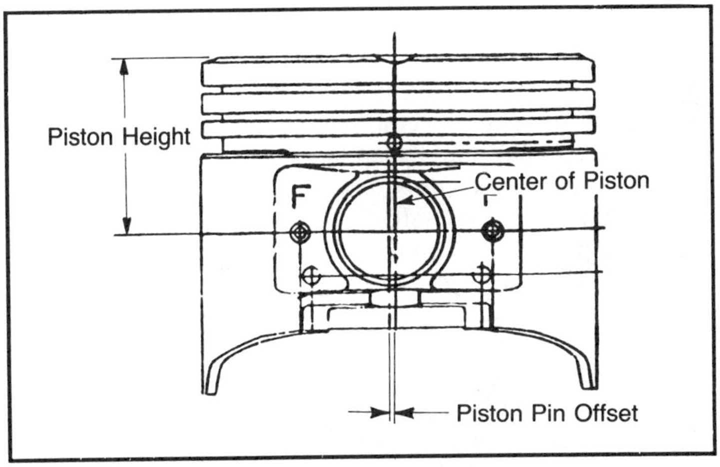

Production pistons and many replacement pistons have piston pin offset. The offset ranges from 0.030-0.060 in. This minimizes piston noise and is indirectly the reason why production/replacement pistons are reversed for racing applications. Racing pistons have the pin on-center.

mended. For turbocharged engines the compression ratio should be kept below 8.0-to-1. In either case the octane of the gasoline used is critical. Also, if either of these engine types is used on the street or in dual-purpose applications, the ring gap should be opened up several thousandths-of-an-inch. If you are having a custom piston made for either of these two applications, don't put the top ring as close to the combustion chamber as you would with a normally aspirated engine. One-sixteenth rings are better choices than the pressure-backed Headland, Dykes or gas port rings. For racing, rings are another story, which is covered in more detail on page 45.

PISTON PINS

The stock A engine piston pin is 0.9842-in. in diameter with a length of 2.99 in. to 3.00 in. This stock pin has a weight of 154.6 grams. Most of the stock pins are pressed. The 340's pins were all floating. The 273's also used full-floating pins, as did the 318 up to 1972. All 360 and 318 production engines from 1973 are equipped with pressed pins. The clearance of the pin in the piston should be 0.0015 in. (See Chapter 3 p. 28 for rod clearances).

For racing purposes, the stock pin is almost too heavy. However, it is durable and you don't want to take a chance of the pin failing. As a general rule, it's best to use the pin that comes with the piston from the manufacturer. This is especially true when it comes to bracket racing.

For a race engine using custom lightweight pistons, you can specify a lighter, stronger pin. Generally, this means shorter in length and smaller in diameter. At this point, the application of the engine is very important. Drag racing and NASCAR engines can't use the same pin. A drag engine can use a 100-120 gram pin. This pin is approximately 2.50-in. long with a diameter of 0.9272 in. A NASCAR engine should use a longer pin. The length should be 2.90 in. with a diameter that's the same as the drag engine. These pins weigh 135 to 145 grams. Note how close the stock pin is to the custom NASCAR pin in length. The slightly lower weight of the race pin is obtained by using a thinner wall thickness and smaller diameter. These race pins are made of a special steel material for added strength. However, this material is very hard on the pistons and rods. These steel pins accelerate the wear on these related parts tremendously, so check them often.

Obviously these special race pins have to be coordinated with the other parts of the short-block. The lock grooves in the pistons need to be cut for this special length. The really short pins can run out of pin tower in some piston designs, so it's important that the custom piston manufacturer be consulted about the pin length. Don't take an existing piston and re-cut the pin locks for shorter pins without consulting the manufacturer first.

A smaller diameter pin has to have the rod changed to accept it properly. A production rod can be bushed down to the proper size. Race steel rods or aluminum rods are generally custom-made and can have the smaller pin hole bored in originally. Otherwise consult the rod manufacturer for the best solution. The small bore diameter must also be machined into the piston. Pistons can't be bushed down! The piston manufacturer must be told originally what pin diameter is desired in the custom pistons.

Pin Height—The pin height should always be measured. Pin height is the distance from the centerline of the pin to the top of the piston or to the flat of a domed piston (not top of dome). This is one of the items that can be changed in a custom piston. If not measured, it can cause problems with piston-to-head clearance, valve-to-piston clearance and/or compression ratio.

Pin Locks—Pin locks are only used with full-floating piston pins. There are several different designs. The production pistons that used floating pins (such as the 340) used a type of single *Tru-Arc*. For all-out racing you want to use either a double Tru-Arc or a thick *Spiral Lock*.

The Spiral Lock is 0.072-in. thick. Don't re-use these once they have been installed in the engine. With each rebuild use a fresh set. A double Tru-Arc system actually consists of two single

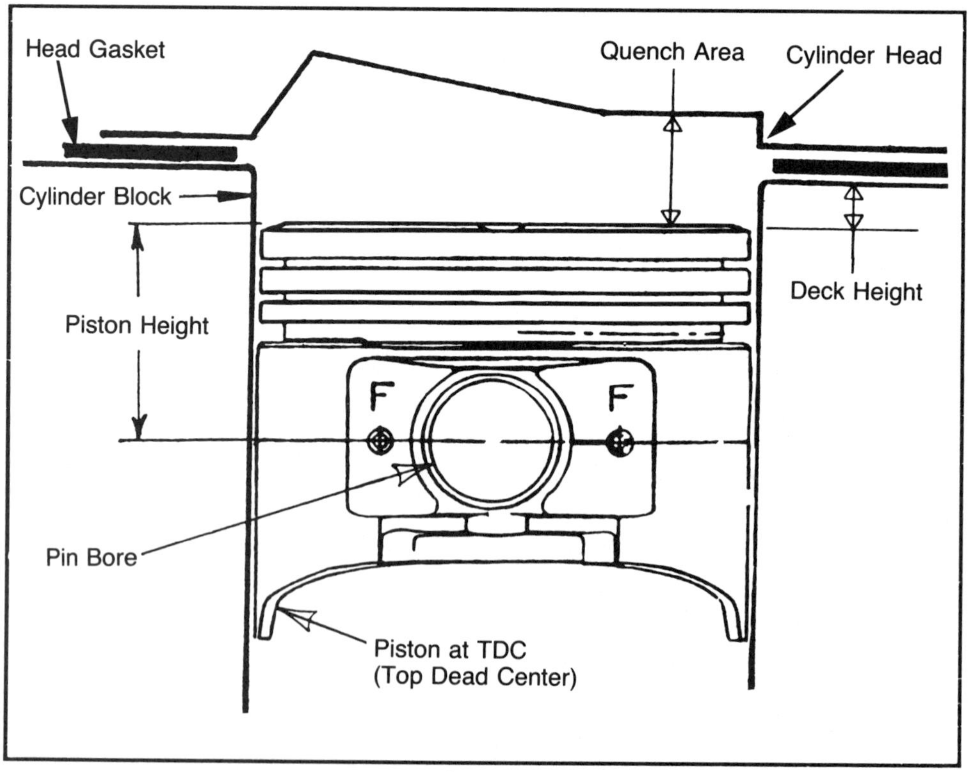

In race engines, all the items shown affect the piston. One of the main design parameters of the piston is its resulting compression ratio in the engine. The piston design has to take into account the piston height, deck height and head gasket thickness directly but also the quench area indirectly.

Tru-Arc's installed back-to-back in the same groove. They are installed so that the flat side of the rings (or the side with the sharp edge) are facing outward. This means that one flat side is next to the pin while the other flat side is next to the cylinder wall or outside of the piston. The gaps in the rings should be 180 deg. apart and aligned in the vertical direction.

With either type of pin lock, the pin end clearance should be set very tight. This clearance should be in the 0.000 to 0.006 in. area. This will require a lot of custom fitting. A true "zero" pin clearance will make it very difficult to get the last lock in place and it must be fully seated to work properly.

Which Pins To Use?—Most aftermarket forged racing pistons are designed for use with full-floating pins, meaning the pin lock grooves are cut in the piston. These same pistons can be used with pressed pins, if desired. If you have forged pistons and they have lock grooves, should you run full-floating pins? The answer to this question is based on rpm. For high rpm, full-floating pins are used. Pressed pins are used in low rpm applications. For the A engines, the low vs. high split is in the 6,500-7,000 rpm range.

You could look at it another way. Use pressed pins with hydraulic cams and floating pins with mechanical cams. Also, keep in mind that floating pins will work well in lower-speed operation, but they won't work better and are *not* worth spending extra money for. Before deciding to go to full-floating pins, be sure to check the rod itself. The floating pins must have a bronze bushing in the small end of the rod. Many rods already have the bushing, so no problem. If it's a pressed-pin rod without a bushing, you should check with your local machine shop to see if that particular rod has enough stock (material) in the small end to accept the bushing without weakening the rod. Running a floating pin steel-on-steel (no bushing) in the rod is *not* recommended. The pin and rod both tend to scuff. This isn't very durable. A pressed pin is much better than a scuffed floating pin.

PIN TIPS

Once the pistons fit the block, they can be installed on the rods to make some rough checks. If the piston is a full-floating design, assembly and disassembly are quite simple. Pressed pins are more difficult. Because the piston is machined after these rough checks are taken, the pin will have to be pressed back out of the rod. This is not advisable, so I'm going to give you a pressed-pin trick to save time and hassle. Take an old, used piston pin to a machine shop and have them turn it 0.002 in. to 0.003 in. undersize across 2/3 of it's length. This removes the press-fit, so the rod and piston can be assembled and disassembled easily by sliding this special pin in and out. Remember to put this special pin back in the toolbox before final assembly.

BASIC PREP

After digesting all of this piston theory, you've probably gone out and selected a set of pistons. Now what are you going to do? Throw them in the engine, right? Hold on! There is still a great deal of work to be done before the pistons can be installed into a finely tuned performance engine. First, check to make sure they are the right pistons. Manufacturers *do* make mistakes! Measure the bore size, pin height, piston weight, pin specs, pin lock grooves, pin oiling and piston ring grooves. Compare these measured answers to the order sheet. Keep in mind, many pistons are ordered strictly by part number, so many of the checks provide valuable information.

Always write the measurements down on your engine build-up sheet. Never start machining the piston before you make the proper measurements. That's a good way to scrap a set of expensive pistons. The first, and most important measurement, is the *piston clearance*, which is the clearance of the piston to the cylinder wall.

Piston-To-Wall—Determining this clearance is really part of the block prep, but it has to be measured and the block bored and honed before the piston will fit properly to make the rest of the measurements to machine the piston. Generally, the pistons are measured for diameter 0.500-in. below and perpendicular to the pin unless the

This picture shows two different styles of piston pins. The one on the right is the production pin while the one on the left is the race design. Note that the race pin is shorter and has a thinner wall thickness making it lighter. Pistons have to be specifically designed for this style pin. They are not interchangeable.

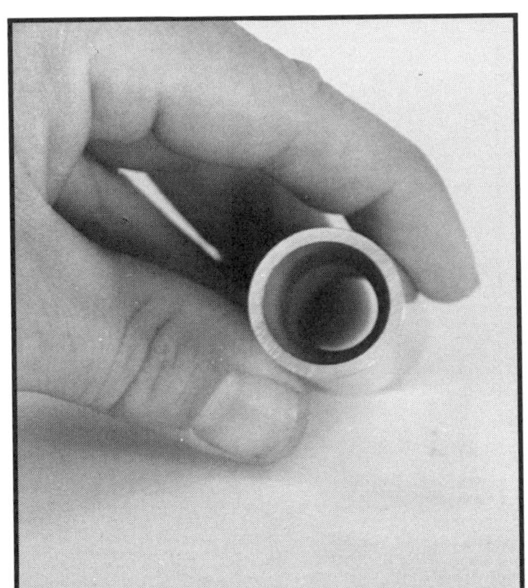

On the outside, this pin is straight like any other, but it is tapered on the inside. It's thinner on the ends and thicker in the middle.

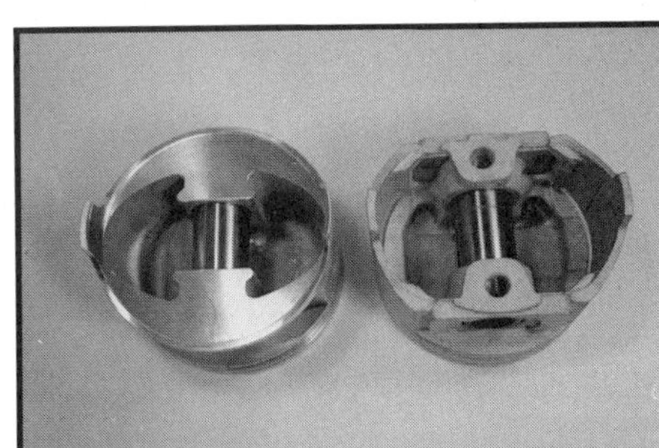

The bottom side of the two pistons (race left and production) shows the different approaches to piston pin oiling. The production piston has two large holes while the race piston appears to have none.

manufacturer stipulates otherwise. The piston-to-wall clearance is calculated by comparing the diameter of the piston to the cylinder bore diameter. If you want to *increase* this clearance, you must hone the block. Keep in mind that most 0.030-in. over pistons are actually 0.030-in. over minus the desired piston clearance. For example, if the piston manufacturer specifies a piston clearance of 0.007 in. to be run on that particular 0.030-in. over piston then the piston should measure 0.023-in. over. This is done to assure that the piston has the proper operating clearance when installed in a 0.030-in. overbore. In a custom-made piston, the manufacturer will ask for the actual bore size desired. In other words it's not a 0.030-in. over 340 cubic-inch Chrysler small-block piston, it's a 4.070-in. bore piston. This means the cylinder bore size would be 4.078 in. (assuming you want 0.008 clearance). It's still *not* a good idea to assume any of these numbers. Measure and be sure! An error of only a couple thousandths (0.002 in.) can change a recommended clearance of 0.007 in. to an actual clearance of 0.005 in. This will cause scuffing, overheating and loss of power. A production piston (cast) and most replacement (cast) pistons are designed to run at 0.002 in. to 0.005 in. piston-to-wall clearances. The forged racing pistons are designed to work best at 0.007 in. to 0.010 in. Some of the "West Coast" rigid-skirt pistons need over 0.012 in. clearance before they'll work properly, but I wouldn't recommend these types of piston skirts. If the piston clearance is too tight, be sure to open it up before running the engine.

On the other end of the scale, some engine builders add *Teflon buttons* to pistons when there is excessive clearance. This can add as much as 100 grams to the weight of the piston. In many instances the buttons end up in the oil pan. I wouldn't recommend them in a racing (gasoline) engine. Larger pistons are a much better solution for power and durability.

The piston and rod assemblies can now be installed into the engine, usually without the rings, so the *deck height*, *valve-to-piston clearance* and *piston-to-head clearance* can be checked. These are the first rough measurements. Eventually, you'll have to use at least one ring on the piston to hold it straight to get the desired valve-to-piston clearance.

Deck Height—Measure the deck height first. Obviously, this is done with the cylinder heads off the engine. Using a dial-indicator gauge mounted in a bridge, measure the distance from the deck surface to the top of the piston immediately above the piston pin centerline. This height is considered the *deck height*. If the top of the piston is above the deck, the reading is *positive*. If it's below, it's *negative*. It's pretty straightforward with a flat-top piston but with a domed piston, the deck height is measured to the flat or lower part of the piston. The deck height has a lot to do with the engine's compression ratio and with the piston-to-head clearance. This piston-to-head clearance will be your next measurement.

Piston-To-Head—There are two ways to look at the piston-to-head clearance. The first is by direct measurement. The second is the sum of the deck height plus the head gasket thickness. The second method is only an approximation and works best with a flat piston. A domed piston has to fit inside the combustion chamber so there are many other things for it to hit than just the bottom of the head. The high-compression '68-'71 340 production engines had a flat piston with a positive deck height higher than the gasket thickness. Therefore, the outside top surface of those pistons were chamfered to fit cleanly inside the combustion chamber for adequate piston-to-head clearance. The deck height-plus-gasket thickness method is a quick check to see if there will be a clearance problem. If there is, then chamfering could be one solution. Domed pistons can pose a more complicated problem.

In racing, the piston can't hit the cylinder head. If it does, something will usually fail. Head gaskets, piston pins, pistons themselves and rods or bearings are all likely victims. To prevent this from happening, make sure there is adequate piston-to-head clearance. With a steel connecting rod, this clearance is 0.055 in. minimum. An

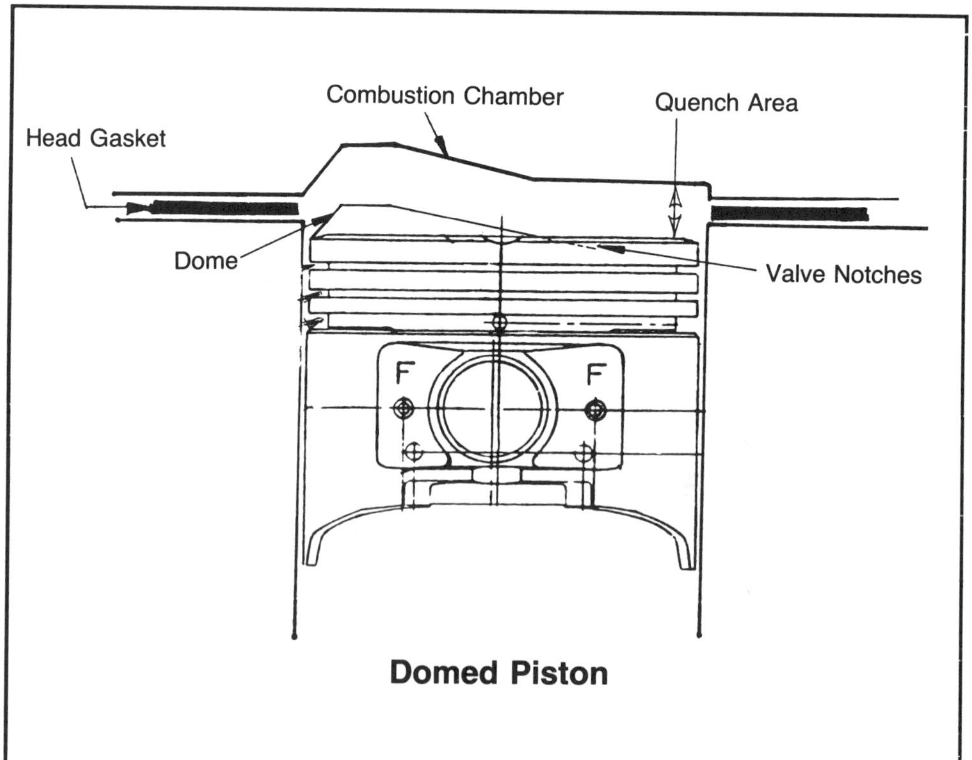

Domed pistons need to be checked more closely than flat pistons. Since the dome fits inside the combustion chamber, clearances to the cylinder head have to be checked in three different directions in addition to the quench area. Clearances to spark plugs can be critical.

aluminum rod engine should have slightly more: 0.060 in. minimum.

To measure, rotate the piston to TDC and install the head. Then, flip the engine over on the stand, and place a dial-indicator on the end of the rod bolt after removing the nuts and cap. Push the piston and rod up until the piston contacts the cylinder head. The dial-indicator should read the proper piston-to-head clearance. If the reading is recorded with the gasket to be used, then the reading represents the total clearance. If checked without a gasket, be sure to add the gasket thickness to the reading to get the total piston-to-head clearance. A clearance of 0.055-in. is recommended.

Valve-To-Piston—The next clearance measured in the block is the valve-to-piston clearance. If this clearance is inadequate, then notches will have to be added to the pistons, or the existing notches deepened. To check valve-to-piston clearance accurately, the camshaft must first be installed at the proper centerline (see Chapter 6 page 72 and Chapter 7 page 93). The clearance can be checked in two ways: by dial-indicator gauge or by using clay on the top of the piston. Clay is the easiest and most popular method, especially for beginners, but is not as accurate as a dial-indicator. It can be used for a preliminary check, however. To check with clay, add a lump to the top of the piston in the area that the valve comes closest to. Install a "used" gasket (the same thickness as the one to be used) and the head, and torque down at least two bolts around the chamber that is being checked. Rotate the crankshaft through two full revolutions. Remove the head, making sure that the clay stays on top of the piston. Stick a steel scale into the thin spots on the clay and record the reading. The minimum clearance for the intake valve is 0.100 in., while the exhaust should be 0.120 in. These numbers are recommended for use in drag racing with a clutch and manual transmission or Lenco. With an automatic transmission you can use 0.090-in. clearance on the intake and 0.100 in. on the exhaust. These last numbers can be used on an oval track engine also if the driver is careful not to over-rev the engine.

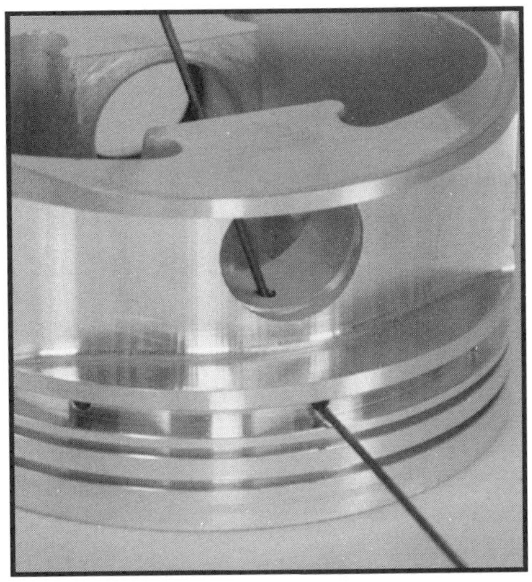

Forged and racing pistons tend to oil the piston pin from the back side of the oil ring groove. The passage is drilled at two different angles—one straight on from the ring groove and one at an angle from the pin bore.

These valve-to-piston clearance numbers are given as "zero" lash numbers. That makes the numbers and method the same for mechanical or hydraulic cams. With mechanical cams you could use a "net" clearance number. This would be the clearance with the cam's valve lash included. For example—instead of 0.090-in. clearance, you could call it 0.060 in. with a 0.030-in. valve lash. I wouldn't recommend this method because it's more confusing and offers more opportunities for an error.

Basic Machining—You can now remove the pistons from the block and disconnect the rods. Take the pistons to the machine shop to have the valve notches cut. The machine shop should know that all A engine valves have a 72 deg. stock valve angle or 18 deg. off of vertical. This is the axis that the cutter must be set to. The diameter of the cutter should always be 0.100 in. larger than the valve used, i.e. the 2.02 intake valve should have a 2.120-in. cutter. This allows 0.050 in. radial clearance for valve deflections, piston rock and machining with tolerance stackups. But keep in mind that in an A engine, not all pistons can be notched the same way. Relative to the front of the engine, four pistons must have the exhaust notch (smaller of two) first or on the left while the other four will

This is a high-ratio forged racing piston. Note that the edges have been smoothed as part of the basic prep.

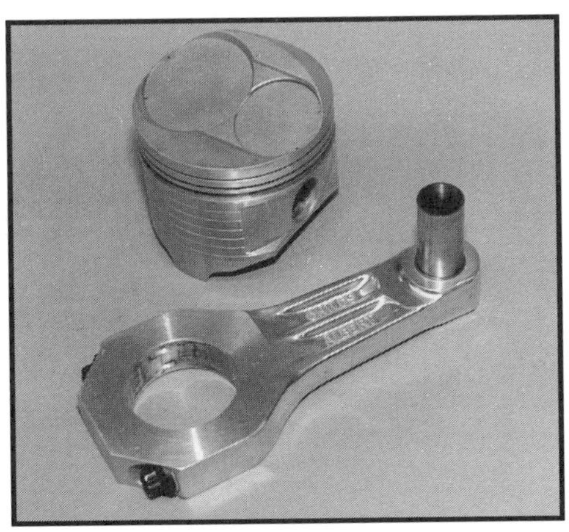

J&E pistons coupled with Childs & Albert connecting rods were used by Dick Landy to build a bulletproof 340 to run in Pro Stock back in 1980. The rod measurement from center to center is 5.7 in. (Photo courtesy Hot Rod Magazine)

Venolia now offers a reverse dome piston that is flush with the top of the block providing a quench area that forces the intake charge into the combustion chamber. It is an ideal piston for engines requiring 9-to-1 compression ratios. (Photo courtesy Venolia pistons)

have to have it second or on the right. Pistons 1-4-5-8 have the exhaust notch first, while pistons 2-3-6-7 have the intake notch first. Also keep in mind that angle-milling of the cylinder head or off-setting the head on the block by moving the dowels will affect the location and angles of the valve notches.

After the notches are cut, remove all sharp edges or corners on the top of the piston with fine sandpaper. The edges of the notches are always sharp and should be sanded lightly to form a slight radius. The top outside edge of the piston can be sanded lightly to remove its sharp edge but don't radius or chamfer it very much because compression ratio will drop slightly, especially with a flat piston. If you are filing and sanding an overall diameter chamfer to gain piston-to-head clearance, then you will have to accept the slight loss of compression ratio. On a domed piston, round-off the edges of the dome as much as possible, and increase the *plug relief*. In many cases, the relief is so small that it will actually hit the plug. Looking from the top, the plug clearance or relief should be increased until it's 3/8 in. or 1/2 in. in radius. Once machined into the piston, its edges should likewise be radiused.

The final check of the piston is to install the piston rings into their proper grooves and measure the vertical clearance in the groove. At this point the piston can be installed into the engine. However, there are several special considerations about rings that you should be familiar with.

PISTON RINGS

Rings are critical to high-performance engine building. There are basically five different designs of piston rings. They are *not* interchangeable at all. The pistons must be cut for the specific ring design selected. Once a certain ring groove is cut, then the ring design can't be changed, only replaced with similar parts. Check the grooves on custom-made pistons to make sure they match the rings. Most people forget this. There is only one exception to this rule. There is also a way around it in some instances. I will cover these exceptions later.

RING SELECTION

Which rings to use? The answer depends entirely on the application. Use the following list as a guideline. The first suggestion under each heading is the first choice, while the next listing is the alternate.

HIGH RPM DRAG USE
(over 8,000 rpm) Dykes or 1/16 in.

SUPER STOCK
Headlands or 1/16 in.

STOCK ELIMINATOR
Dykes or 1/16 in. w/spacer and gapless second ring.

BRACKET RACING—1/16 in.

OVAL TRACK
Dykes or 1/16 in.

STREET
5/64 in. moly, low tension using stock or replacement pistons.

The 0.043 in. was omitted because it is most commonly used with *gas ports*. Gas ports are 14 equally spaced holes drilled vertically down from the top of the piston to the inside top ring groove. The holes are 0.040 in.-0.045 in. in diameter. The holes allow combustion chamber gas to force the ring out against the cylinder wall to seal it. They work, but beware of rapid cylinder bore wear, which can get expensive, especially if the engine will be turning high revs for long periods of time.

RING TYPES

There are five basic types of rings: 5/64 in., 1/16 in., 0.043 in., Dykes and Headland. Then there are several variations of each style. The *5/64-in.* ring has been used in all production A engines since 1964. The *1/16-in.* ring is a thinner version of the production ring, and is used for performance engines. It's easy to install, durable and seals well at both low *and* high engine speeds. The *0.043-in.* ring is even thinner than the 1/16, and is used only for racing in conjunction with gas ports. The *Dykes* ring has an L-shaped cross-section. It's a pressure-backed ring, used only for racing engines that run maximum rpm most of the time (such as oval track). The *Headland* ring also has an L-shaped cross-section and resembles a large, thicker Dykes ring. They differ in that the Headland is installed right at the top of the piston. Headland rings are popular choices for Super Stock, but the novice engine builder might find them hard to work with. Finally, there is another type of ring that has been around for years, but is only recently gaining in popularity. It is called the *gapless ring*. There are several versions of this ring on the market, but basically it is a 1/16-in. ring with notched ends. The notches allow the ring to fit together in the ring groove without an actual gap.

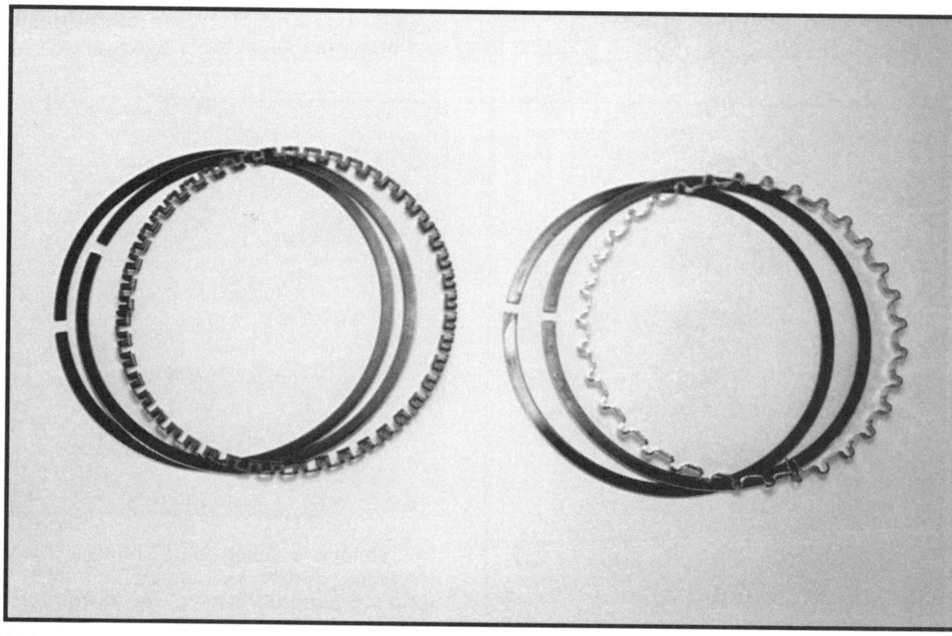

This photo shows two different oil ring styles. Note that the rails are the same but the expanders are different.

Roundness—Rings need to be checked for several items. One of the first is *roundness* in the cylinder bore. To check for this, install the ring in its final-honed bore with the honing plate bolted on. Place a light below the ring in the crankshaft area and look at the area where the ring touches the cylinder wall, all the way around. You may use an old flat-top piston upside-down in the center of the bore to block light and to highlight any gaps at the edge. The ring isn't going to seal in any areas where light shines through. This is important when the actual ring is selected for a given bore size. Equal-to-bore size, 0.005-in. and 0.010-in. oversize rings should all seal to a given bore size. Any ring larger than 0.010-in. oversize will become oval in the bore, and only seal in four places. If it's *light-checked*, you'll see those four areas. If you install the rings this way, there will be a lot of oil *blow-by* and the engine will be down on power.

For example, if the engine is bored 0.030-in. oversize, then an equal-to-bore size ring would read 0.030 in.; a 0.005-in. over ring would have an actual size of 0.035 in.; and a 0.010-in. ring would actually be 0.040 in. All of these would seal properly. Any ring size larger than this wouldn't seal the bore properly, even if you could gap it correctly.

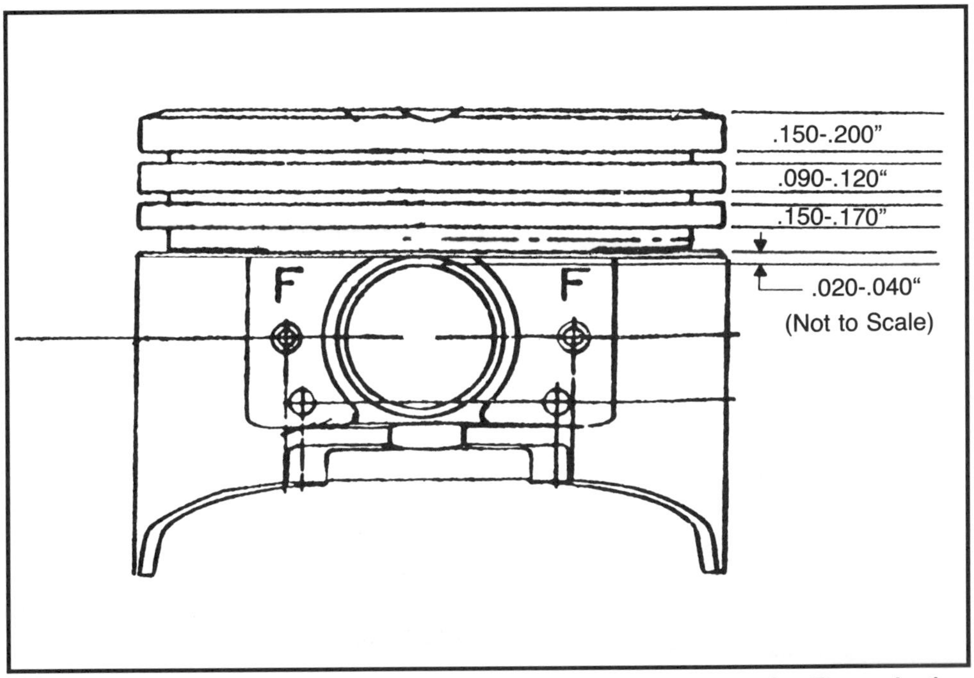

The numbers shown are for straight 1/16 rings as the best overall racing choice. The production engines are based on the straight 5/64 ring. The smaller clearance number could be considered a minimum. However, the 0.043 straight ring can run slightly closer, but not to the pin. If the engine is to have a blower or use nitrous, the support between the rings should be widened.

Ring Gap—The next thing to check on piston rings is the *ring end gap*. The 0.005-in. and 0.010-in. oversize rings on a given bore are used to cut down the ring end gap. For racing, I don't recommend using 0.010-in. oversize rings. Rather, I recommend 0.005-in. oversize rings. This means that you should use 0.005-in. rings in a stock bore, 0.035-in. over rings in a 0.030-in. overbore, and 0.065-in. rings in a 0.060-in. overbore.

To get the desired ring gap, the ring has to be filed. The reason I recommended the 0.005-in. ring is because it provides the extra material needed for filing, without the extra load and its resulting friction increase. But before you start filing rings, you should decide what gaps are needed. All three rings (top, second and oil) are usually gapped differently for maximum performance. As a guideline, gap the top rings between 0.015-0.017 in. and the second rings between 0.010-0.012 in. These numbers are for an approximate 4-in. bore. Larger bores should have a slightly higher ring gap, while smaller bores should be slightly lower.

When filing, be sure to hold the ring flat relative to the file and file from the outside inward, *never* from the inside

It's hard to show the difference between the production 5/64 in. compression rings and the high-performance 1/16 in. rings. This photo shows eight of each with the 5/64 in. rings at left.

This is a typical piston ring gapper. It grinds the ends of the ring to get the desired clearance, and keeps that gap perpendicular to the body of the ring.

out. If you use a special rotary ring file, always rotate in the same direction to accomplish this "outside-in" filing action.

Another consideration relating to ring gap is *ring location*. As a general rule, the closer the ring is located to the combustion chamber, the wider the gap will be. This is critical to the top ring but doesn't really apply to the second ring.

Vertical Ring Clearance—A ring must seal to both the cylinder wall and the piston. The *vertical clearance* in the ring groove determines how well the ring will seal to the piston. Chrysler recommends a vertical clearance for the top ring of 0.0005 in. to 0.0015 in. for production engines. For serious racing, go no more than 0.0010 in. The second ring can have more clearance, 0.0015 in. to 0.0025 in.

CHRYSLER PARTS LIST

PISTONS

310 Forged 9.0 Compression (0.030-in. oversize)	P4452750
318 Forged 11.5 Compression (0.030-in. oversize)	P4120281
340 Forged 11.5 Compression (0.030-in. oversize)	P3690825
360 Forged 11.5 Compression (0.030-in. oversize)	P4007042

NOTE: Above pistons use 1/16 race ring sets.

RINGS

5/64 318 c.i.d. Set	
Std.	P4349844
0.030-in. oversize	P4349845
0.060-in.	P4349846
5/64 340 c.i.d. Set	
0.030-in. oversize	P4349848
0.060-in. oversize	P4349849
5/64 360 c.i.d. Set	
0.030-in. oversize	P4349851
0.060-in. oversize	P4349852

NOTE: All rings feature moly-coated top rings and medium tension. 5/64 is the stock ring groove so they can be used on stock pistons.

RACING RINGS

1/16 318 c.i.d. Set	
0.030-in. oversize	P4349824
1/16 340 c.i.d. Set	
0.005-in. oversize	P4349825
0.030-in. oversize	P4349826
0.060-in. oversize	P4349827
1/16 360 c.i.d. Set	
Std. Size	P4349828
0.030-in. oversize	P4349829
0.060-in. oversize	P4349830

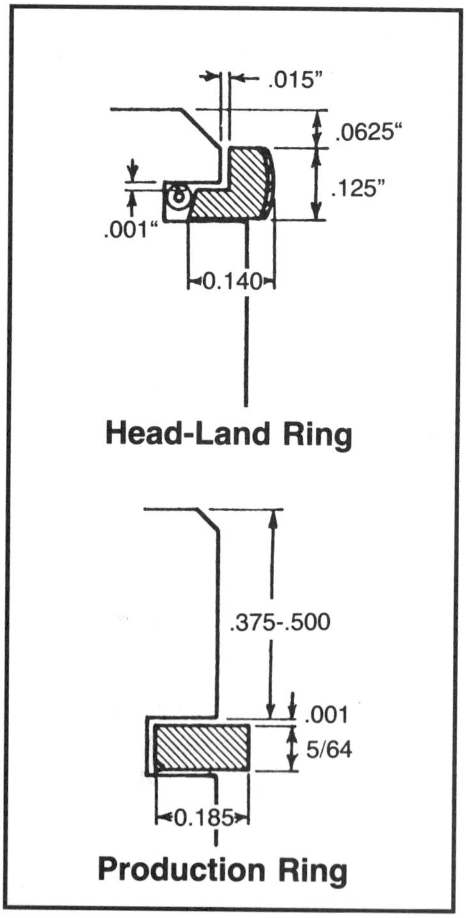

The Head-Land ring requires the top ring groove be cut totally different from any other ring set-up. The 5/64 shown is typical of production engines. The 1/16 ring groove is narrower and closer to the top. The Head-Land dimensions shown as those suggested by Sealed Power (the manufacturer.)

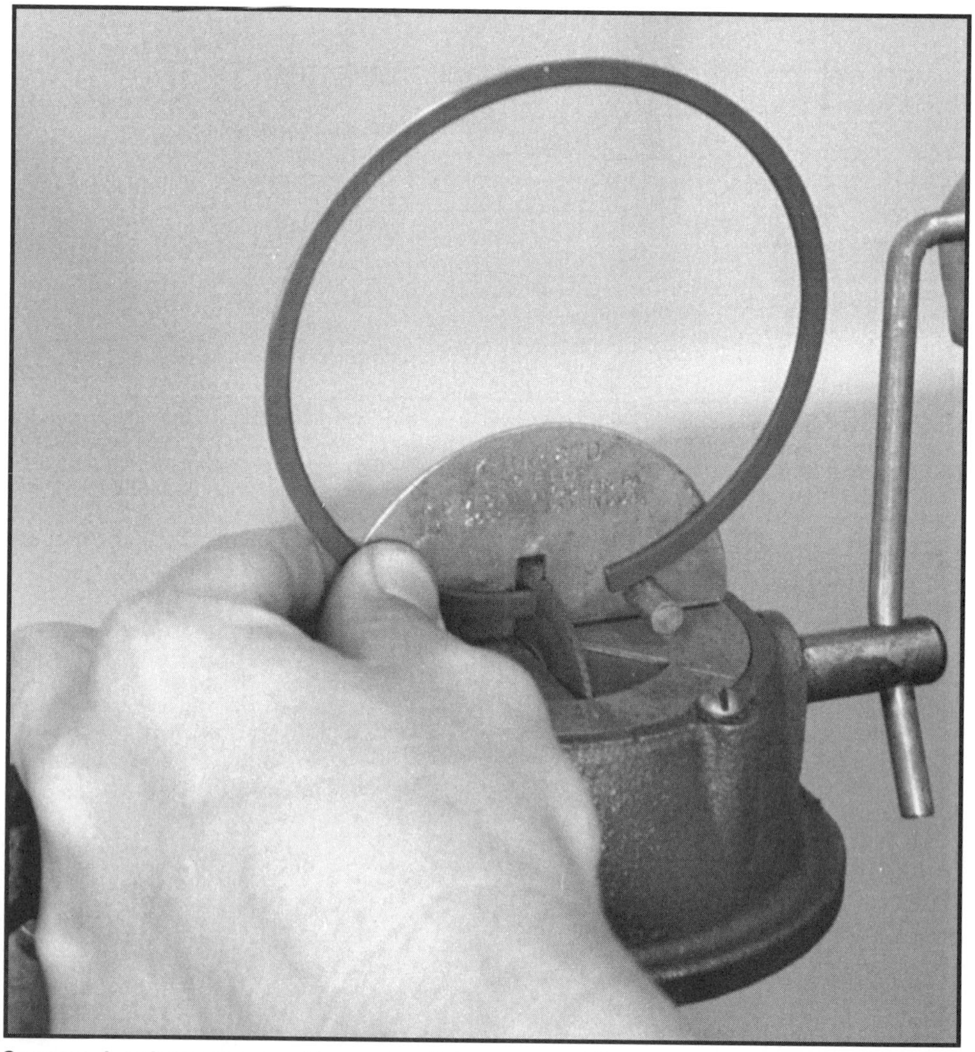

Compression rings are ground (gapped), oil rings are not. Ring must be held square to grinder to obtain properly shaped gap, i.e. parallel to each end and perpendicular to the cylinder wall.

Check the lower ring groove surface for *flatness*. In a used piston, check the ring groove for wear, by looking at the ring from the edge of the ring groove. If it is worn, the clearance reading will vary and be inaccurate.

COMPETITION RINGS

I have covered some of the general specifications of all types of rings. Now, I'd like to look at facts that pertain to racing rings only. Ideally, a racing ring should meet the following requirements:

- High ring placement in the bore for the top ring.
- Small vertical clearance in the ring groove for the top and second rings.
- Tight end gaps.
- Light ring tension (esp. on oil ring).
- A plasma-moly coating.
- Three rings per piston.

High Placement—The high placement requirement for the top ring applies primarily to the 1/16-in. and 0.043-in. rings. The 5/64 in. is not a race ring. The top surface of the 1/16-in. or 0.043-in. ring should be 0.070 in. to 0.090 in. down from the top of the piston. The Headland ring is obviously at the top of the piston.

Ring Tension—A ring with light tension means that the ring doesn't push as hard against the cylinder wall as a stock ring. This feature cuts down on engine friction and helps increase the engine's net output. However, don't select rings too loose, or they won't seal well and the engine will burn oil. For drag racing, most light tension rings work fine. For oval and endurance racing, a more moderate tension is recommended.

Coatings—Plasma-moly coating is a surface coating that is applied to the face of the ring—the part of the ring that seals against the cylinder bore. The "moly" is an excellent sealing agent, and is slippery. It's applied to most race ring designs and is popular on the Headland, 1/16-in. and some 5/64-in. rings. Make sure the ring tips don't touch during operation, or the moly will rub off and the sealing will go away. Of all coatings, plasma-moly is the best one.

Floating pins require pin locks or snap rings. This photo shows Tru-Arcs which require snap ring pliers to install. Piston groove is cut to fit specific snap ring or lock.

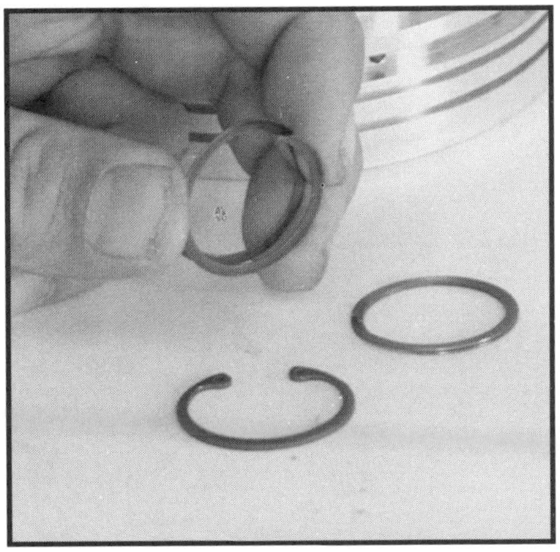

LEFT—The second style of pin lock is the Spiral Lock. Ring to upper right is also a Spiral Lock. They can be installed without special tools but you will need a screwdriver and a little practice.

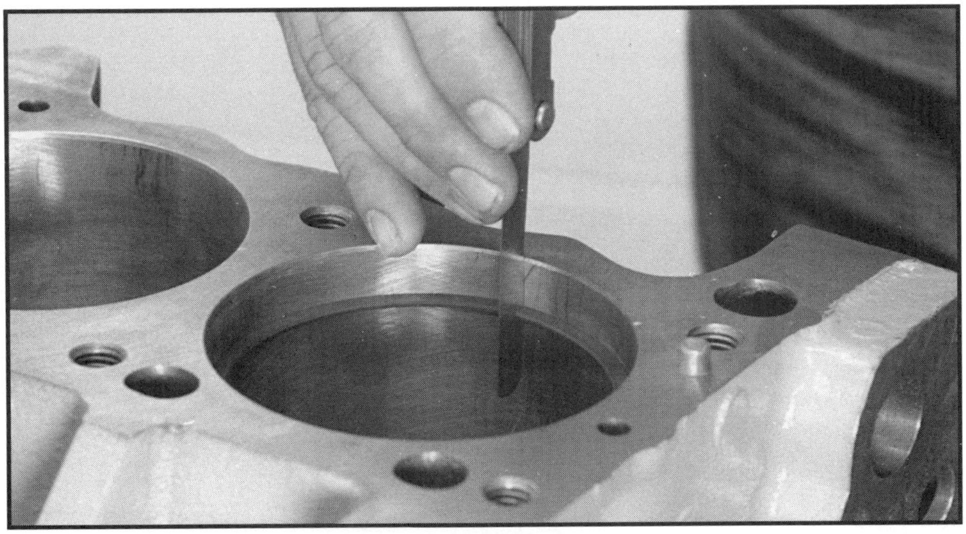

The free-standing gap is unimportant. Ring end gap must be measured in the proper cylinder bore and in the actual ring travel area.

5 CYLINDER HEADS

In the previous four Chapters I have covered the block, crank, rods and pistons separately. Assembled together, they constitute the basic foundation of our engine, the short-block. They must work properly together to extract maximum power and reliability from the engine. Cylinder heads directly affect the magnitude of this output, which is known as *horsepower*. Actually, the cylinder head *assembly* has the greatest influence on an engine's horsepower output. The cylinder head by itself is where the most potential lies to make power. But remember, great heads on a great short-block will produce the most power. Use these same great heads on a poor short-block and you will end up with a weak engine, one that makes *less* power than a stock one. Cylinder heads are given a lot of attention, and too often the short-block is taken for granted. But if the short-block isn't right then the heads can't deliver. At that point all your effort and money is down the drain. In truth, the short-block holds the key to the cylinder head's output. The short-block isn't exciting and gets no credit for the high output of the total engine assembly. Typically the cylinder heads, cam or carburetion get *all* the credit!

In reality, the heads must work together with the cam and carburetion as well as the short-block. When you are building an engine, you don't want to build ten sets of heads to go with it. Selecting a set of heads to suit a particular application, then preparing them properly are the key factors to high horsepower output. To do this, you must first determine what the engine is going to be used for, and consider vehicle-related factors such as how fast the vehicle has to go, how heavy it is, what transmission is going to be used, the type of racing and whether or not it will be used for other purposes. Once these questions are answered, the heads can be selected and modified to complement the short-block assembly.

IDENTIFICATION

There are two basic types of production A engine cylinder heads: the small-port 318 and the big-port 360. The small-port 318 obviously includes the 273 while the 360 also includes the 340. This may appear fairly simple, but it gets more complicated.

Since 1978, the big-port 360 head has been used on the 318 as part of the E48 high-performance four-barrel or *police* option. Then there's the 340 T/A in 1970, which used a specially machined 360 head. There are also big- and small-valve versions of the 340/360 cylinder heads.

CASTING NUMBERS

The best way to identify a cylinder head is to check the *casting number*. It's usually located on top of the cylinder head, on the intake runner. To check for it, the valve cover must be removed. It's a seven-digit number such as 3418915. Most racers tend to shorten this to just the last three digits—in this case "915." While casting numbers are good for identification, don't use them as a guide for purchasing heads. Dealers and speed shops generally don't have the casting numbers on file. To buy the head it is best to have the *part number*. The other problem with casting numbers is that two different heads can have the same casting number. The 340 T/A and the 360 did! There are also big- and small-valve (2.02-in. and 1.88-in. intakes) versions of the 340 casting that were produced from '71-'73.

What the casting number does indicate is whether the *port configuration* and *size* are the same, and whether or

A typical A engine production cylinder head (intake side). This one is a new-style 318. Note how much the outside vertical wall cuts into the port's area (just in-board from the gasket face.)

not the size and shape of the combustion chambers are equal. On the other hand, it doesn't say how the heads were machined. The most obvious example is the *valve size*. Installing the *valve seat* in the head is done as a machining operation to the casting. The 1.88-in. intake seat could be the standard size but a larger 2.02-in. valve could have been used by machining a little more material from the seat area. This is exactly how the 1.88-in. intake 360 heads and the 2.02-in. intake 340 heads were built in 1971. The 340 T/A was built the same way out of the 360, but with an additional twist. The 318/340/360 heads in the early 70's had a hole machined vertically next to the outside of the intake port to allow the intake pushrods to get to the intake rocker arm. The 340 T/A heads had offset intake rocker arms so that this hole could be moved away from the intake port itself. This operation was done to the 360 casting in order to create the 340 T/A. So the casting numbers stay the same, but the actual piece changes.

Valve Sizes—Another method to identify A engine cylinder heads is to measure the *valve sizes* (outside diameters). All the small-port heads used 1.75-in. intakes/1.50-in. exhausts. The big-port heads used the 2.02-in. intake/1.60-in. exhaust up to '71 and also a small intake valve version (standard on the 360 in '71) at 1.88-in. From '72 on, the big-port head used 1.88-in. intakes and 1.60-in. exhausts.

Probably the most important cylinder head to identify is the A engine W-2. It was introduced in 1976 for racing purposes but was never used in production. It has large oval intake ports. It also is the only A engine head that doesn't have a heat-crossover passage. I will discuss the W-2 in more detail later on in this Chapter.

SELECTION

Once you are able to distinguish between the different heads available from the aftermarket, swap meets, or junkyards (or you can order most of these heads directly from Mopar Performance), you must then decide which are best for your engine. Evaluate all types of heads from a performance standpoint. For stock applications of the 273 and 318, the small-port head is best. For modified 273/318 engines used in racing, the 360 (small intake valve) heads are recommended. For the '72-'73 340 and 360 in stock trim, the stock 360 head is best. Don't use the small-port heads on the larger 340 or 360 A engines. The large valve (2.02 in.) versions of the big-port head are best for modified 340/360 engines that are raced. For all-out racing, the W-2 is best.

In class-style drag racing, the rules of the sanctioning body must be closely considered. In NHRA Stock and Super Stock classes, the head can't be changed from stock. In bracket racing the major consideration should be cost. I will close in on these variables as I move through the basic prep of the cylinder head and into the many aspects of this highly technical area.

BASIC PREP

The first step in preparing cylinder heads for racing or high-performance applications is to inspect them closely. After finding the casting number and measuring the outside diameters of the valves, check the block side of the heads with a straightedge to make sure they are square, or flat. Next, inspect the *valve guides* and the *valve seats* for cracks, especially in used heads. Pay special attention to the exhaust side.

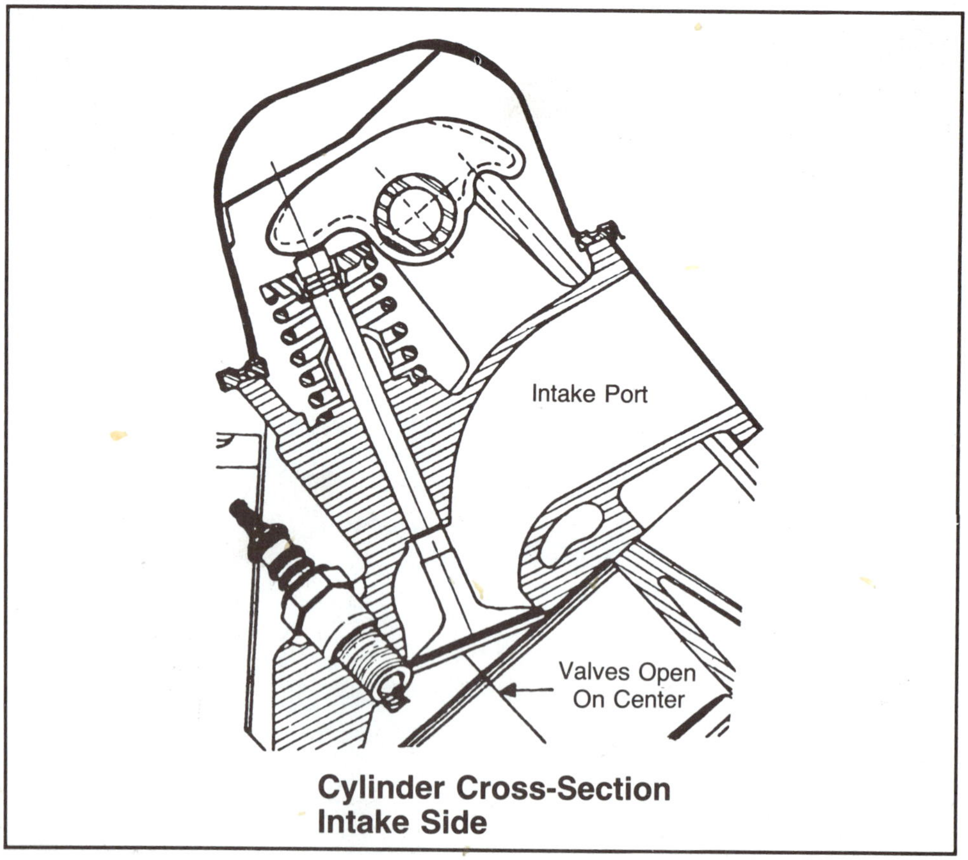

Cylinder head cross-section through the intake port. Note the hydraulic valvetrain. Also note high, angled position of the spark plug. (Courtesy Mopar Performance)

This is a typical A Engine head from the exhaust side. Note that the air pump (emissions) holes just below each port are plugged.

If the heads are used, they should be hot-tanked and deburred unless used in NHRA Stock or Super Stock, in which case they are best left unmarked. This is because NHRA and IHRA do not like machine marks on stock cylinder heads. For all other applications, the heads should be deburred and the various bolt holes chamfered. *NOTE*: If the head is to be milled, then the deburring and chamfering should be done *after* the milling process.

MILLING

The first major consideration on the head is how much to *mill* or machine off, if it needs to be machined at all. The machining is done by a professional machine shop, but the question is, how much should be milled?

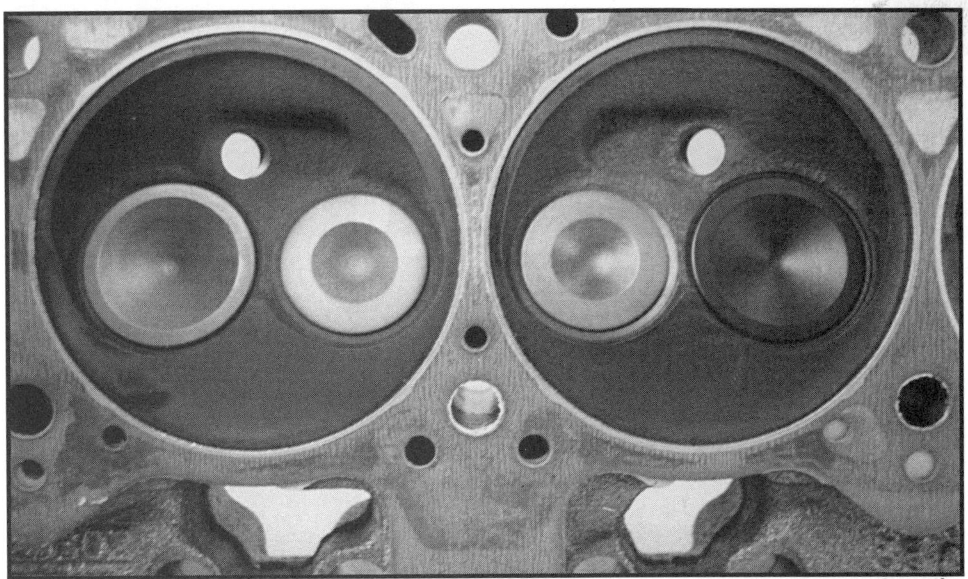

This is the 318 combustion chamber with stock valves on the seat. Note the distance from the edge of the valve to the outside edge of the chamber.

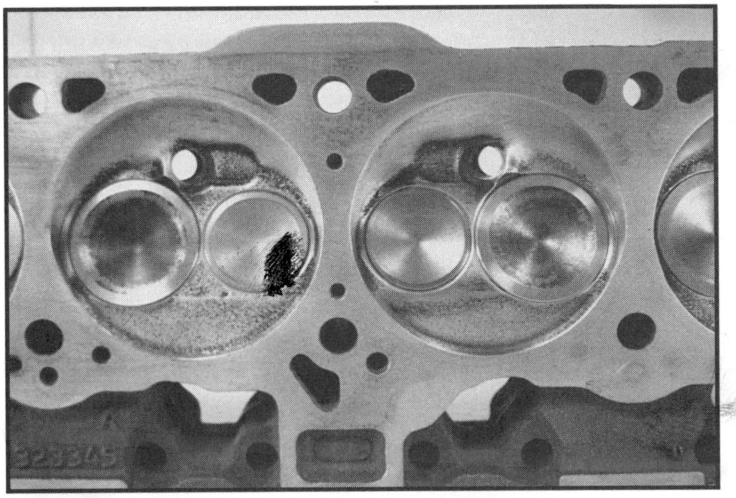

This is a 340/360 combustion chamber. Compare the valve-to-valve clearance and note the closeness of the combustion chamber's edge to the edge of the valves. Compare these features to the 318's shown in the previous photo.

This is a new style 318 ('86 and newer) chamber shape. Note the point in the chamber between the valves pointing down. This is part of the new swirl-port technology that was originally developed on 4-cylinder engines and for emissions reasons. It can also be useful in racing when used properly.

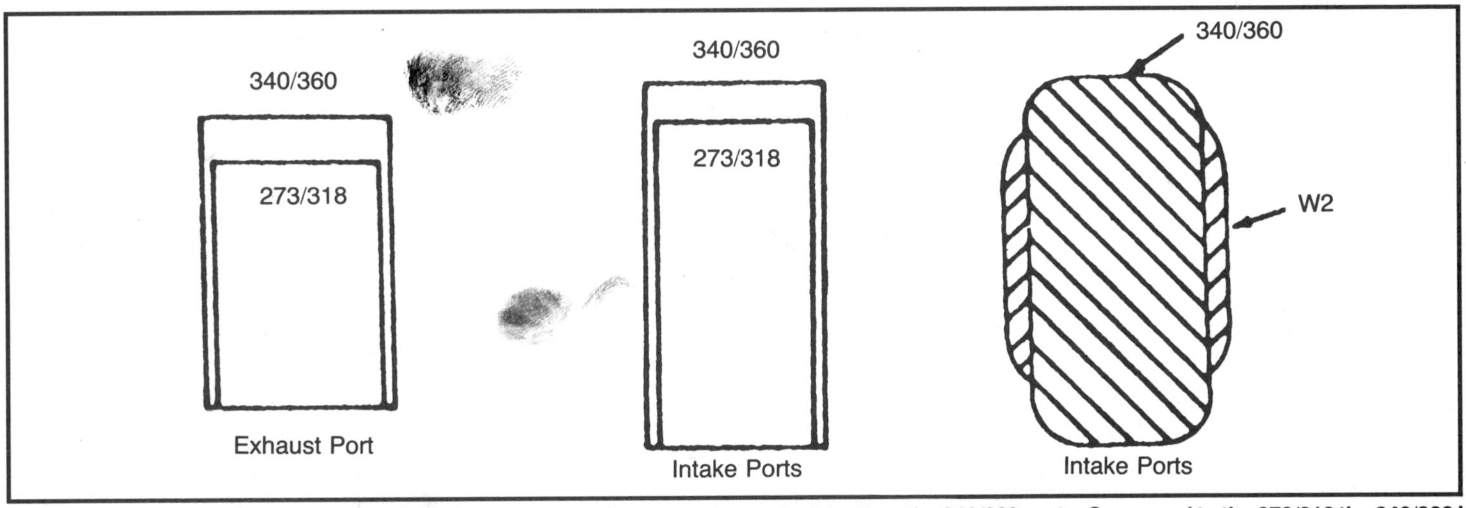

The relationship of various ports is shown. Note that the oval-port W-2 is much wider than the 340/360 ports. Compared to the 273/318 the 340/360 is taller and wider. (Courtesy Mopar Performance)

The maximum A engine heads can be milled is 0.060 in. If more than this is needed, then the block might have to be milled or domed pistons might need to be used. Engines that are milled excessively are generally hard to seal (gaskets, manifold fit, etc.) and more expensive to build (more work checking clearances and additional machine shop time).

In most cases, heads are only machined by a few thousandths-of-an-inch. But you'll still have to supply a specific number to the machinist. To do that, you'll have to measure a couple of areas. The first is how even and level (flat) the block-side is, taken with a straightedge. The second measurement is recorded from the volume of the existing combustion chambers. This is commonly referred to as *cc-ing the head*.

CC-ing—To cc a head you will need a 100-cc burette, a flat, clear plate (usually plexiglass and approx. 1/4-in. thick) with a hole off to one side, some cc-ing fluid (parts cleaning solvent is a good choice) and a light grease, such as petroleum jelly.

First, turn the head upside-down (deck surface up) and support it above the work bench so that it's almost horizontal. One end or edge should be slightly higher than the other. Next, insert an intake valve and an exhaust valve in one of the center combustion chambers. Later on, after all valve work is done, all chambers can be cc'd to check that final volumes are within 3/10 of a cc. But for now, just record the volume of one chamber to calculate compression ratio to determine the amount of head milling necessary. Seal the valves with the light grease, then insert a spark plug. Apply a thin ring of the light grease around the outside of the combustion chamber, then carefully lay the flat plate over the combustion chamber so that the hole is on the *high* side. Fill the chamber with fluid from the burette. The amount used indicates the cc volume of the combustion chamber. As a general guide, the stock 340/360 chambers run in the 68 to 75 cc area. A racing 340/360 head will typically run in the 63 to 66 cc range. The 273/318 heads run about 3 to 5 cc's smaller than the 340/360 heads, equal-to-equal.

Compression Ratio—Once the head is milled flat or straight, further milling is done to increase the *compression ratio* of the engine. More compression will make more power *if* you have the proper fuel. So the decision comes down to the type of fuel used. I will cover fuel and its effects upon the engine and compression ratio in more detail in Chapter 7 on *Blueprinting*, but for the moment use the following general guidelines:

Use a machinist's straightedge and feeler gauges to check for head squareness. Clearance is measured in the middle. The straightedge is run down the center of the head and also on both diagonals. If it's not flat or square, then it must be milled.

- Best Pump Gas—9.5-to-1 compression ratio maximum.
- Cheap Racing Gas—12.0-to-1 compression ratio maximum.
- Top Grade Racing Gas—13.5 to 14-to-1 compression ratio.

Most amateur racers use pump gas so I'll use this for an example. The best pump gas (usually no higher than 92 octane) dictates that the compression ratio be 9.5-to-1 or less. Now that the final compression ratio (9.5 in in our example) has been determined, you need to calculate what it's going to take to get your engine to this ratio.

First, you'll need to know the *exact* bore and stroke of the engine. This is used to determine the *displacement* of one cylinder. As an easy example, suppose you have a flat piston that's 0.030 in. below the deck at TDC (top dead center) and that the head gasket thickness is also 0.030 in. This is information that can be easily measured, or obtained from your engine building records. Assume that the cylinder

Milling requires careful set-up and measurement before any actual machining is done. Here, Robert Landy from DLI gets set to mill off in increments. You should ask your machinist to do the same. The last cut on the head should be a light one. (Photo by Michael Lutfy)

head's combustion chamber measures 73 cc's. With the engine *bore size* number (not piston diameter), calculate the volume above the piston at TDC and the volume of the gasket. In this example, add these two volumes to the combustion chamber volume in the head to find the *total combustion chamber volume*. This number is added to the cylinder size (volume) and the answer is divided by the total chamber volume. This final answer is

The first step to cc-ing the cylinder head is to seal the valves (intake and exhaust) to the head. A light grease applied to the valve seat will be adequate.

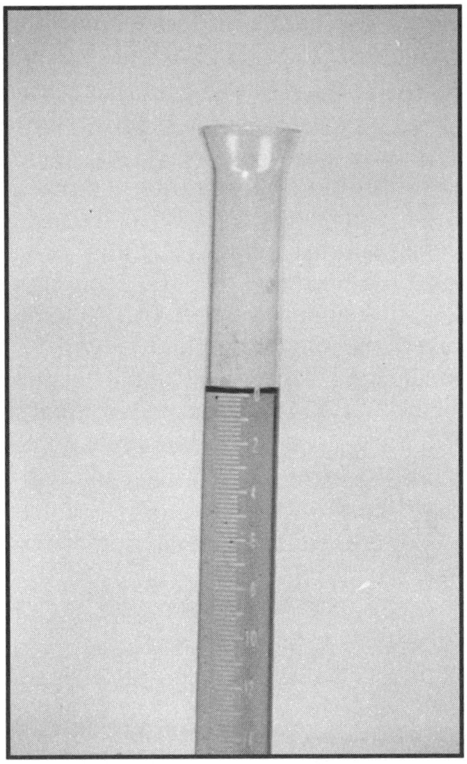

The next step is to fill and zero the burette. It's best to use a colored cc-ing fluid so that it's easier to read.

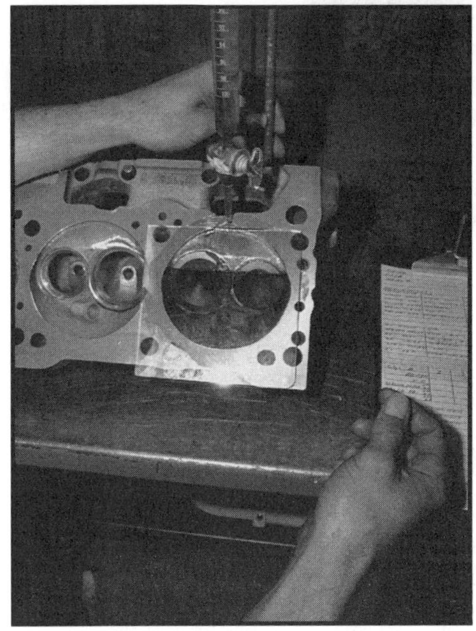

You should use a burette with a 100 cc capacity. Once set up, the chamber is filled with the cc-ing fluid, careful to avoid any air bubbles. Read the burette for the combustion chamber volume. Don't fill the hole in the cover plate itself. Pay careful attention to the volume after you have stopped filling the chamber. It will give you a good indication if there has been any leak.

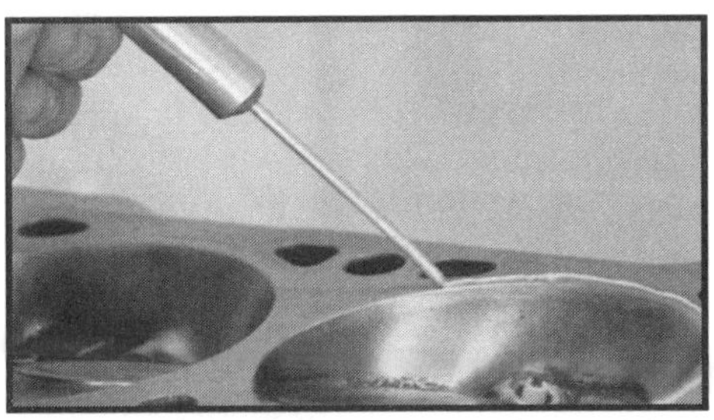

With the valves seated (no springs required) and the spark plug installed, the head is set up inverted-combustion chamber up. You want to have one side of the chamber higher than the other. You don't want it flat or level. Then you run a small bead of light grease completely around the chamber.

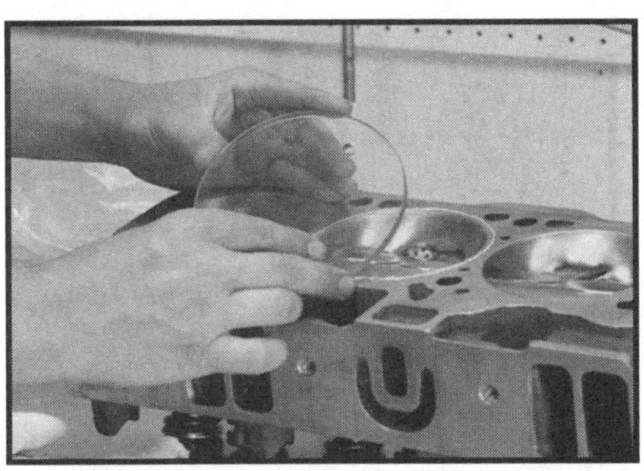

Next you place the thick, flat clear plate (1/4 inch plexiglass) over the chamber. The plate can be round or square as long as it fully covers the chamber. The small, drilled hole should be to the high side of the chamber.

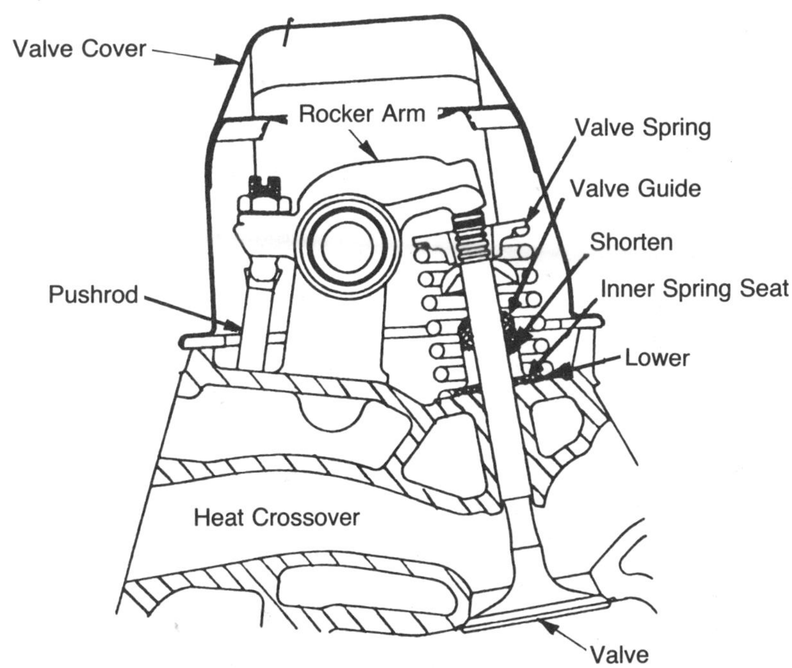

Cylinder Head Cross-Section

Cylinder head cross-section through the exhaust port and heat crossover. All production A engine heads have a heat crossover, the W-2 doesn't. Note the mechanical valvetrain shown. Also note waterjacket on either side of exhaust valve stem above the exhaust port. Race cams and mechanical valvetrains, will likely lift the valve higher and use dual valve springs (a single spring is shown). The extra lift may cause shortening of the valve guide to maintain keeper-to-guide clearance at max lift and the inner spring seat is machined flat with the outer for the dual spring. Any other spring seat machining must be done lightly and carefully because of the waterjacket. (Courtesy Mopar Performance)

the engine's *final compression ratio.* Now do you see how important it is to keep good records? Imagine how hard it would be if you didn't!

Getting back to the above example, assume the compression ratio figured out to be 9.0-to-1. Since 9.5 is desired, the chamber is too big. Milling the head is the best and easiest method to increase compression ratio. Can you tell the machinist you want him to mill the head .5 compression ratio? Probably not! To get a number, go back to the equation for calculating compression ratio and use it to calculate the desired chamber volume for 9.5-to-1 compression ratio. I will assume that for the above example, the chamber needs to be 6.0 cc smaller in total volume. Since everything else in the engine is a constant at this point, the 6 cc's have to come out of the cylinder head (not out of the gasket and not from above the piston). So this means that the original volume of 73 cc's must be reduced to 67 cc's. You might be able to relay this information to the machinist, but I suggest you go one more step.

There is a relationship between the cylinder head's combustion chamber size and the amount that is to be milled off, i.e. a *thickness-to-volume ratio.* There are two numbers that are generally used as a guideline. For the big-chamber heads (340/360), every 0.0048 in. milled off reduces the chamber volume by 1 cc. For the small-chamber heads (273/318), 0.0053 in. milled off subtracts 1 cc.

Let's assume the head in the example is a 340 head, so you would use the 0.0048 in. number. Therefore, the number to give the machine shop is 0.0048 in. multiplied by 6 (cc), which is 0.0288 in. Round it off to 0.029 in. (29 thousandths-of-an inch). That is the number to give to the machinist.

One other thing to consider: If the *deck* surface of the head is machined, the *intake* must be machined as well so the manifold will fit properly. This numerical relationship is the same for *all* A engine heads. Basically, for every 0.010 in. milled off the head (or block), mill 0.0095 in. off the intake surface. In the above example, this means the intake surface should be milled 0.0275 in. (0.0095 in. x 2.9).

Spring Seats—Check the valve *spring seats* in the head. Most stock heads are machined for only a single spring, meaning the spring seat has a large inside diameter (I.D.) and a *step*. If dual-springs are used, the steps have to be removed and the I.D. decreased so the dual-springs will fit. In some cases, the dual-springs' I.D will be small enough so that the whole outside diameter (O.D.) of the valve guides will have to be machined smaller.

Valve Guides—The next items to check are the *valve guides.* Valve guides are critical to power output and durability. If the engine is to be seriously raced, or if it is used, then *bronzewell* guides should be installed. This is even recommended for new heads with good guides, because the engine will make more power and sustain it longer. Durabililty increases because of the bronze-to-steel wear characteristics, and the spiral design of the bronzewall guide and its lubrication qualities. When honed to an exact clearance and combined with a good valve job, bronzewell guides locate valves precisely over the *valve seat.* If the valve head is perfectly round and the seat in the head is perfectly centered (a definition of a good valve job) then the valve will seal well and therefore produce good power. If the valve head moves around because of excessive clearance in the guide, then the valve head will contact one side of the seat and only partially seal. Also, the localized contact will overstress the seat and cause it to wear quickly into an oval shape. The resulting oval seat and round valve will seal very poorly, and the power loss will be extreme. It can be worse than a bent valve.

Valve stem sizes are also related to valve guides. The standard A engine uses 3/8-in. valve stems. For all-out professional class racing, it is some-

ANGLE-MILLING

For the last few years, *angle-milling* has been a hot topic in drag racing. Where it's allowed, most small-blocks use angle-milled heads. This isn't a bracket racer tip—it's much too expensive and complicated. It's usually done in highly modified classes. Basically, angle-milling is taking more material off one side of the head than the other. The "angle" is between the intake and exhaust *sides* of the head *not* between one end and the other.

Let's explain this in more detail by giving an example. Usually the head is milled flat or parallel to its existing surface. For example, if 0.030 in. was milled off the head regularly, then this would mean that you would mill 0.030 in. off both the top (intake) and 0.030 in. off the bottom side of the head. On the other hand, if you milled 0.020 in. off the intake side and 0.040 in. off the exhaust side, the head would be angle-milled. The nature of angle-milling requires that some amount be milled off both sides although typically the amount removed from the intake side is very small-only a few thousandths. If this angle-milling is done to the head, then the head bolt bosses must be refaced, the head bolt bores redrilled and the manifold surface remilled. It's usually best to use two intake gaskets on this type of set-up. Also the valve notch angles and locations will have to be recalculated and the pistons renotched. Why would we go to this much trouble? Angle-milling effectively *raises* the intake port. Raising the intake port is known to make more horsepower (flow more air). It's less expensive and more reliable to angle-mill than it is to raise the port by welding.

The cutter is installed in the special seat and guide machine. It will pilot on the valve guide itself.

times best to use smaller valve stems. The most commonly used valve stem sizes are 11/32 in. and 5/16 in. The 5/16 in. was used on the 426 Hemi from '64 to '71. The 11/32 in. wasn't used in production. These valves are usually made of stainless steel or titanium. However, oversized valve stems should not be used to compensate for worn guides. You'll only be prolonging the problem, not correcting it.

Measuring Guides—Whether or not bronzewell guides are used, the *height* of the valve guide and the *keeper-to-guide* clearance need to be measured. This clearance should be 0.100 in. at maximum cam lift.

To measure, insert valves into the intake and exhaust guides in one of the

WORN VALVE SEATS

The standard heads and the cast-iron W-2s use the machined surface in the cast iron of the head as the valve seat. After 3 or so valve jobs, the seats sink and wear. Oversize valves can fix this but there are limits on how big you can go. Another solution is to replace the seats. Generally this is expensive and therefore not recommended. These valve seat inserts (the existing cast-iron seat is machined out) require very accurate machining and must be centered precisely above the guide. Doing this much machining on the head may induce stresses that will cause the seat to move slightly and require a fresh valve job to get the performance back. It's critical how this style of seat fits into the port. Sharp edges and corners will hurt flow and cause power losses. With all this in mind it's often cheaper and easier to buy a new head. However, replacement seats do allow you to save an old head that's very valuable, no longer available, etc. Carefully done by experts it should be as good as new. There is an exception to this.

High-octane leaded fuel is hard to find today. Most older high-performance A engines (the 340 especially) were built with high-compression engines—10.5-to-1 and higher. Getting gas for these cars is very difficult. When you do find high octane out of a pump, it's usually unleaded. The older engines need leaded gas to keep the exhaust valve seats lubricated. Unleaded gas will cause the seats to wear out quickly. Newer engines ('72 and up) were designed to run unleaded fuel and as such have hardened exhaust seats. The solution to this is to replace the exhaust seats with hardened seats made of Stellite material. They only need to be added on the exhaust side. All these early high-performance engines were either 340's or 360's and they all use a 1.60-in. exhaust valve. That means that only a 1.60-in. size Stellite seat (PN P4452186) needs to be considered. Once installed and the new seat grounded, the head will work as well on unleaded as the current engines do! Then it becomes a question of compression ratio versus octane level.

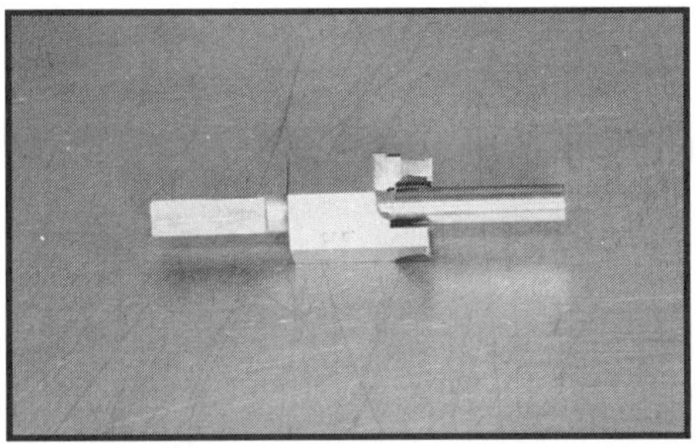

To cut, shorten or narrow the valve guide on the head for special racing/performance purposes, a special cutter is required.

Spring seats are cut in a similar manner on the same machine but a different cutter is used. Usually this is only done to lower the inner spring seat to the same height as the outer for use with dual valve springs.

Although the seat and guide machine looks like a big drill press, it is much more sophisticated. Bridgeports and other similar machines can be used but are slower and require more care in operation.

The cutter has narrowed the guide and shortened it at the same time. It's also concentric with the guide itself.

chambers. Then install the keepers and retainers to be used *without* the spring. Preload the retainer and valve upward by hand, then measure and record the height between the valve seat and the retainer. This clearance would also be the *installed height* of the spring. Next, push the valve downward, holding the retainer in place on the valve until the keeper touches the guide. Measure the distance from the valve seat to the retainer again. Subtracting these two numbers gives the keeper-to-guide clearance when the valve is closed or *seated*. It's a big number, around 3/4 in. If you subtract the valve lift of the cam from this measured clearance, the result is the keeper-to-guide clearance at the *maximum lift* of the cam. This number should be equal to or greater than 0.100 in. If it is less than 0.100 in., then the guide will have to be shortened.

For example, a typical A engine installed height is 1.69 in. Assume that the keepers hit the guide at 1.00 in. This means that the clearance is 0.690 in. in the seated (closed) valve position. At this point, there are two ways to look at it. One, you could install any cam up to 0.590-in. lift (figured by subtracting 0.100 in. from 0.690 in.) and have adequate clearance. Two, if you select a 0.620-in. lift cam then the guide would have to be shortened 0.030 in. (0.690 in. − 0.620 in. = 0.070 in. which is 0.030 in. less than the 0.100 in. minimum required).

Valve Seals—Valve seals should be considered next. The *PC-style* seal requires simple machining on the top of the guide, but can accept the same clearance between the keepers and guide. An *umbrella* seal requires no change to the guide itself but does need more clearance than 0.100 in. The 0.100 in. is adequate if measured with the seal in place. Remember that umbrella seals cannot be used with dual springs because the inside diameter of the inner spring is too small. Therefore, single springs will have to used, which limits the maximum lift of the camshaft to around 0.500 in. This lower lift usually allows for the extra clearance required.

Details—The heads are nearly ready for the machinist, except for a few more details. First is the *oil feed hole* to the valvetrain. The oil that lubricates the valvetrain comes up from the block, through a drilled passage in the head and rocker stand and onto the rocker shaft. If this hole is not drilled through, is plugged or restricted, then the valvetrain won't be oiled properly.

Next, set the heads on the block

AIR PUMP HOLES

Most of the newer production heads have air pump holes in the exhaust ports. If the heads are to be used for performance or racing purposes, these holes should be plugged. Threading the holes and installing a small Allen screw is probably best with the heads off the engine. If left open, they will leak exhaust when headers are installed.

There are two exhaust flange bolt patterns on the W-2 cast-iron heads. You will only use one set of holes. On the holes that you are not going to use, you should plug them with small Allen screws. Be sure to use sealer on the threads.

A finished machined valve guide. Note that the second guide up or rearward from the front is shorter, and narrower than the other guides. Also note that there isn't any chamfer on the top of this guide.

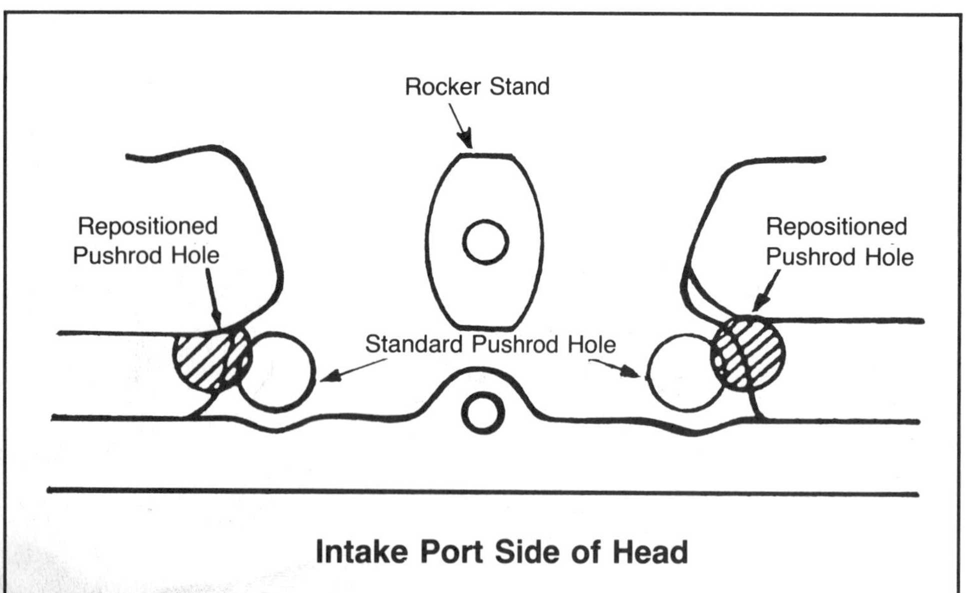

Pushrod holes on 340 T/A heads (1970) were offset from the intake port. Better porting can be done as a result, left and right offset rocker arms must be used. Note that the moved pushrod hole "breaks out" of the head casting. (Courtesy Mopar Performance)

DLI installs bronzewell guides, then checks the clearance and hones them for an exact fit. (Photo by Michael Lutfy)

without a gasket and make sure they sit flat against the block's deck. If they don't, the *dowel pins* may be too tall and are holding the head up off the deck surface. Or, the dowel holes in the head may not be drilled deep enough or may have a restriction in them. Either way, a machine shop can fix it. As long as the dowel locates the head adequately, it's more cost-effective to shorten the dowel.

While the head is on the empty block, check the combustion chambers' alignment over the cylinder bore by using a light and looking up the cylinder bores. All four chambers on one bank would need to be observed before any judgment can be made. Look closely for sharp edges of the chamber hanging over the cylinder bore. If all the chambers appear shifted in one direction, then *offset dowels* can be used to bring the head(s) back into alignment. This is rarely required, however.

With the pistons installed in the block, install the heads and check piston-to-head clearance as outlined in Chapter 4. Since the block itself should be milled at this point, this check should be quite accurate. The proper piston-to-head clearance is 0.055 in. (minimum). Why check this now before sending the heads to the machine shop? The 273/318 engines have smaller combustion chambers, which "hang over" the bores. These wouldn't be affected by head milling. However, the production 340 piston has a *positive deck height*, meaning the piston top is above the deck. As such, the piston fits inside the combustion

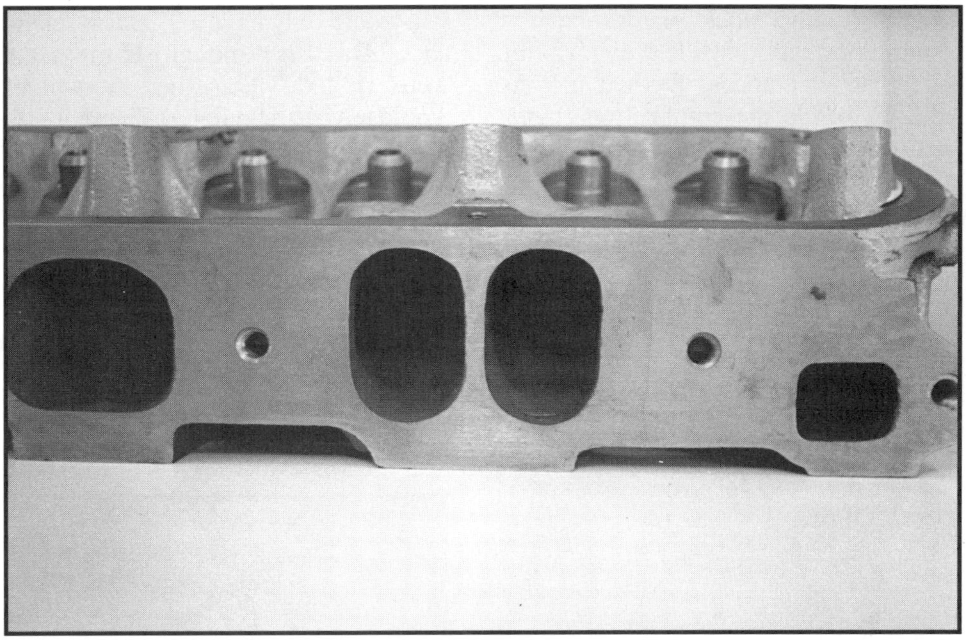

A cast-iron, oval-port W-2 version of the A engine head. The space on the left is black because there is no heat crossover. This is a race W-2 head because of the flat, cut-off rocker pedestals.

The exhaust side of the W-2 head is similar to production heads. However, note the flat, short, rocker stands. Also notice the height of the top of the exhaust port relative to the bolt holes compared to production port.

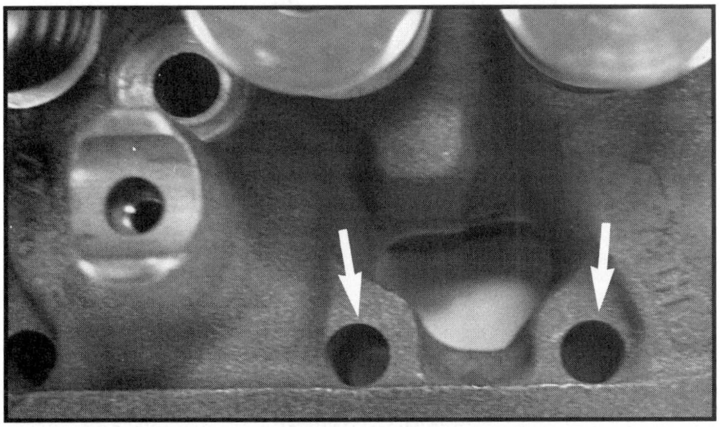

All pushrod holes in A engines up to 1986 are small.

The '86 and newer 318 head had a much larger pushrod hole.

chamber. The stock pistons have a chamfer on their outside diameters for adequate clearances. Domed pistons also fit inside the combustion chamber. In either of these cases, milling the heads decreases the amount of piston-to-head clearance. For example, if you have a 0.065-in. clearance now and you are planning on milling 0.030 in. off the head, then you would only have a 0.035-in. clearance. This isn't enough! If the heads were milled, then the short-block would have to be redone to increase the clearance. It's best to adjust the head milling before it's done.

Pushrod Holes—The final check on the heads are the *pushrod holes* (for clearance) machined into the head. There are a few important items to note. First, if the engine is a real 340 T/A, then the pushrod holes are *offset*. This means a stock 340/360 head can't be used without remachining these clearance holes. Also, 340 T/A heads can't be used on a stock 318/340/360 unless the offset valvetrain from the T/A is also used. The '85 and newer 318 engine uses an enormous pushrod hole for clearance for the production hydraulic roller camshaft. Standard 318 heads cannot be used on these roller cam engines unless the pushrods are clearanced in the head by enlarging the stock hole. However, the material in older heads may be inadequate for drilling this new type of hole. And the W-2 is in a class by itself with even more highly offset rockers than the T/A. Keep in mind that the reason the pushrod holes were moved on the T/A and W-2 was to allow for bigger ports and better porting capabilities. If you are planning to port the head, this could be a big feature for a real T/A or W-2 head or problem with stock 340/360 heads if you break through.

SPECIAL HEADS

Before I leave the area of cylinder heads and castings, there are several types of heads that should be discussed. They are quite unique, and some are difficult to find. However, they all have special applications and features.

360 Police Head—The special 360 police head was introduced in March

ECONO HEAD

The Econo version of the W-2 cast-iron cylinder head (P4120043) for A engines was sold out in 1985, so it is no longer available from the factory. You may find them in salvage yards or swap meets. However, the standard W-2 head (P3870812) can be substituted with some modifications. The valvetrain is the only difference between the two heads. Therefore, the performance stays the same.

The difference in valvetrain is in the location of the rocker shaft. The Econo W-2 uses the shaft in the production location with cast-in pedestals, while the standard W-2 (race version) has special rocker stands that move the shaft up and away from the valve tip. This requires longer valves.

There are two ways to use the standard W-2 head in the Econo head applications. One is to use the shafts 2.02 and 1.60 dia. "long" valves with the Race (P4120589). With this setup, 'B' engine valve springs are used because the installed height is 1.86-1.90 in. This is not the "econo" approach but generates the best valve geometry, best valvetrain oiling and is the easiest to do. It's a bolt-on job and requires no machining.

The second method involves reworking the Race W-2 rocker stands (P4120102). By reworking the stands you use standard 2.02 in. intake and 1.60 in. exhaust valves (standard length) and 340 T/A rocker shaft (P3577086). Reworking the stands consists of machining 0.130 in. off the bottom while removing the existing step, and machining (drilling) the attaching hole (3/8 in. dia.) back on-center, which moves it 0.110 in.

These operations should be done by a machine shop. It is also recommended, for better durability, that the attaching screw hole in the bottom of the stand be welded up before it is redrilled back on center. With this modified stand the T/A rocker shaft *must* be used.

The first step of a valve job is to measure the seat for runout. A special type of dial-indicator is used. It will be used at each step in the seat grinding operation. Indicator is piloted on the shaft located in guide. Guides should always be replaced or repaired first before the valve seat is machined.

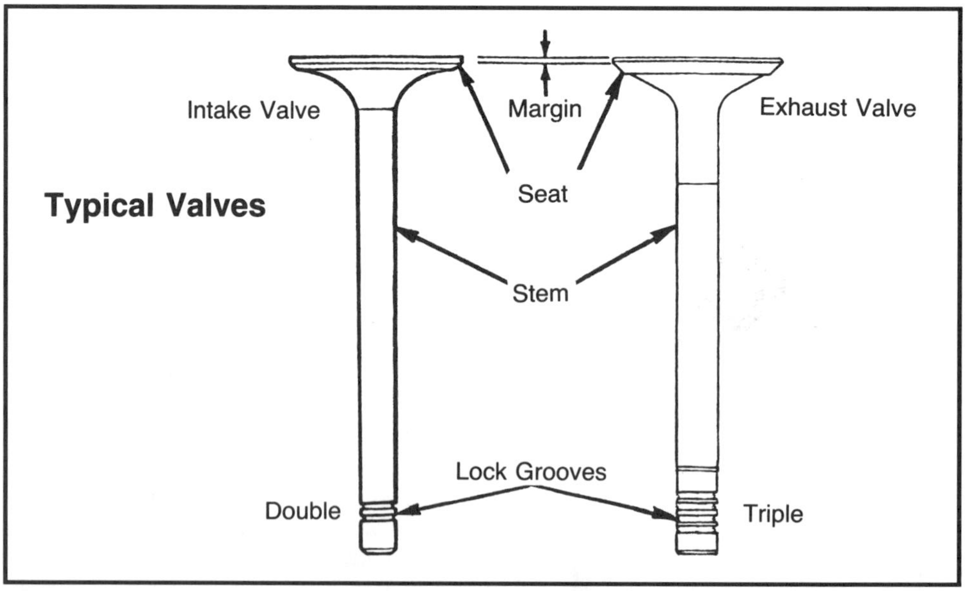

Valves and their various features. Note lock grooves on exhaust valve are triple while intake valve uses double. Occasionally, the intake uses a single groove in production. All race valves are single groove (intake and exhaust). In A engines, usually two stems sizes are used—3/8 in. and 11/32 in.

of 1977. This "new" casting added more water jacket holes in the lower edge (spark plug side) of the cylinder head. These extra holes are for sand removal only. They do not affect water flow or cooling. For identification purposes, the older police heads had 4 D-shaped holes while the newer heads have 9 triangular-shaped holes. The new head uses a new production gasket (PN:4027329). This gasket can be used with the older heads but the older gaskets cannot be used on the newer heads.

W-2—The W-2 cast-iron A engine heads were introduced in September 1976. They will out-perform any of the production cylinder heads built to date. There are two basic versions, Race and Econo. The Econo head was introduced in 1977 and carried up until 1985. The only real difference between these two heads are the rocker shaft *pedestals*. The Econo head has them cast in while the Race head machines them off and uses a separate stand. The Econo head is so named because of the valvetrain used, and not because of its characteristics. The cast-in pedestals allow most of the 340 T/A valvetrain to be used except for the intake rockers. For a list of specific features of both types of W-2 heads, see the sidebars.

VALVE SEAT PREP

The preparation of the *valve seat* and *valve face*, generally lumped together into the category of *valve job*, is one of the most important areas in the engine building process that relates directly to performance. All the high-tech, high-buck hardware plugged into the engine won't do any good unless the valves seal properly. Leaking valves cost horsepower. But sealing is only part of the story. *Flow* is the other part. The type of valve job affects the flow capacity of the ports.

A good valve job starts with good guides as mentioned earlier. The valve

W-2 HEAD FEATURES

- **PORTS**—Large, oval-shaped intake ports with added material around the port to allow for optional intake porting. High, round exhaust port biased away from the centroid of the cylinder. Both ports feature improved airflow and better breathing capabilities.

- **COMBUSTION CHAMBER**— The combustion chambers are the same as 340, same spark plug location, basic volume and shape.

- **SPARK PLUG**—Plug size same as standard 340 (N9Y or N63Y).

- **HEAT CROSSOVER PASSAGE**— None.

- **VALVE SIZES**—Same as 1971 340; 2.02-in. intake and 1.60-in. exhaust.

- **VALVE SEATS**—Cut for above valve size. Semi-finished to allow the trick race valve seat to be used. This holds true for both the Race and Econo heads.

- **INTAKE MANIFOLD**—Special manifolds must be used. Larger oval ports require intake manifold bolts to be spread apart. A 340 standard manifold could be used by welding, porting and remachining the intake surface and bolt holes.

- **EXHAUST MANIFOLD**—Can use stock 340-360 exhaust manifolds and headers. Some grinding will be required for best flow through the stock manifolds. Use W-2 exhaust gasket as template. The W-2 also has provision for 4-bolt header flanges. The exhaust flange bolt holes that are not used must be plugged with studs on the W-2 cylinder heads.

- **VALVETRAIN**—The intake pushrods are inclined at an angle to allow maximum intake porting. The highly inclined pushrods require special offset bushed rocker arms for the intake valves, (Right PN: P4120200 Left PN: P4120201). These are special for the W-2. The exhaust rocker arm is PN: P4120202. *Bushed* rockers have an actual bushing that bears on the rocker shaft. These rockers are recommended for racing. The Econo W-2 can use the unbushed rockers. Intake PN: P3870825 right/P3870826 left, and the exhaust rocker is PN: P3870827. These rockers are offset the same as the ones above but are unbushed.

  The special W-2 rocker arms require the special rocker arm shaft (PN: P4120589). This rocker shaft incorporates special oiling features that are required for high engine speed, racing applications. The Econo head should use the 340 T/A rocker shaft PN: P3577086.

  The rocker shaft supports have been milled off the W-2 heads except for the Econo head and require the special supports. These stands are made of aluminum (PN: P4120098), or steel (PN: P4120102). See Econo head sidebar for Race head conversion.

- **CASTING FEATURES**—The exhaust valve guides are "wet" for better exhaust valve cooling. The rocker shaft pedestals have added material and a large radius next to the head for added strength. The exhaust valve guides are machined lower for high lift cams and better valve seal clearance. The valve spring seat has a large diameter for Hemi-style valve springs. Extra material is added to exhaust side at each end of the head in the head bolt area for added strength and stiffness. The casting number on both Race and Econo W-2 heads is 3870810.

- **PERFORMANCE**—From a power standpoint, both the Race W-2 and the Econo W-2 make the same horsepower. Both heads offer a substantial power increase over the standard 340-360 heads.

- **ALUMINUM CASTINGS**—There have been two aluminum W-2 heads—one from Batten and one from Diamond. The Batten version was introduced first and came in two versions—a standard W-2 big-oval and a big-square port (on the intake sides). The Diamond head is the newer head. It also came in two versions—a *quench* chamber and a *non-quench* chamber. In both cases the Diamond head uses the big-square port on the intake side. Both the Diamond and Batten heads use the W-2 race valvetrain. Both use valve guide inserts (separate from the head) and special intake and exhaust valve seats since valves cannot run directly on the aluminum like they do on cast iron. The biggest advantages of aluminum cylinder heads are their lightweight and ability to be repaired.

seat in the head must be round and concentric with the guide. Round means that the runout should be less than 0.001 in. The valve head must also be round and concentric about the stem. These two features will guarantee good sealing. To attack the flow aspects you have to deal with the *shape* of the seat itself. Shape is defined in terms of *angles*, *widths* and *radii*. There are different styles of valve jobs for different applications—bracket, competition, stock or three-angle. I will start with the competition valve job because it's the most complicated and detailed. Keep in mind that there are two aspects to all valve jobs—*head seat* and *valve face*; both for intake and exhaust. The specs that work best in competition are as follows:

HEAD SEAT

Intake

- Seat itself: 45 deg., 0.085-0.090 in. wide
- Top cut: 30 deg.
- Bottom cut: 60-70 deg. or approach angle 0.100-in. wide
- Vertical cut: Blending to port or 90 deg.

Exhaust

- Seat itself: 45 deg. 0.065-0.070 in. wide
- Top cut: 30 deg. cut as narrow as possible
- Bottom cut: 60-70 deg.
- Vertical cut: Blending to port or 90 deg. angle

Chrysler prefers 70 deg. for the bottom cut but many valve shops prefer 60 to 65 deg. The orbital grinder used to grind seats tends to chatter at higher angles near 70 deg. The same holds true for the last cut from the valve seat

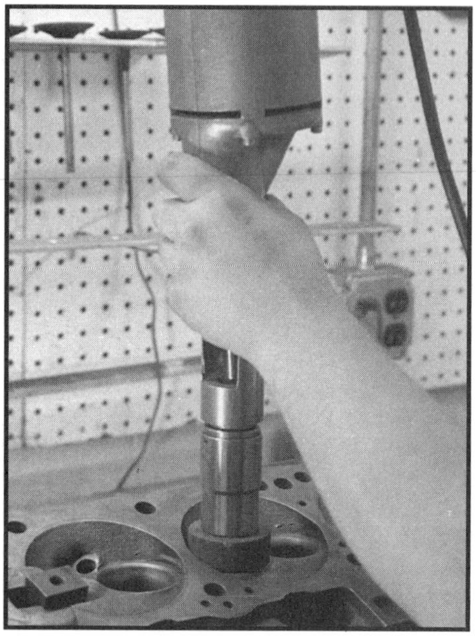

The seat grinder and a large, pre-angled stone cutter grinds the seat. Depending on the type of valve job needed, the process can include several steps with different styles of grinders, stones, and/or cutters. It also pilots on the shaft in the guide.

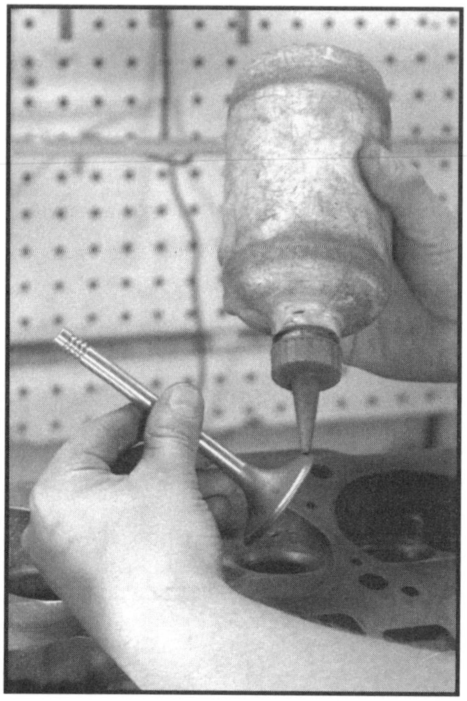

The last step in the valve job is to lap-in the valves. It's best to use a fine lapping compound applied directly to the valve seat.

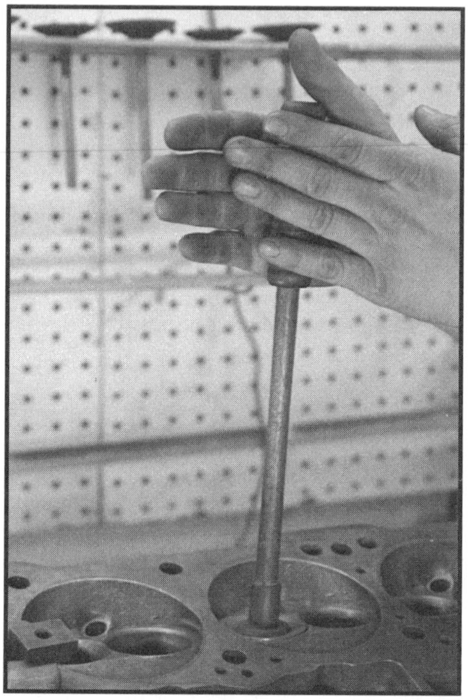

Once the lapping compound is applied, the valve can be lapped-in by hand. It should only be done lightly to mate the two surfaces.

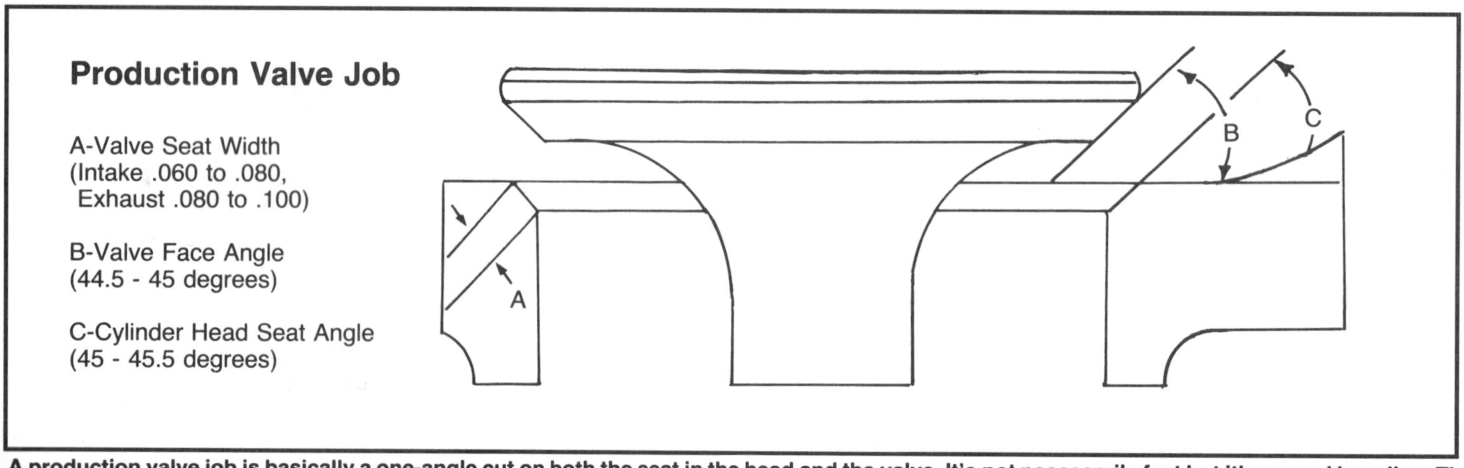

A production valve job is basically a one-angle cut on both the seat in the head and the valve. It's not necessarily fast but it's a good baseline. The rest of the seat is roughed in by a multi-angled cutter rather than ground in as the seats are.

itself to the port. This is more properly described as *blending*, going from the bottom cut and blending into the port. This is usually done very carefully with a hand grinder. If you have the equipment to grind at 90 deg., then a vertical cut 0.100-in. wide will also work fine.

VALVE FACE

Intake

- Seat itself: 45 deg., 0.085-0.090 in. wide
- Margin: 90 deg., 0.050 in. wide, sharp/square top
- Back cut: 30 deg. as underhead shape allows

Exhaust

- Seat itself: 45 deg., 0.060-0.080 in. wide
- Margin: 90 deg., 0.080-0.090 in. wide
- Top radius: clipped very small
- Back cut: 30 deg. as underhead shape allows.

Bracket Racing—A valve job for bracket racing is similar to the all-out competition job, with some differences. Usually the bottom cut is 70 deg. and is blended into the port smoothly. Since bracket engines typically use stock parts, then *back-cutting* is more likely. The stock valves have an underhead shape that lend themselves to back-cutting. Bracket engines are not likely to use the more expensive race valves that don't require back-cutting.

Stock—A *stock* valve job is generally called a *three-angle valve job*. This is all that's allowed in NHRA Stock and Super Stock classes. In this type of valve job, the top cut is 45 deg. and the bottom cut is 70 deg. That's all. In some cases, the top cut is omitted.

The top cut may only be up to 0.250 in. larger than the outside diameter of the valve head. The bottom cut can be up to 70 deg. but the *depth* (width indirectly) is limited to 0.250 in. measured from the top of the valve seat to the bottom of the bottom cut.

The above specifications serve as a general guide. The shape of the actual valve used on the valve seat can affect the ultimate performance. However the above specs will be very close. If you have a flow bench and lots of time you might find improvements. If you plan on using a unique valve shape, it's best to have it flowed first!

Short-Side Radius—One of the most popular terms in valve seat language is *short-side radius*. This relates to the floor of the port as it turns and passes the valve seat upstream (or away from the chamber) of the valve seat. The idea is to get as much short-side radius as possible. The W-2 heads excel in this area. Keeping the valve high in the chamber and maintaining the highest short-side on the port will flow the best and therefore make the most power. Bigger valves (larger diameter) move the valve seat outward, which also helps the short-side radius. Sinking the valve is never recommended because it hurts performance. One of the reasons for this is that it decreases the short-side radius.

Shrouding—A typical combustion chamber *shrouds* the valves, meaning

The valve should be inspected and measured before and after grinding.

A valve grinding machine is required to grind the valve face. It's a very specialized machine. Note the valve head just below the lamp.

The underhead area of the valve can be polished by putting the stem in a drill press and using a polishing cloth.

that it encloses them. Bigger valves make this shrouding worse. A large cylinder bore helps alleviate this condition and the related engine performance by unshrouding the valve. Relieving the combustion chamber around the outside diameter of the valves will also help.

VALVES

The valves in the A engine small-block Mopar *open on center*. Technically, this means that as the valve head travels along the centerline of the valve stem moving up to maximum lift, the maximum O.D. of the valve crosses the plane of the block's deck surface at the cylinder bore's centerline. This does not mean much until you realize what this does for performance. It allows the A engine to use a larger valve than other small-blocks—all other things being equal. For example, if you compare a Chevy and Mopar small-block, each with 4.0 in. bores, then the A engine can install the larger valves without notching the cylinder bores or offsetting the guides, which can be expensive. At the same valve size as other small-blocks, such as 2.02-in. intake and 1.60-in. exhaust, the valves are less shrouded as they approach maximum lift and therefore flow more *real* air—as installed on the engine. In most cases, the extra valve size is used on the intake side rather than the exhaust.

VALVE SIZES

There are three *intake* valve sizes used in the production A engines: 1.78 in., 1.88 in. and 2.02 in. There are

Generally the stem and keeper grooves are polished by hand, if required.

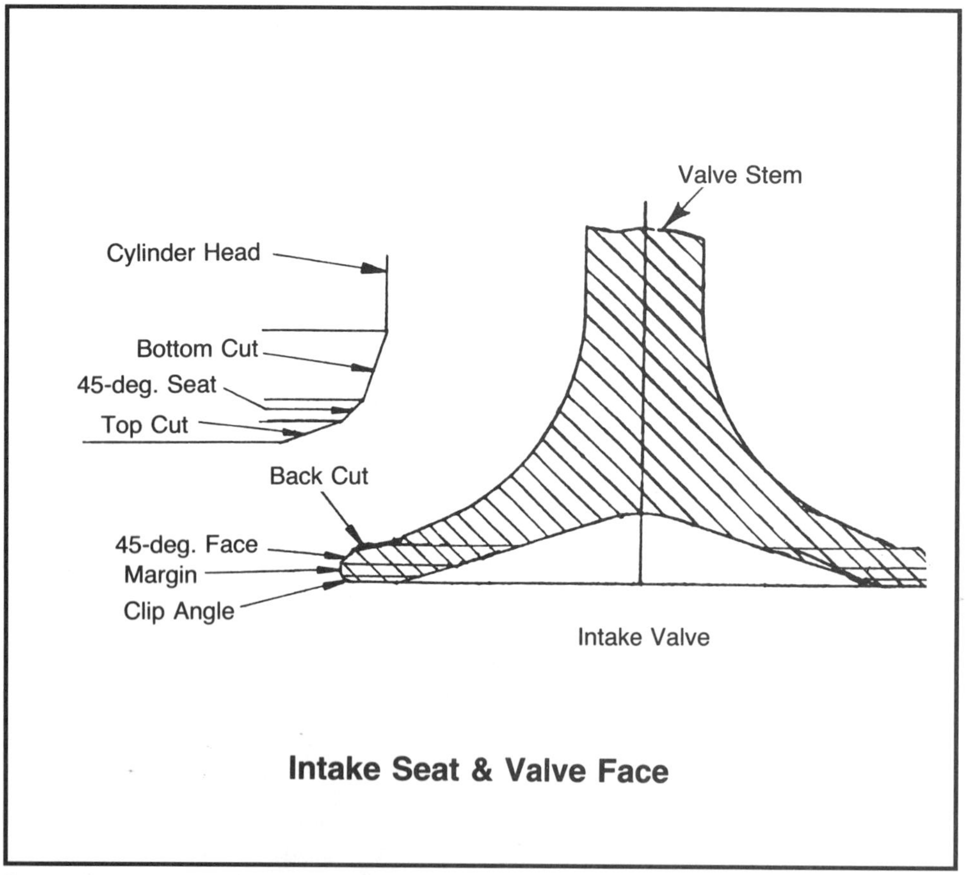

Competition valve job features three angles on both the valve and the seat in the head. The preparation of the valve face and seat in the head is critical to the sealing of the valve and engine output. A wider seat is more durable but flows slightly less. Sealing is extremely critical on the intake side. A race valve may not require a back-cut but note the ski-jump typical of stock valves shown on right side. The clip angle, sometimes called top cut, is not always used and is shown for reference only.

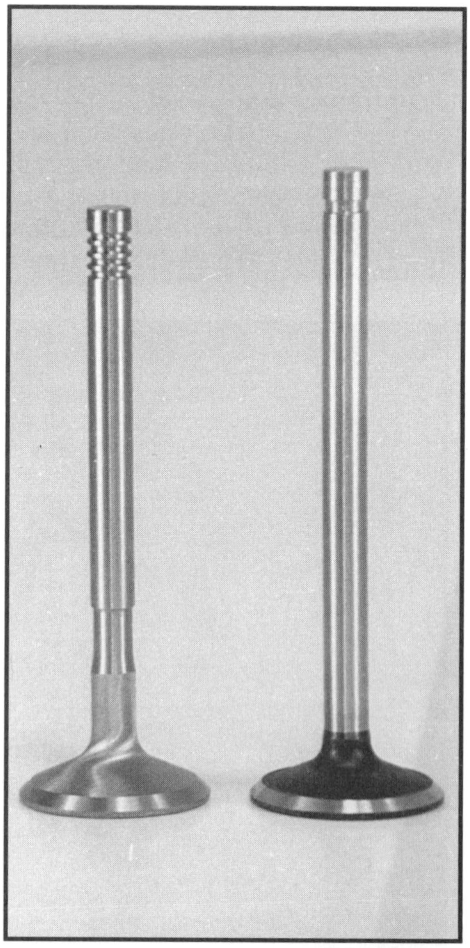

Exhaust valves are varied. Note that both of these valves are more tulip-shaped than the nailhead intake valves.

only two exhaust valve sizes: 1.50 in. and 1.60 in. All these valves use a 3/8-in. stem diameter and have the same overall length. For racing there are valves that change all of these characteristics. I will look at each of the individual characteristics of the valve separately.

Head Diameter—The most easily identified feature of the valve is the *head diameter*. A 2.08 in. valve is the largest, generally used in the larger bore 340's. It's too big for the 318 and is not needed. Most of the exhaust valves stay at 1.60 in. Chrysler has used a 1.65-in. O.D. exhaust valve, but it is usually cut down to 1.61 in. to 1.62 in. by the time it's installed in a race engine. The bigger valve generally works better with the stock-style exhaust. On the W-2, the exhaust port itself flows so well that the bigger O.D. valve isn't needed.

Overall Length—Overall length of the valve is the next easiest feature to observe. Although any length is possible, there are basically two lengths used on the A engine: *stock* and *long*. Long valves are basically the same as stock valves except that they are approximately 0.250 in. longer. The *keeper grooves* also move up. These valves are used in the highest output versions of the Mopar small-block but long valves don't make more power. So why use them? Because they increase the *installed height* of the valve spring so bigger, stiffer valve springs can be used. The bigger springs are needed to work with the high-lift cams, which in turn work with modified heads that flow more air at high lifts. That all adds up to more horsepower potential.

Valve Stem Diameter—The *valve stem diameter* is hard to visually notice but is easy to measure with *dial-vernier calipers* or a *micrometer*. There are only three stem diameters used—3/8 in., 11/32 in. and 5/16 in. The 3/8-in. diameter is used in all the production A engine heads and the cast-iron W-2s. The 11/32 in. and 5/16 in. sizes are only used in racing. To use these special valve stem sizes, the valve guide must be replaced or a different bronzewell guide used that will reach the smaller inside diameter. These valves are more exotic and more expensive than the 3/8 in. stemmed valves. They are usually made of special material, such as stainless steel or titanium.

Keeper Grooves—Valves come with three different styles of keeper grooves; *single, double* and *triple groove*. The single groove is used in all racing applications. The double and triple groove valves are only used in

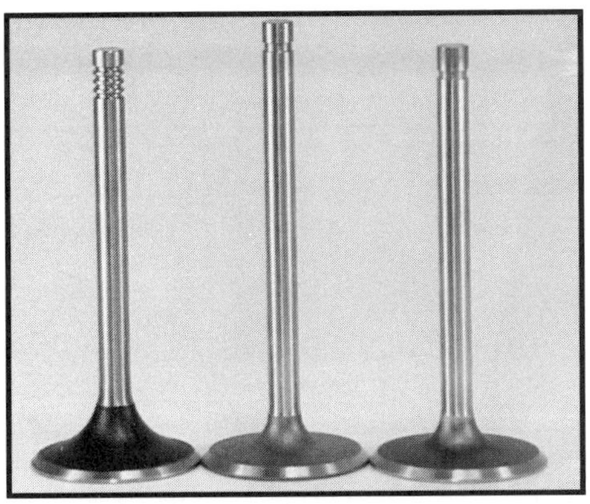

Intake valves come in many shapes and sizes. Note that the valve on the left is more tulip-shaped while the one on the right is more a nailhead. Also observe that the middle valve is longer than the other two. It's for the race W-2 set-up that uses longer valves.

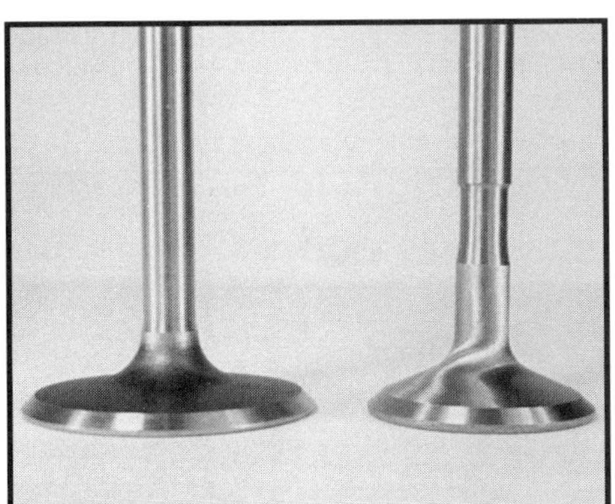

The intake valve on the left is a nailhead. The exhaust valve on the right is a semi-tulip. Concentrate on the valve's underhead angle to tell one from the other. The exhaust valve is also swirl-polished. Note that both valves are not back cut and that neither has a ski-jump.

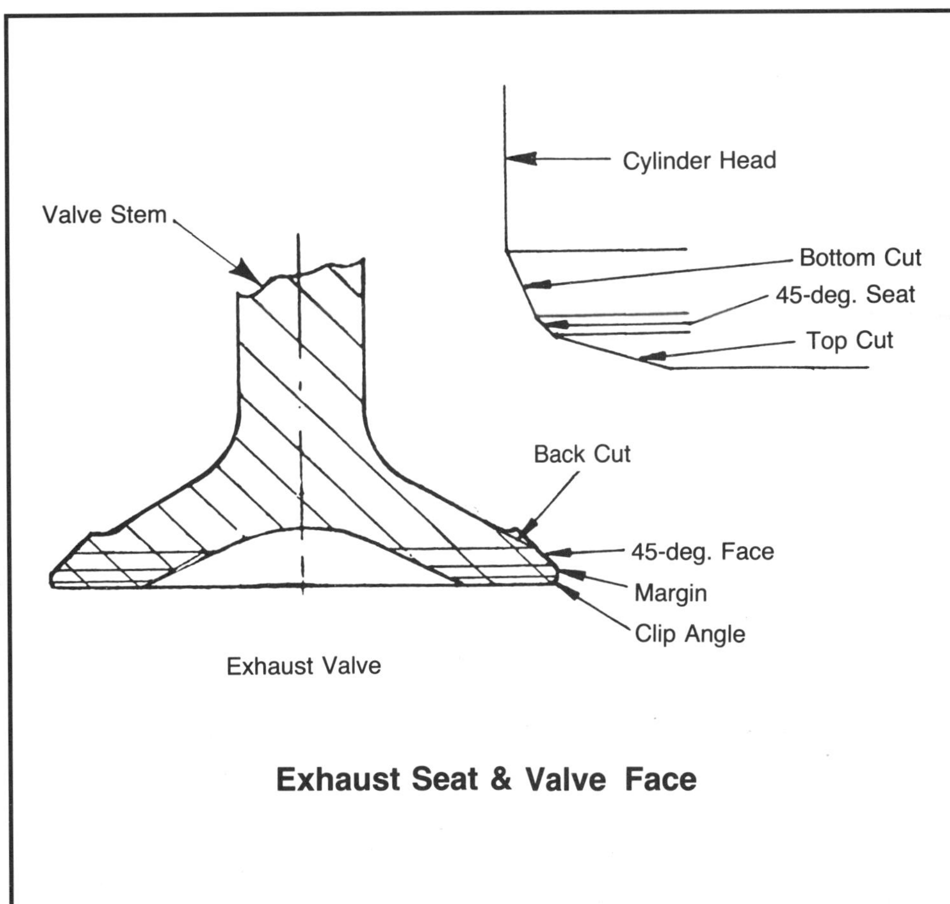

Exhaust valves tend to be more tulip-shaped than intakes. Race valves may not need to be back-cut, i.e. the ski-jump shown on the left isn't there. The clip angle is shown for reference only and is not used in many valve jobs.

production or as stock replacements. However, single groove valves are also used in production. The triple groove *lock* valves are generally found on the exhaust valve, while the single and double grooves are used on the intake. As long as the proper keepers are selected to go with the number of grooves and the stem diameter, there shouldn't be any problems. Hardened keepers are recommended for racing, by the way.

Valve Shape—The *underhead* valve shape is the hardest feature of the valve to describe. There are two basic shapes—*tulip* and *nailhead*. The nailhead looks like a flat disc attached to the stem with *zero radius*, or a square corner. The tulip has a very large radius and gradually blends into the stem, which may extend up from the seat 1.5 in. or more. Actually, no valve is a true nailhead. Most valves are mixtures of the two styles—part nailhead and part tulip. Generally the full tulip valve isn't used in wedge cylinder heads, including the A engines. It tends to block the port or restrict the port's flow at low lifts. Tulips do flow well at high lifts but are best used in hemispherical combustion chambers like the 426 Hemi.

Clipping—Clipping the valve is a process that shortens the valve's overall length by machining off a small amount of the tip. This is *not* recommended because it reduces the strength of the valve tip, making it more susceptible to wear. It also affects the valvetrain geometry, which will usually cause problems. If the tip is rehardened after it is machined then the valve should be fine and the wear properties will be restored. Double-check your geometry before clipping even if you can get the valves rehardened.

PORT FLOW

A good flow bench is expensive, so a specialist will probably flow the heads. The question is, why flow the heads at all? Just flowing the heads alone may yield an interesting number or chart but it doesn't increase power output. A flow bench is most useful when making changes to the heads. That way, a change can be tested to see whether it produces a gain or not. Relative comparisons of changes can be useful in developing an engine.

There are many tricks to making changes based on the flow numbers. Keep in mind that gaining flow is not always the most important factor. In many cases, lower flow with more area under the curve makes more power.

Valve Location—The valve location on the A engine is very favorable for higher port flows as the valves open

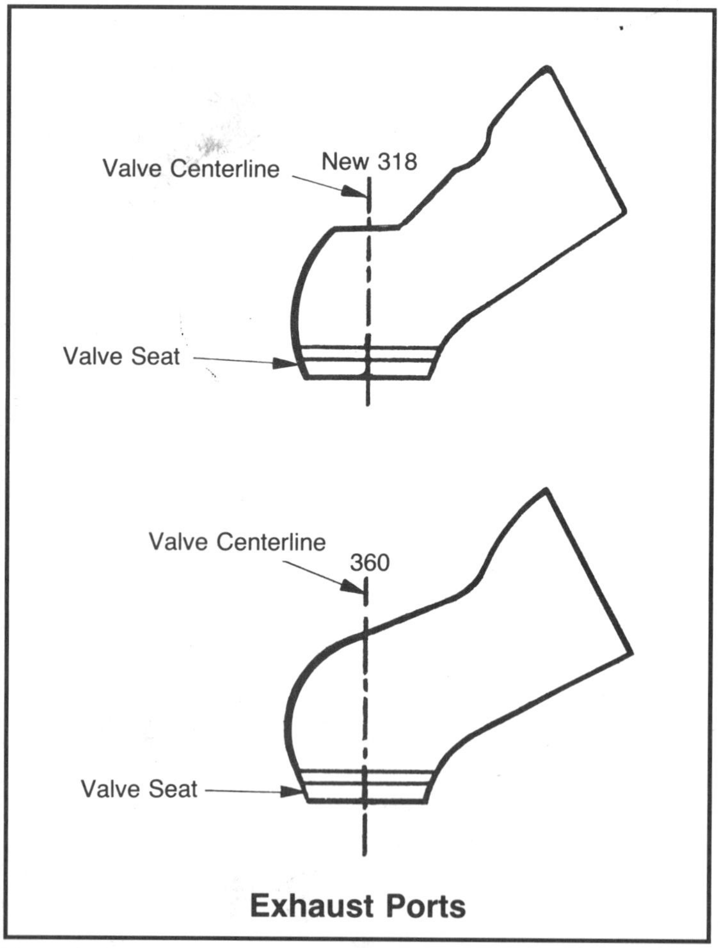

Side view of exhaust ports comparing the 318 to the 360. The 360 is larger but the 318 has the floor "raised" and the long-side radius doesn't go away from the valve centerline like it does on the 360 and most other small-blocks. These are very good features for high-flow ports and racing.

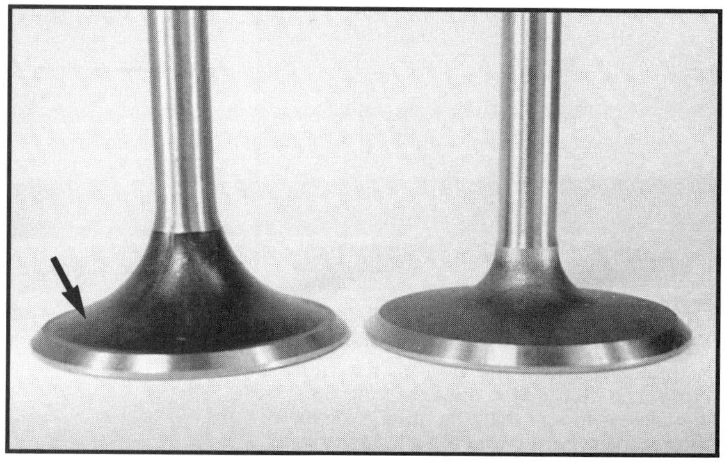

The extremes of basic valve design are illustrated in this photo. The valve on the left is tulip-shaped while the valve on the right is nailhead-shaped. The valve on the left has a small ski-jump (arrow). The valve on the right has none.

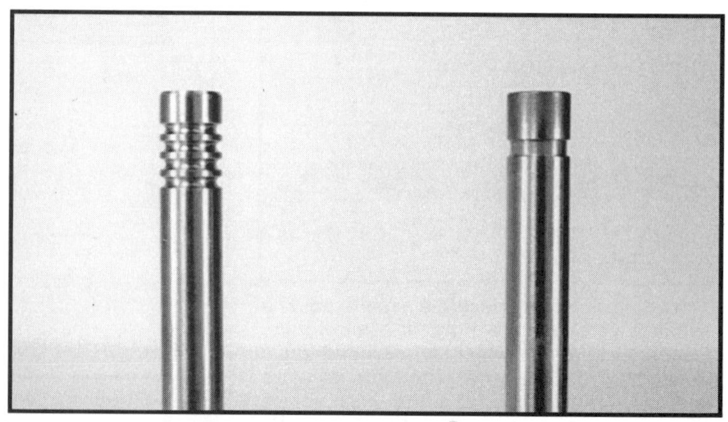

The extremes of valve stem locks are illustrated here. The valve on the left is a triple lock groove while the one on the right is a single groove.

on center. As mentioned earlier, the underhead valve shape affects flow; the tulip-shape is best at high lifts and the nailhead is best at low lifts. The ideal valve shape would be a combination of the two. The subtleties of the valve job will affect flow, but a flow bench should be used to find these slight changes in angles and widths. Be sure to always do all seat development with the valve shape that's going to be used in the real engine.

In many cases, it's advantageous to flow the intake manifold with the intake port. It's harder to do but gives better overall data and evaluates the intake at the same time. *Port matching*, matching the intake manifold runners to the intake ports in the head (and to a lesser extent matching the headers to the exhaust ports), also helps the ports flow more cubic-feet per minute (cfm).

Reversion—For several years, *reversion* was a popular topic when discussing flow theory and engine performance. Reversion occurs when exhaust gas flows in the intake port at overlap. But reversion doesn't seem to play such a critical role today. Now, *port volumes* and *lengths* seem to be more of a factor.

Flow Tips—If the intake manifold isn't flowed with the head then an *ideal entrance* should be added at the entrance of the port into the head casting. It only needs to be added to the port that is being flowed. This entrance should then be kept constant for all the heads of this type that you flow and all comparisons that you make.

In wedge-shaped combustion chambers like the A engine, the valves ideally prefer to be *up* in the chamber for the best flow and horsepower. Sinking the valve will reduce flow and power. So it's a good idea to develop a *valve depth* measurement for the valves used and the modifications made. It eliminates one more variable.

The *heat crossover passage* used in all production heads also hurts flow. It can be blocked off so that it doesn't affect the carburetor and intake manifold but it will hurt the exhaust flow slightly. It's not worth bothering with in most cases. The W-2 doesn't have a heat crossover passage for this reason. This is a feature that must come in the casting itself. It can't be added or milled away. Also, don't try to weld it closed.

Offsetting Guides—Offsetting the valve guides can be very expensive, but offers an advantage to a professional racer. Generally, with a 4.04-in. bore and the 2.02-in. intake/ 1.60-in.exhaust valve sizes used in a high-performance A engine, clearances around the valve are close but not a problem. Putting in an oversized intake makes it very close. A port will flow better if the valve centerlines are moved apart. If this is done, then move the exhaust toward the cylinder wall first. Only consider this with W-2 heads. This trick is being used by the Chevy small-block engine builders because their valves *don't* open on center. This limits their valve size and shrouds their valves more. Offsetting the guides is their method of trying to copy the A engine. Obviously the A engine can use the same technique to an even greater advantage.

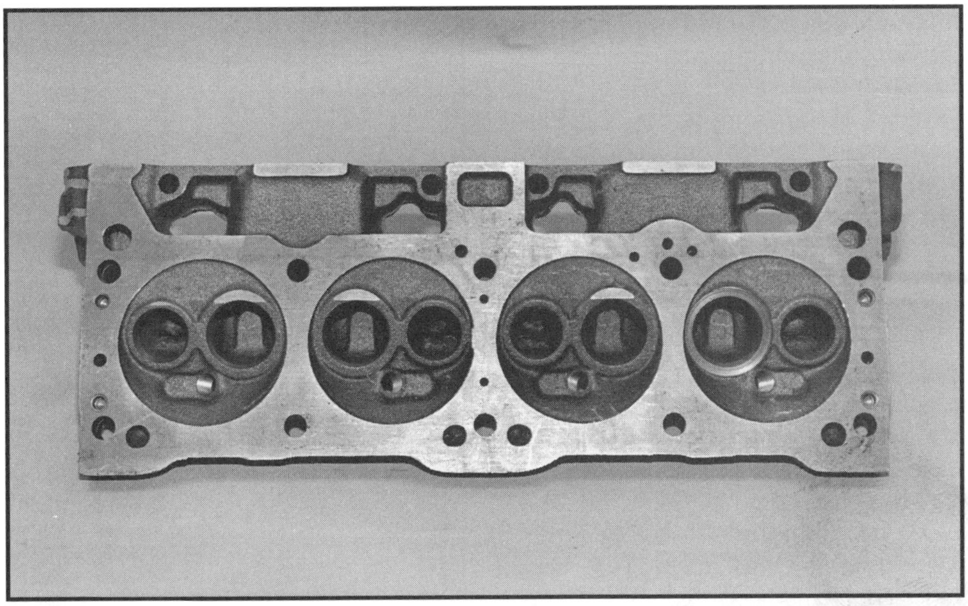

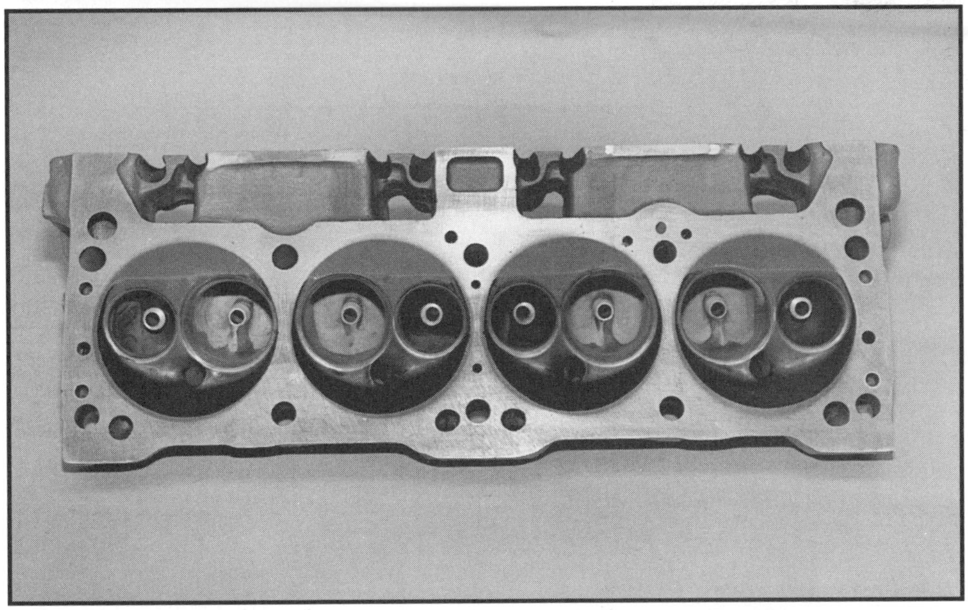

These two photos illustrate what a little porting and machining can do to improve flow. These are both stock 340 head castings. The one on top is untouched, while the other has been worked over by Dick Landy and his guys at DLI. The DLI head has larger combustion chambers, .25-in. in diameter and 1.3 cc's bigger in volume. (Photos courtesy Hot Rod Magazine)

HEAD PORTING

Cylinder head porting is removing metal from the sides of the port in order to change its shape. It's hard work and quite time-consuming. It also requires a lot of attention to detail. The majority of the porting must be done by hand using a high-speed hand grinder. Technically speaking, *polishing*, where the cast surface is machined to a smooth, shiny surface, is not porting because polishing doesn't change the shape of the port. It does "smooth" the flow, but it really makes it look pretty more than anything else. Most amateurs don't make this distinction and consider porting and polishing to be one and the same.

When it comes to porting cylinder heads, there are all different levels, from mild to completely altering the port configuration. There are home kits for skilled enthusiasts to "clean-up" the ports or slightly change the ports' configuration. But for racing, it's best to have a professional head shop do the porting, and that means it is going to cost quite a bit.

Many people try to port heads themselves, and when I'm asked the best way to go about it, I always say that everyone should port a set of heads themselves at least once. That way they understand how much work is involved and are not so likely to complain about the expense when they have it done by someone else the next time. But let's not take the porting operation too lightly. A good rule to follow is don't port it yourself unless you can afford to "trash" one head. You will likely nick a seat, or break into the water jacket or develop a crack. On a set of race heads, this could spell disaster especially in terms of money. The production 318 and 360 heads are still being built and are very common. Porting an aluminum W-2 or a 340 T/A head might be best left to a professional.

PORTING TIPS

As a general rule, the wall finish on the intake ports should be somewhat rough, not shiny smooth. The port flow will also increase if the *bowls* are enlarged. The bowl is the area in the port immediately above the valve seat (as opposed to the combustion chamber, which is *downstream*). This works on both the intake and exhaust sides. Porting them larger than stock is a plus, but don't make them bigger than the seat itself!

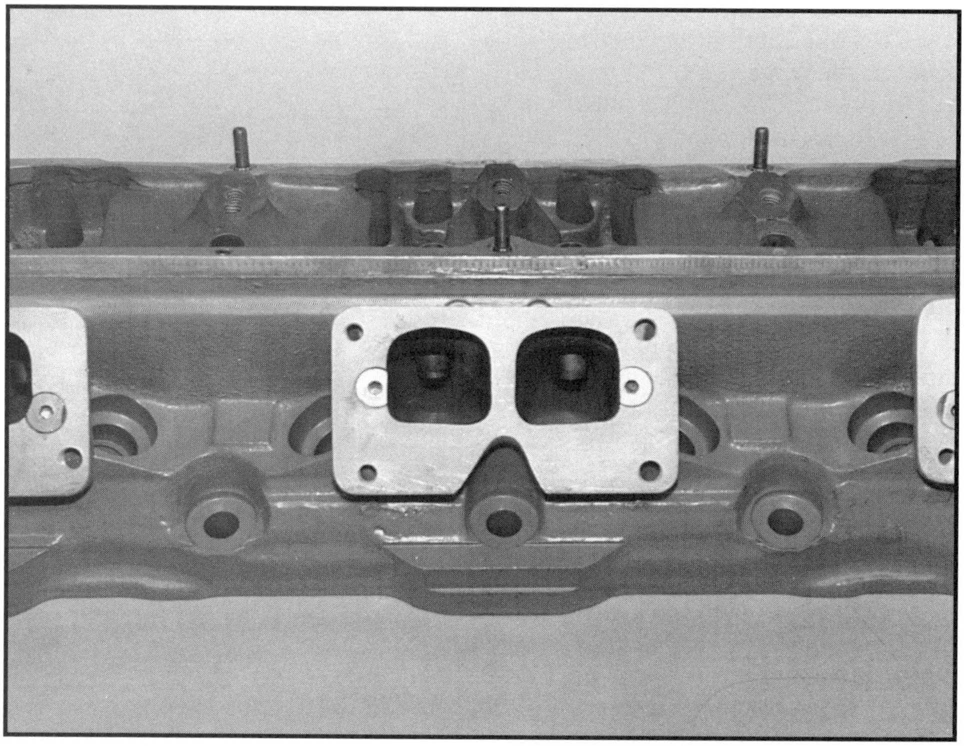

Head porting is difficult, time-consuming and easy to mess up, therefore it's best to leave the job to professionals. DLI went all-out with these 340 exhaust ports. Shown on bottom is the stock head. In the DLI version (top), using a magnesium tooling plate, the exhaust ports were enlarged and then welded inside to achieve better airflow. The plates are bolted in the stock header flange holes and new holes in the plates are used to mount big-tube Hooker headers. (Photos courtesy of Hot Rod Magazine)

Don't try this at home! Stock intake ports (top) were heavily modified by Dick Landy Industries (bottom). DLI relied on Plastic Mastic for the roof of the port. Additionally, Landy used a magnesium tooling plate to form the point where the valve cover bolts to the head. (Photos courtesy Hot Rod Magazine)

Also, the valve guide bosses can be narrowed, and in some cases shortened. However, shortening the guide cuts down its overall length, which gives the valve stem less support. Then the valve head will move more, the valve angle will change quicker and the guide will wear out much more quickly. Caution must be used here. Additionally, if you make the guide too thin, it may crack.

The last "trick" relating to porting is the port's *volume* in cc's. A related consideration is the port's length. It's hard to do much about the port's length in the head itself. This is usually done in the manifold. What the racer is looking to modify is the overall *runner length*, which is the sum of the runner in the head and the runner in the intake manifold. Though this volume is also the total runner volume, the head can be looked at separately.

The volume of the port in the head is the amount of cc's required to fill the port up to the manifold face with the valve on the seat. For the 340/360 intake port, the volume is approximately 160 cc's. The exhaust on these big-port heads is around 75 cc's. The small-port 318 heads are 140 cc's on the intake side and 60 cc's on the exhaust side. Obviously, the lengths of the port runners in these two heads are the same. One aspect to keep in mind relative to port volumes is that bigger is not always better, even in all-out race engines. Small engines (318, 305, 292, etc.) prefer a small runner (less volume) for high velocity in the manifold and ports. The bigger engines (340, 355, 360, 372, etc.) with a tunnel ram and 8-barrel induction systems, prefer larger ports. A single 4-barrel package tends to fall somewhere in-between.

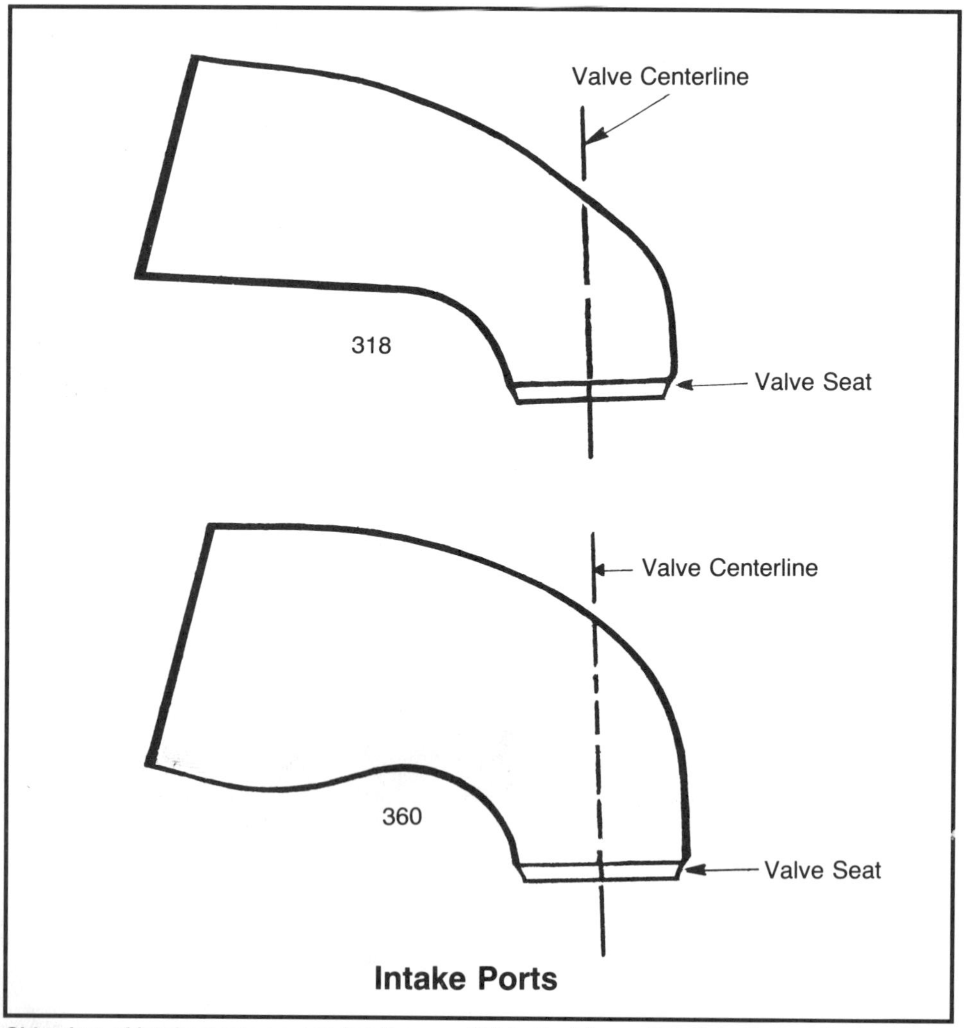

Side view of intake ports comparing the new 318 to the bigger 360. Note that while the 360 is larger, the new 318 is higher and smoother.

HEAD GASKETS

There are many different head gaskets for the A engine family, but I'll limit my discussion to those used in racing. Basically there are the stock, production composite gasket, the steel shim, the Teflon-blue and the latest O-ring gasket, both from Fel-Pro.

Many racers like the *steel shim*. It is strong but isn't very flexible. This means that it doesn't seal to the highs and lows, nor does it move with the head and block as they bend under

CHRYSLER PARTS LIST

CYLINDER HEADS	
W-2 Cast-Iron Casting	P3870812
W-5 Alum. Casting	P4452924

NOTE: The W-2 heads require special valvetrain and intake manifold.

HEAD ASSEMBLIES	
Ported Bracket Head Assy.	P4286864
Race Ported W-2 Assy.	P4349084
318 Ported Swirl Port	P4452748
HEAD BOLTS	
STD Heads	P4120471
Package (includes bolts & washers)	
Milled Heads	P4007666
Bolt & Spacer Pkg. For Ported W-2 Heads	P4007712
STUDS	
W-2 Stud Kit	P4120542
GASKETS	
Steel Shim Set	P4120094
Telfon-coated Set	P4120093
O-Ring style Set	P4349557
VALVES (Standard Length)	
1.88-in. Intake	P4452017
2.02-in. Intake (3 groove)	P3690230
2.08-in. Intake (1 groove)	P4007940
1.60-in. Exhaust (3 groove)	P3690231
VALVES (3/8-in. Stem)-Long Stem	
2.02-in. Intake (1 groove)	P3614584
2.08-in. Intake (1 groove)	P4007941
1.60-in. Exhaust (1 groove)	P3614589
VALVES (11/32-in. Stem)-Long	
2.19-in. Intake (1 groove)	P4120144
1.65-in. Exhaust (1 groove)	P4120145

dynamic stress. Typically, these gaskets work for very short periods, but that's not a problem because racers constantly rebuild engines and replace gaskets all the time.

The A engine head gaskets should be chosen according to the *compression ratio*. At low compression ratios there's no problem. Below 12.0-to-1, use the *Teflon-blue* gasket made by Fel-Pro. It's a composite gasket, approximately 0.040-in. thick and works well up to its limit.

In the "old" days, if the compression was over 12.0-to-1, then the block or head had to be *O-ringed*. Today, there is an alternative—the *O-ring* gasket made by Fel-Pro. It looks similar to the blue gasket but it has an O-ring built into the gasket around each cylinder. It seals very well at high compression ratios.

If you want to O-ring your block or head, have the machinist cut a small groove around each cylinder bore or combustion chamber. The O-ring itself is made of stainless steel wire, 0.032-in. thick. The grooves cut are 0.032-in. wide and 0.017-in. deep.

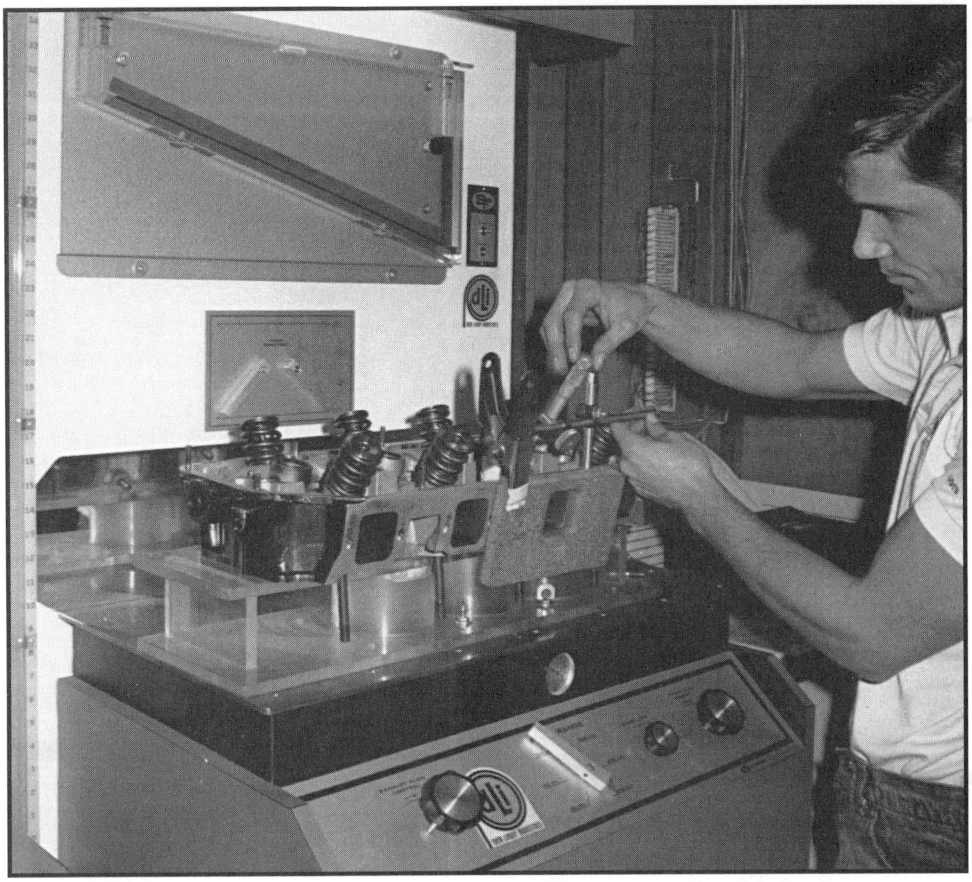

A good flow bench can help determine whether changes work or not. Robert Landy is prepping a 340 head for flow-testing on a SuperFlow unit. (Photo by Michael Lutfy)

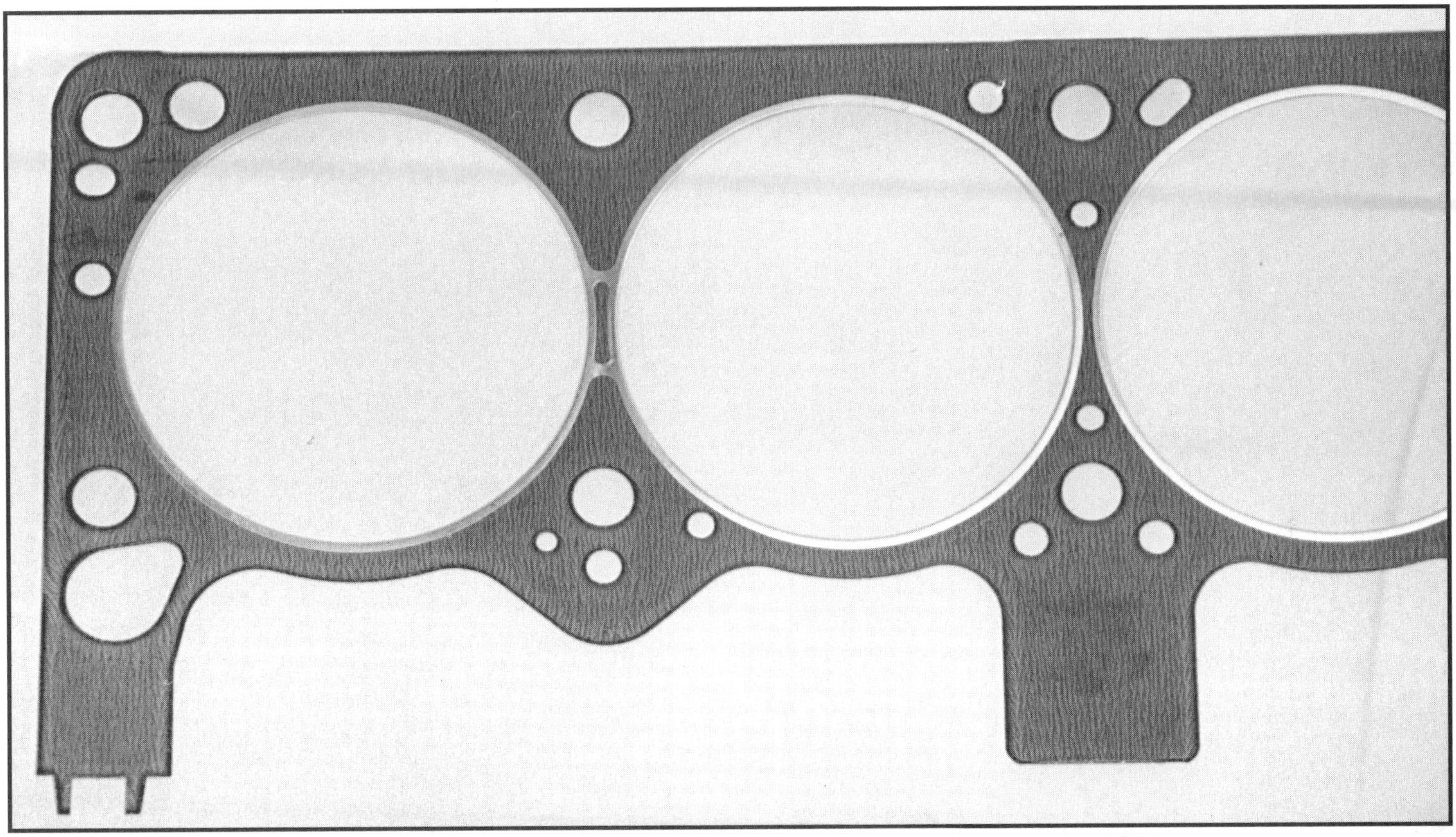

Typical A engine head gasket is shown. There are steel shim, composite, teflon and O-ring style gaskets available.

6 CAMSHAFTS & VALVETRAIN

CAMSHAFTS

There has been more written about camshafts and their related parts than virtually any other engine performance part. A lot of this theory can be difficult to understand. With this in mind, I will cover what may be old ground to some of you, but I need to make sure that we're thinking along the same lines before I get into theory.

The camshaft is the most popular performance item in an engine. Cylinder heads can deliver a more substantial increase in horsepower, but cams are changed more frequently. They are not the easiest item to change in a V-8 but the selection is almost infinite. From one standpoint, the vast selection makes the choice very difficult, but on the other hand, you shouldn't have much trouble finding the perfect camshaft suited for your application. The problem will be selecting the best one and then choosing the valvetrain to go with it. Before you can select a camshaft properly, you need to learn the terminology used and be familiar with the basic principles of camshaft design, theory and function.

TERMS

All of the A engine V-8's are *4-stroke* engines. This means two things. One, is that it takes two complete revolutions of the crankshaft to complete one cycle. Two is that one cycle is divided into four phases; *intake*, *compression*, *power* and *exhaust*. Each phase has a length of 180 deg. of crankshaft rotation. The camshaft in a 4-stroke engine rotates at 1/2 the crankshaft speed. In other words, the cam drive has a 1:2 ratio or the crank rotates twice for each revolution of the cam. This is an important factor, because camshaft durations are typically given and discussed in *crankshaft degrees*. Now for the characteristics of the camshaft itself.

Duration—The *duration* of the cam is how long (in crank degrees) that the valve is held open relative to some common *baseline*. Obviously there are two potential durations for a cam—one for the intakes and one for the exhausts. They don't have to be the same and in some cases, it is best to have them different. The original 340 cam was a 268/276-deg. design. The first number is always the intake duration, in this case 268 deg. The second number is the exhaust duration (276 deg.). In many cases the intake and exhaust lobes are the same, such as 284/284 deg. It can be listed separately or lumped together and called a 284 deg. cam. The problem with duration is establishing the *common baseline*. Virtually every cam grinder uses his own method or baseline. Therefore, it can get very confusing if one tries to compare one manufacturer to another. This means that a cam with 284-deg. of duration made by Chrysler is not the same size as a 284 cam from Crane, Isky, Lunati, Competition Cams, etc. You also don't know whether it's bigger or smaller.

In the early 70's, a new baseline was introduced to solve this dilemma. It

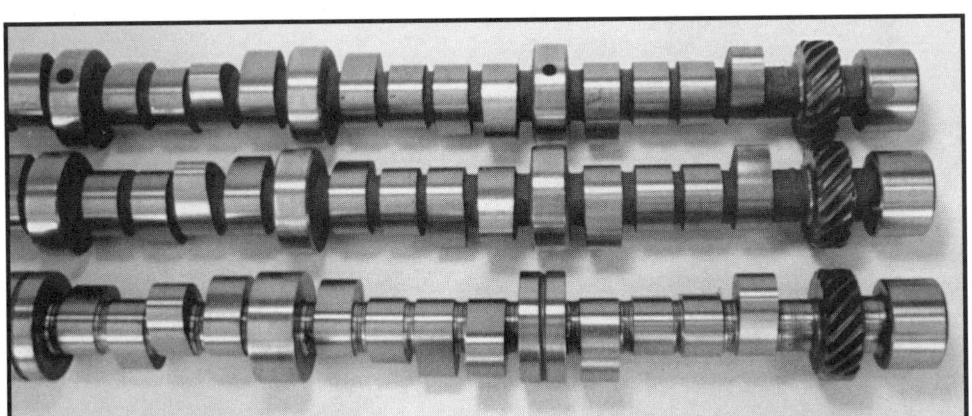

Camshafts look alike at first glance. The top is a small hydraulic, the middle one is mechanical, while the bottom one is a high-lift roller.

was called *duration-at-fifty* (or at 0.050 in.). Initially, this was done to compare the very large cams (330 to 370 deg.) by the standard method of rating them.

These cams were so big that they all started in about the same position but they certainly were not all the same size (duration). For example, in the early 70's Chrysler designed two cams that both had 340-deg. of duration but one had a duration at 0.050 in. of 286 while the other had a duration at 0.050 in. of 296. Obviously, the old method was inadequate to define these really big race cams.

Duration-at-fifty was hailed as a great breakthrough. All the aftermarket manufacturers started using it on their cams to decrease customer confusion and help compare one cam to another—no matter who manufactured it. Then problems arose. Customers that had been using 284-deg. cams now had to buy 210-deg. cams. These appeared to be smaller than stock, which averaged around 260 deg. Technical engineers, (such as yours truly) would spend all of their time explaining why there was a difference and why 210 was really better than 260. So it was back to the old method. The duration-at-fifty method is still used on the giant cams, especially big rollers, but most of the small and intermediate cams are back to an advertised duration. Another problem with using duration-at-fifty for comparisons of smaller cams is that the 210-deg. cam would have *zero overlap*. In small camshafts and cams selected for dual-purpose applications, the overlap is one of the most important numbers, a number which the duration-at-fifty method eliminates. As a result, the old advertised duration method is better for comparisons of smaller camshafts. A small cam is one with 0.600-in. lift or less.

Lift—The lift of the cam is much easier to explain and understand than duration. Simply stated, lift is how far the valve is moved. Generally, it's a more reliable way of comparing camshafts of various manufacturers. A 0.500-in. lift cam is bigger than a 0.450-in. lift cam. It's also true that the higher lift cams will make more power than the lower lift versions, at least when comparing cams with 0.620-in. lift and less. For example a 0.500-in. lift cam will make more power than a 0.450-in. lift design. Though this method seems simple and straightforward, it isn't very popular with racers and performance engine builders.

There is one point relative to cam lift that seems to confuse the average engine builder. Similar to duration, there are two lift numbers—*lobe lift* and *valve lift*. Lobe lift is the lift of the cam at the lobe itself. It's harder to measure and rarely used to describe a cam. The valve lift is the lift of the cam at the valve. All the numbers for lift listed above are valve lift numbers. This is also what cam manufacturers list in their catalogs. Technically speaking, valve lift is really *designed* valve lift. Design valve lift is defined as the lobe lift multiplied by the engine's design rocker arm ratio. The A engine has a design rocker arm ratio of 1.50. Chrysler uses this method on their production cams, while aftermarket manufacturers, such as Crane, Crower and Mopar Performance, use it as well for their camshaft designs. However, confusion sets in when it comes to *actual* valve lift. The actual lift has to be measured *at the valve*. There are many experimental errors involved, such as deflections, geometry etc., but the biggest one is the rocker arm ratio. As stated, the design ratio is 1.50, but in practice, the actual ratio is closer to 1.46 to 1.48. This makes the cams measure small at the valve. For example, on a 0.500-in. lift cam, the lobe lift can be calculated to be 0.333-in. (0.500 divided by 1.50). Using 1.46 as the actual rocker ratio, the 0.333-in. lobe lift cam yields 0.486-in. lift at the valve (0.333-in. multiplied by 1.46). So when you measure the cam at the valve, you may think that it is a smaller one, but it isn't really. Using this example, it's obvious that valve lift is one good method to compare cams.

Lobe Centers—By definition this is the *centerline of the cam lobe*. This definition is often over-simplified. On the A engine, as well as all other ohv V-8's, there are two lobe centers per cam: one for the intake side and one for the exhaust. Lobe center is stated in degrees relative to top-dead center (TDC). Typically, all these centerlines are listed for the #1 cylinder's valve relative to TDC of the #1 cylinder. A commonly used lobe center is 110 deg. This means that the center of the lobe (where maximum lift occurs) is located 110 deg. after TDC. Unless stated otherwise by the manufacturer, the exhaust lobe center is the same as the intake's, in this case 110 deg. Lobe centers are the physical relationship of intake lobe to the exhaust lobe, which is ground into the cam by the manufacturer. Lobe centers can also be considered as the sum of the intake and exhaust centerlines divided by two.

A related aspect to lobe centers is the *cam's centerline*. The centerline of the camshaft is the exact relationship between the centerline of the lobe and

A engines can accept a variety of cam drives. From left to right—roller chain, silent chain and Cloyes matched, multi-keyway, roller chain set-up. The Cloyes matched roller chain set-up has five different keyways which allows for different cam centerlines without other hardware.

TDC of the piston. There are two centerlines but usually only the intake centerline is used for discussion. Unlike the lobe centers, which don't change once the cam is ground, the cam's centerline or installation centerline can be changed in either direction. At the same time, once the intake center is determined, then the exhaust centerline is defined. For example, if the cam is ground on 110 deg. lobe centers and the installation centerline is 110, then the exhaust centerline is also 110 deg. If this same cam's intake centerline is changed to 105 deg., then the exhaust centerline has to be 115 deg.

Overlap—The overlap of the cam is defined as the amount of time (measured in degrees) that both the intake and exhaust valves are open at the same time. Obviously, overlap is a function of the duration baseline. Overlap is very important in small camshafts, especially hydraulic ones. Once the cams get into the 0.550-in. lift range and up, the durations of the profiles get large enough that the overlap is in the 90 to 100 deg. area. These are so large that the overlap effect is virtually eliminated. So don't use overlap as a method to compare big camshafts—only small ones. It is listed in most specification sheets as an aid to calculating the various events and centerlines.

Valve Lash—Commonly used on mechanical camshafts, valve lash is the clearance between the valve stem tip and the tip of the rocker arm, measured when the tappet is on the cam's base circle (usually 180 cam deg. away from the max lift point). Typical valve lash settings are 0.028-in. intake and 0.032-in. exhaust. For example, this means that there is 0.028 in. of clearance between the intake valve tip and the rocker arm. These valve lash numbers are usually given "cold." They are also *positive*, meaning a real clearance or gap exists. In hydraulic cams with adjustable rockers, the clearance (valve lash) is negative, which means it's *preloaded*. The proper valve lash, and whether it is a hot or cold setting, is given with the manufacturer's specs on the camshaft.

Base Circle—The cam's base circle is the lowest point (closest to the camshaft's rotational center) on the lobe. The base circle typically takes up the majority of the actual lobe in degrees of rotation. The *lobe lift* is measured against the base circle as its zero point. With a 300-deg. cam, the base circle takes up over 320 deg. with the ramps taking the remainder.

Keys—The centerline of the A engine cam is located by a key and a keyway. The best way to change the cam's centerline is to use *offset keys*. They are available in 1-, 2-, 3-, 4- and 5-deg. offsets (double for crank degrees).

Casting Numbers—Most production cams have a *casting number*, or at least part of one. But unlike heads and blocks, these casting numbers don't identify the cam. They would tell you that it was an A engine cam but you should already know that from other information that's easier to spot. In production (stock) all the A engine cams are ground from the same casting. Therefore, you cannot tell a cam from a 340 cubic-inch engine from one used in a 318 by its casting number alone.

Visual Description—All A engine cams look the same to the eye. They all have five *support bearings* or *journals*. Each cam has sixteen lobes, four between each set of journals. The distributor drive gear is located at the rear of the cam next to the smallest "bearing" journal. The keyway is located on a short snout at the front from the cam, on the opposite end from the drive gear.

THEORY

Camshaft theories abound in high-performance circles, more so than any other part. It's easy to get lost in all this theory, but some of it can be useful.

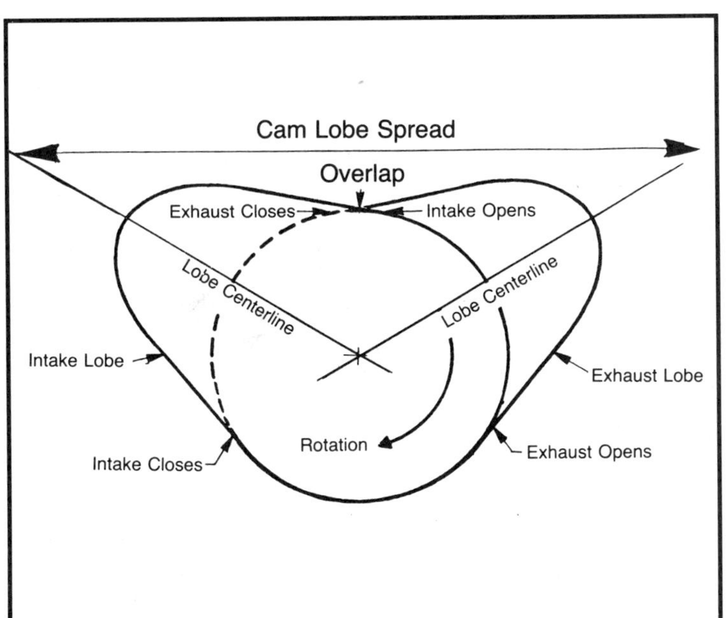

Typical cam features. Lobe spread on a race engine is 104-108 deg. while it's 108-114 on production engines.

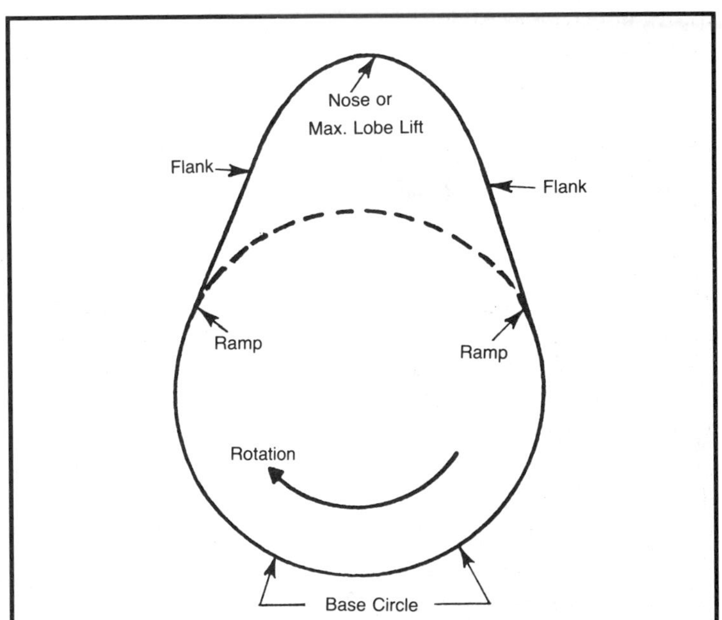

General cam lobe terms. The important features are the max lobe lift (nose) and the base circle.

The problem with the theory I'm going to cover next is that it only applies to big cams and very high-output engines, and doesn't work with smaller cams. However, I do offer specific cam selections and recommendations in the *Tables* section located in the back of this book.

Ramps—The *ramps* are the part of the cam profile that connect the base circle to the actual profile, the part that would be considered inside the duration number. In other words, the ramp is outside the duration number. There are opening and closing ramps. There are also unique hydraulic and mechanical tappet ramps. There are many cams that have very little in the way of ramps. They are very small or short. This style of cam tends to be hard on the valvetrain. Cams used for NHRA/ IHRA Stock Eliminator competition, (sometimes called "cheater" cams), are almost rampless and tend to be somewhat limited in rpm.

Area—Cam designers like to talk about the area under the curve. This is the area between the actual *lift profile* for the cam and the baseline, which represents the cam's *base circle*. For example, take two different race cams with the same lift of 0.650 in. and the same duration at 0.050 in. of 280 deg. Assume that camshaft A has a duration at 0.100-in. lift of 255 deg. Then, you find that cam B has 260 deg. at 0.100-in. In this case, cam B would have the greater area under the curve. Stock Eliminator cams are built using this same "area under the curve" principle.

Valve Float—Valve float occurs when the valve position is no longer defined by the cam profile. In other words, it occurs when there is clearance in the valvetrain parts with the tappet past the ramp of the profile. Valve float has three basic levels—*false motion*, *valve bounce* and *valve float*. Most racers define valve float as when the valves bounce so badly that the engine shuts down. With a hydraulic cam, valve float will cause *lifter pump-up*, which holds the valve off the seat a small amount for 360 deg., which shuts down the engine. With mechanical cams, the power falls way off and in some cases, the valves hit the pistons and bend.

Full valve float is the most extreme case. Inside the engine, the tappet bounces from the opening side of the profile over the nose past the closing side, all the way to the base circle, in some cases without touching the cam profile itself. False motion is the mildest case. It usually occurs on the closing side of the profile first. Valve bounce occurs near the nose (max. lift of the cam) or just to the closing side. It will cost the engine output but may not be noticed by the driver other than it doesn't "pull" on the upper rpm range.

Valve float is also related to valve spring pressure, both closed on the seat and open. It's also affected by the intensity of the cam profile itself. Intensity (harshness) of the cam is related to the area under the curve. Profiles with more "area" are most intense and tend to break pieces, wear out springs and float valves quicker than other milder profiles.

Degreeing—This is another word for centerlining the cam, or installing the cam relative to the crankshaft. There are two basic methods to degreeing. One is to set the centerline of the

On the Cloyes multi-keyed set-up, there are five different symbols on the lower (crank) sprocket which have to line-up with the upper sprocket to give the proper amount of advance or retard.

number one intake lobe at the proper relationship to TDC. The second is to set the intake opening/exhaust closing, intake closing/exhaust opening cycles properly, relative to TDC and BDC. The second method has four measurements and they are taken in the ramp area. The valve isn't moving very fast at these points. On the other hand, the first method only measures one item, the cam's centerline. The measurements are taken in the 0.050 to 0.100 in. range before and after maximum lift. At this point the valve is moving fast and very small movements of the crank (degree wheel) yield big movements in the valve (dial indicator). The centerlining method tends to be more accurate and much less confusing. For more on degreeing camshafts, turn to Chapter 7 on *Blueprinting*.

Advancing/Retarding—Once a cam has been degreed, or installed at the recommended centerline, it can be *advanced* or *retarded* from this point by using offset keys and re-degreeing. Technically speaking, if the cam is put in straight-up (installation centerline equals lobe center or intake and exhaust centerlines as installed are the same), then it is not advanced or retarded at that point. Typically, manufacturers do *not* install straight-up and usually recommend a centerline that is 2 deg. off of straight-up. A common one is to grind the cam on 110 centers and install the intake at 108. The commonly accepted definition of advancing is to advance the centerline (intake) toward TDC. In our example above the cam was advanced 2 deg. (110 to 108). From a performance standpoint, advancing the cam gives more torque. Usually a cam isn't advanced or retarded more than 4-deg.

Clearances—Valve-to-piston clearance should always be checked. The procedure is detailed in Chapter 4 on page 41 and in Chapter 7 on page 98. The intake should have at least 0.090 in. of clearance, while the exhaust should have a minimum of 0.100 in. These clearances can be checked using clay on the top of the piston or with a very light spring with a dial-indicator set-up on the retainer. The second method takes more time, because a measurement is taken at every degree from 10 deg. before TDC to 10 deg. after TDC. The first method is quicker, but the second is more accurate. If you use clay, then increase your measured clearances to 0.100 in. on the intake and 0.120 in. on the exhaust. The exhaust clearance is higher, because it tends to float first. Also keep in mind that the cam's centerline affects the valve-to-piston clearances. If the cam is advanced then there will be less valve-to-piston clearance. Be sure of the cam clearances in the engine assembly at the actual centerline used or planned, not just the manufacturer's recommended one.

Cam Drive—There are basically three types of cam drives: *silent chain*, *roller chain* and *gear drives*. The cam drive connects the crank to the cam. The silent chain is used in most production applications. The roller chain is used in the high-performance production engines and most racing/performance engines. Gear drives tend to be noisy and expensive so they should only be used in "ultimate" engines. Generally the chain (roller) drive is best because it can act as a damper for engine harmonics and vibration that come with high rpm.

Distributor Drive—The rear of an A engine camshaft has a helix gear which turns the distributor. It also turns the oil pump. This intermediate

Gear drives are not common on A engines but are available and offer benefits for certain types of all-out racing.

Although basically the same overall size, the roller chain crank sprocket on the left with its double row of drive teeth looks stronger than the silent chain sprocket shown on the right.

The production "silent chain" is shown at the bottom with the roller chain shown to the top. They are interchangeable as complete set-ups, not as individual pieces. This means that the roller chain cannot be used on silent chain sprockets and vice versa.

shaft consists of a shaft and a gear. The gear mates with the cam gear. The shaft has a hex drive that fits into the top of the oil pump. The top end of the shaft, that is installed in the engine, has a slot for the distributor shaft to fit into. There is a heavy-duty shaft used for racing purposes that is slightly larger in diameter and has a large radius to the hex drive. This makes it stronger and cuts down on the amount of twist. If a roller cam is used, then the distributor drive must be changed from the steel unit used in production to an aluminum bronze unit. Otherwise, it will wear out almost instantly and fill the engine with dirt!

CAM RECOMMENDATIONS

The problem with making cam recommendations is the almost infinite number of cars, engine and parts combinations that it has to be done for. There are lots of cams to choose from, but there are even more engine packages and special applications that it's going into. About the best I can do is recommend cams for *general* and *specific* applications. For general applications, see the accompanying sidebar. For special applications, go to the back of the book under *Tables* and look for the *Performance Packages*. In that section, I've grouped performance possibilities into packages using the basic system used by Mopar Performance. At the end of each package, I suggest the best cam to go with it. I've used 5 basic groups for each engine size, called *Stages*. How the engine is being raced and what parts are being used in the engine obviously affect which cam to use.

VALVETRAIN

The valvetrain is everything that connects the camshaft to the valve. The valves themselves are covered in Chapter 5 on *Cylinder Heads*. The word "valvetrain" includes; tappets (lifters), pushrods, rocker arms, rocker shafts, rocker stands, valve springs, retainers and keepers. There are two basic types of valvetrain—*hydraulic* and *mechanical*. They are almost totally different, but there is a lot of overlap. As I examine each individual piece, you'll realize that the valvetrain plays a more complicated role than just connecting the valves to the camshaft. However, to choose the valvetrain components, the camshaft *must* be selected first. So before going on, be sure to read the previous section on cams and select one. There are very few cases where the valvetrain dictates the cam selection. As such, I will assume that at this point you have already selected your cam.

TAPPETS

The *tappet* or *lifter* is the first thing to consider once the cam is selected, because it rides directly on the cam lobes. There are basically five different types of tappets; *hydraulic*, *flat/mechanical*, *mushroom*, *roller* and *hydraulic roller*. They are not interchangeable, although they all fit in the same basic tappet hole in the block. Which one used is determined by the cam selected.

Hydraulic—The hydraulic tappet was introduced to the A engine family in 1967 on the 318 engine. The earlier 273's had flat-tappet (mechanical) designs. The hydraulic tappet auto-

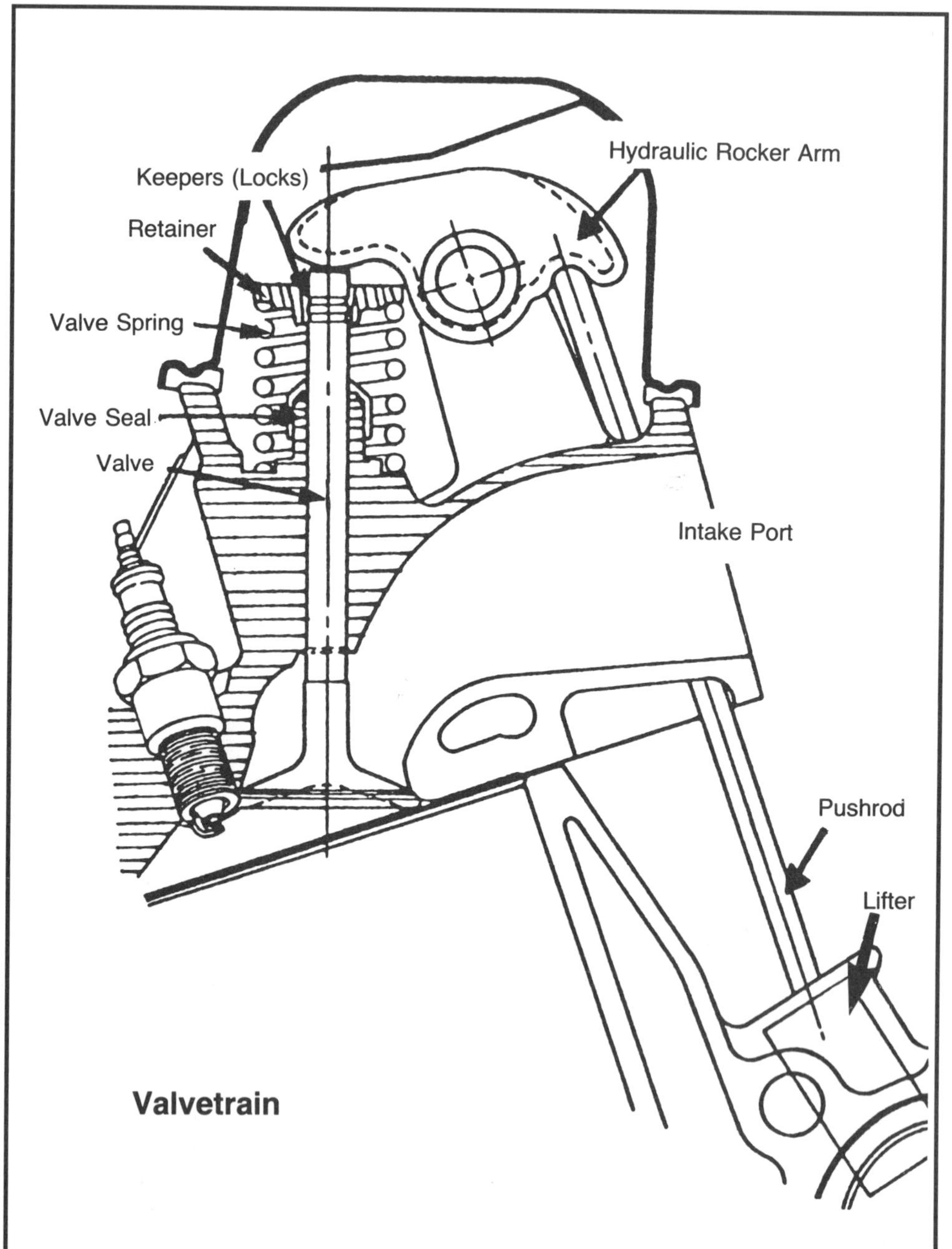

General A engine valvetrain. Note the pushrod angle in the tappet. (Courtesy Mopar Performance).

matically adjusts the valve lash in the valvetrain using the engine's oil pressure. Therefore, it compensates for wear, and there are no adjusting nuts that may come loose due to vibration. The hydraulic tappet itself has a *spherical radius* to its bottom (the part that touches the cam lobe). If you place two tappet bottoms together, you can feel them wobble slightly. If you hold them in the same position up to a light, it is easy to see the light on the edges. Also, there is a small hole in the outside barrel of the tappet that allows the oil pressure from the engine's oil galley to get inside the tappet.

Hydraulic tappets must be used on a hydraulic camshaft. Physically, they could be installed on a flat-tappet cam but the engine wouldn't perform well and the cam/tappet package would wear out quickly. There have been a very limited number of cams ground by the aftermarket manufacturers that would accept both hydraulic and mechanical tappets. Since new tappets are used with any new cam, trying to adapt a hydraulic tappet to a mechanical cam is pointless. Used cams should use the same tappets in the same order that they were used on the cam originally. If not, then new tappets should be used. Used tappets by themselves should be thrown away.

These are two types of cam drive sprockets. The cast-iron roller chain sprocket is on the left and looks stronger. The silent chain sprocket has nylon drive teeth and an aluminum hub.

GENERAL CAM RECOMMENDATIONS

The goal of any camshaft is to extract the most torque and horsepower from the parts of the engine. Obviously, the application is important. Class racing is the easiest to pick for because the rules define everything so closely. The hardest group to pick cams for are the street machines, street rods and dual-purpose vehicles. The specific applications and actual parts used are so varied that it's almost impossible to make an *exact* recommendation. However, I can get close with a lot of "ifs".

There are a lot of engine types between the class racers and street machines that fall into a more general category. From the engine standpoint, there is everything from a high-performance, stock 318 to an all-out, tunnel-ram 8-barrel 355. Horsepower ranges from 150 hp to over 700 hp. If you lump all the non-race cars into the dual-purpose catagory, all you have left are the various race cars/engines.

With any race camshaft, try to balance the intake port and exhaust port efficiencies. Typically, the intake lobe will be different than the exhaust lobe. How much different depends upon the type of racing, the engine itself, the heads and the rod ratio. With a highly efficient exhaust port, one that flows a high percentage of the intake port flow, the camshaft can use the same lobe on the intake and exhaust. I will assume a more standard relationship between the intake and exhaust ports. That narrows our list of variables down to engine and type of racing.

DRAG RACING
(8-barrel carburetion, ported heads, high compression)

372-396 Displacement
296-298 Intake/305-308 Exhaust
110 to 112 deg. centerline

345-370 Displacement
290 Intake/300 Exhaust
108 to 100 deg. centerline

325-345 Displacement
286 Intake/296 Exhaust
108 to 110 deg. centerline

310-325 Displacement
282 Intake/292 Exhaust
106 to 108 deg. centerline

290-310 Displacement
278 Intake/288 Exhaust
106 to 108 deg. centerline

DRAG RACING
(4-barrel, everything else same as above)

340-360 Displacement
288 Intake/298 Exhaust
104 to 106 deg. centerline

300-320 Displacement
282 Intake/292 Exhaust
104 to 106 deg. centerline

NASCAR
(4-barrel)

340-360 Displacement
280-282 Intake/286-288 Exhaust
108 to 110 centerline

ROAD RACE
(4-barrel)

340-360 Displacement
272-278 Intake/278-284 Exhaust
106 to 108 deg. centerline

The above duration numbers are all measured at 0.050 in. for consistency. For the most part it is assumed that they are rollers. The roller-style cam allows for a greater maximum valve velocity than a flat tappet design. This allows the rollers to have more area under the curve at the same duration compared to a flat tappet. If rollers are not legal or they are considered inadequate in reliability, then a mushroon lifter design is the next best choice.

Given the engines and cam numbers listed above, if you used wider centerlines than shown, then the engine's range will be reduced. Also the wider centers will shift the power range toward the high rpm scale. Wide centers work best on larger displacement engines.

Flat-Tappet—The flat-tappet was used in the original 273 in 1964. The original, flat-tappet cams were very mild and yielded less performance in the engine than the newer hydraulics of the last twenty years. As such, I won't dwell on them except to say that they share the same design features as their bigger brothers. Today the flat-tappet (mechanical) cams are larger than the typical hydraulic and are used in serious racing applications. Because this group of cams tends to be larger, they make more power and run higher rpms. Without the extra lift and duration typically used they would not make more power. If a mechanical flat-tappet is compared to a hydraulic tappet, the flat-tappet will be lighter. This gives the flat-tappet an advantage at higher rpms. It also doesn't "pump-up" like a hydraulic tappet will when pushed past its limit. This can be a plus and a minus. A flat-tappet can work at higher rpm, but if the driver makes an error (such as a missed shift, which over-revs the engine) it won't pump-up and shut-off (limit) the engine. As a result, mechanical cams are less forgiving and driver errors will be costly. The break-even point between hydraulic and mechanical cams is 0.515-in. lift. More lift than this and a mechanical should be considered. If less lift is used, then go with a hydraulic. This will most likely change in the future, at least for some applications, and there are always exceptions.

The typical hydraulic tappet is on the left. In the middle is a standard mechanical or flat tappet. The tappet on the right is a mushroom—a special version of a flat tappet.

The flat-tappet itself is not exactly flat. It has a spherical radius but this radius is over twice as large as the hydraulic's so it appears flat.

Mushroom Tappet—The mushroom tappet is a special version of the flat-tappet. Its bottom is larger than the main body (hence the name mushroom), therefore mushroom tappets need to be installed from the bottom, not the top. The larger bottom allows the mushroom cam profiles to be larger and more aggressive than the standard flat-tappet profile. As such, they tend to work well with cams in the 0.630-0.700-in. lift area. There's nothing special about mushroom tappets, except for the upside-down installation. The cam and engine will run fine. However, if you make a mistake (over-rev the engine), then you will probably break the bottom of the mushroom lifter. To prevent this, you should *back-*

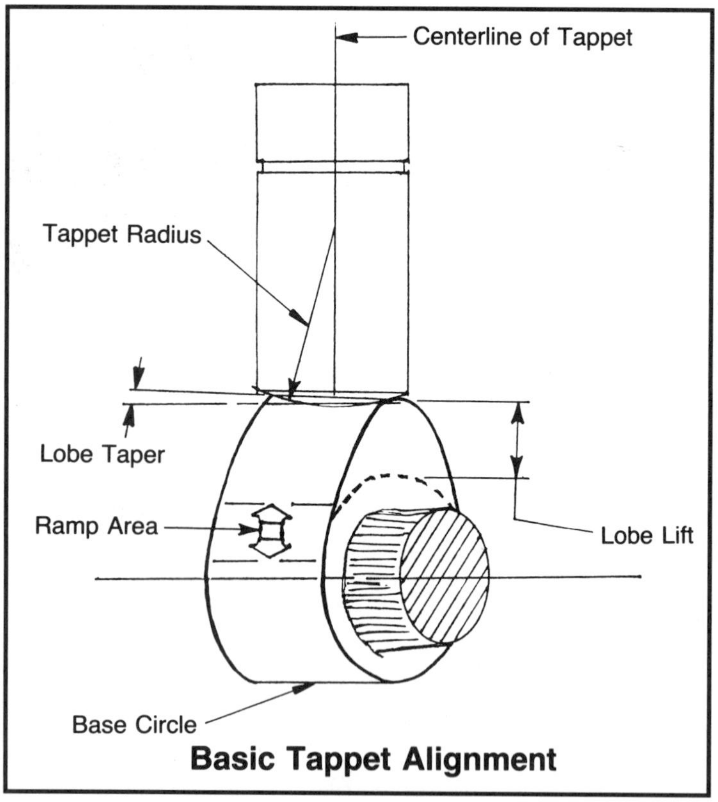

Basic tappet alignment on A engines. The tappet radius gives the tappet a crown. The tappet must be selected to work with the cam based on the cam's lobe taper. A mechanical cam is sometimes called a flat-tappet becuase the tappet radius is so large that it's virtually flat. These cams have much lower lobe tapers to go with the "flat" tappet, Hydraulic tappets usually have a short enough radius that the curve or crown is noticeable.

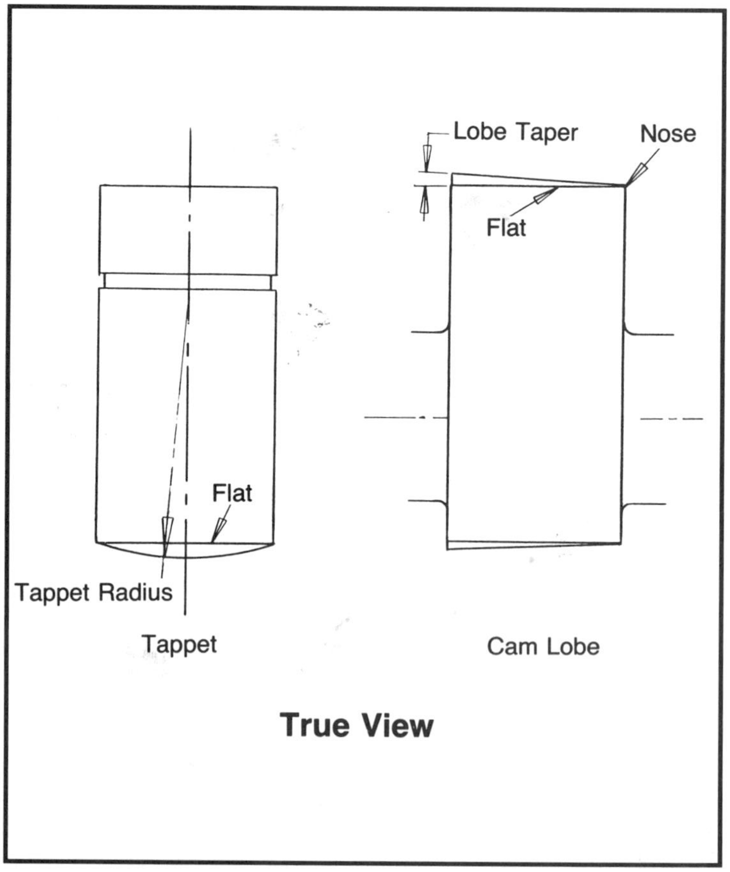

Typical hydraulic tappet (on left) has a definite crown. If in doubt, hold two tappet bottoms together and hold up to the light. The cam lobe has taper but it's harder to measure. You don't want to mix and match parts. When in doubt, get new pieces.

spotface the bottom of the lifter bore. All this does is provide a smooth, flat surface that's perpendicular to the tappet centerline, so it will hit squarely when you "float" it.

Roller Tappets—Roller tappets are similar to mechanical tappets, in that they require adjustable rocker arms. They offer the biggest lift and most aggressive profile potential, therefore they are commonly used for racing with cams that have large lift and duration specs. Roller lifters and cams require more attention than flat-tappets, and they need extra care as well. The bottom of the roller tappet itself is a *roller*. It should be checked to see that the center of the roller is on the center of the tappet body. Check the two sides on a surface plate. They should be within 0.004 in. The other key to the roller tappet is the *guide bar*. The guide bar keeps the tappet from turning sideways. The guide bar is the most common failure in rollers and roller cams and should be checked frequently. There are many different guide bar designs. Some designs get close to the block and require clearance to be ground on the block itself.

Most of the time, the oil to the tappets must be blocked when using roller cams and tappets. This can be done by bushing the tappet bores in the block, which is expensive. As a less expensive alternative, Chrysler recommends drilling out the right side oil galley and inserting a tube (for more information, go back to Chapter 1 on *Cylinder Blocks*). Use only the kit available from Chrysler and follow the directions to the letter. Rollers that have a *high skirt* will uncover the tappet galley and the engine will lose oil pressure. Also, most roller tappets work best at high tappet clearances (around 0.002-in.). This also causes a loss in oil pressure and increases the engine's demand for oil. Both of these problems (real or potential) are solved by installing the oil galley tube.

Hydraulic Rollers—The latest technological breakthrough on the A engine is the hydraulic roller. It was introduced on the A engine in 1985, and is only used on mild engines. It was initially developed to improve fuel economy. Therefore, don't install this production package into an engine ex-

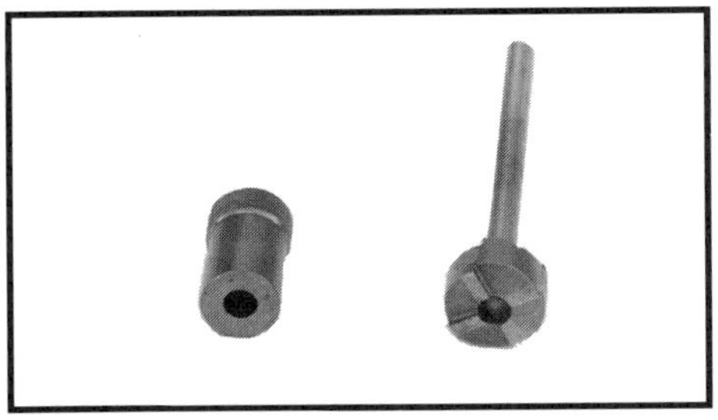

Typical back spot-facing tool. The cutter and shaft are on the right and the pilot bushing is on the left. The pilot bushing is inserted into the tappet bore and the cutter and shaft are attached to an electric drill and pulled up against the bottom of the tappet bore to cut the spot-face. The cutter diameter is larger than the outside diameter of the mushroom lifter.

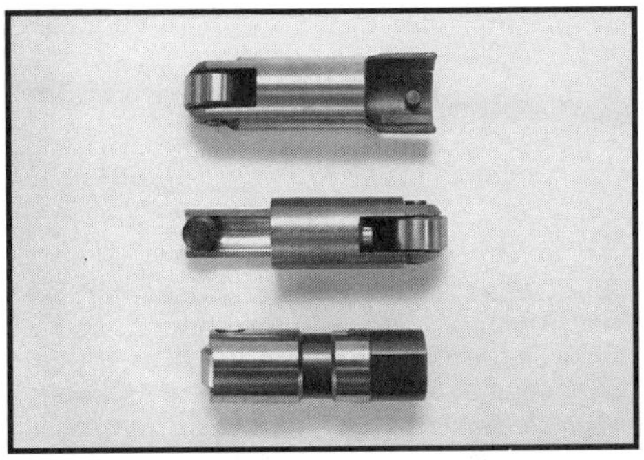

These are various roller tappets. The top two are race mechanical rollers that require a separate guide bar (not shown). The bottom tappet is the production hydraulic roller used in production A engines since 1986. They are not interchangeable.

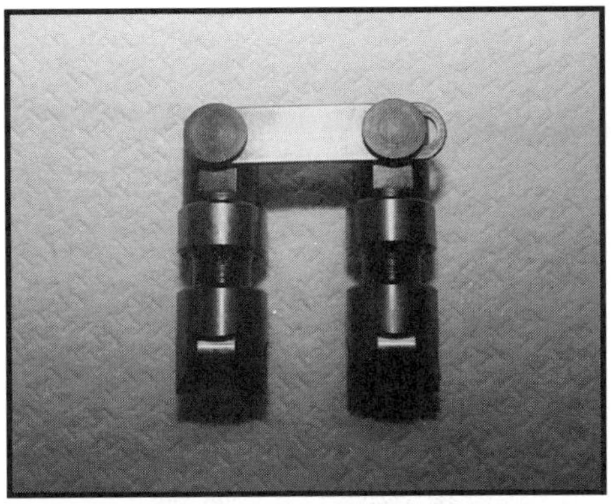

This is a Crane race roller tappet set. Although the guide bar is a separate physical piece, it doesn't come apart once it's assembled by the factory.

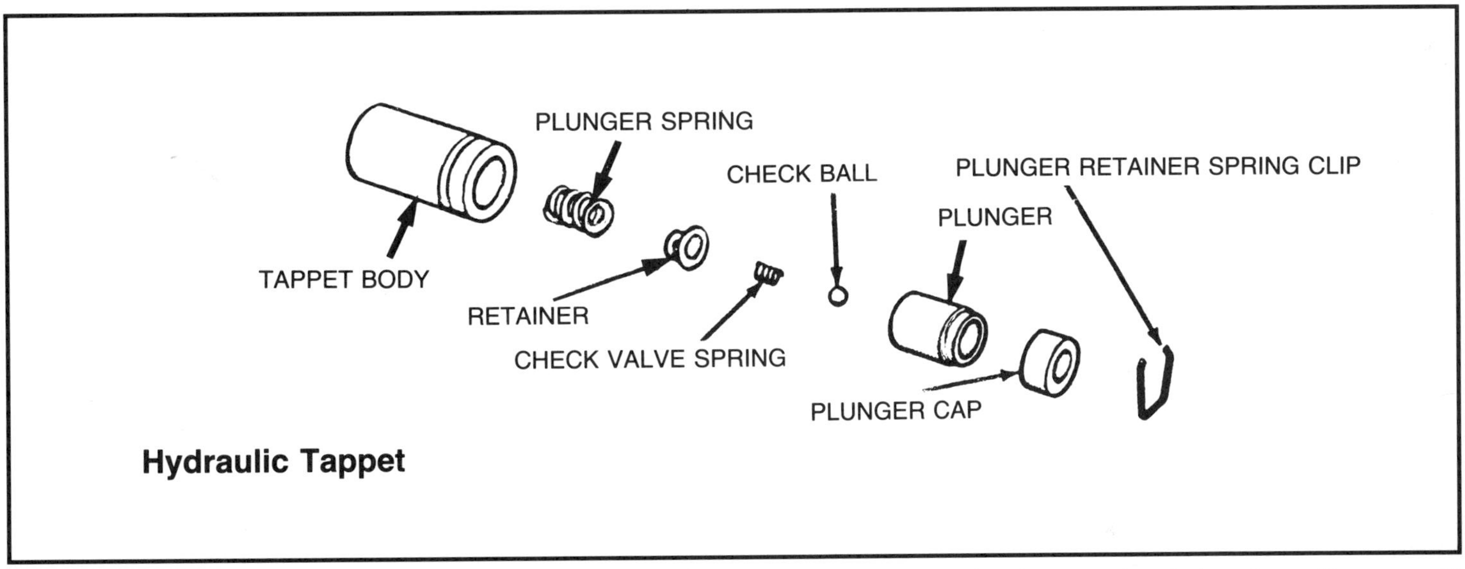

There's more inside a hydraulic tappet than meets the eye. The key pieces are the spring clip and the plunger spring. Spring clip must be removed to disassemble the tappet and can be easily lost. Once disassembled and oil is drained, the plunger/cap can be pushed down against the plunger spring to check operating clearances, adjustment and lash dry setting. (Courtesy Mopar Performance).

pecting more performance. Also, the hydraulic roller package doesn't fit other non-roller production engines. There are head and block changes that rule-out swapping to other engines at this time. Although not yet available, performance cam profiles in the hydraulic roller are coming. To use the hydraulic rollers, you must have the aligning *yoke*, which acts as a guide bar. Also, you'll need the retainer that screws into the block and holds the yoke in position. At this point in their development, the tappets are heavy and float at a relatively low rpm for performance/racing applications.

PUSHRODS

Pushrods connect the tappet to the rocker arm. The most important features of pushrods are how *straight*, *stiff* and how *light* they are. As with lifters, pushrods can be grouped into two basic categories—hydraulic and mechanical. The hydraulic pushrods have a *ball* connection on both ends. The mechanical pushrods have a ball on the tappet end and a socket on the other end at the rocker arm.

Pushrod overall length is another important feature. All the A engine hydraulics are the same length up to 1985. Engines after '85, as well as the newer 318 roller, are the exceptions. The hydraulic roller engine uses much shorter pushrods (over 1/2 in. shorter than standard). All hydraulic pushrods use the same diameter.

Mechanical pushrods were used in production on the early 273's and the 1970 340 T/A. The 273 used a mechanical cam, but the 340 T/A was unique because it was actually a hydraulic cam. These production mechanical pushrods were slightly larger in outside diameter (5/16 in.) than the hydraulics (9/32 in.). They are also slightly shorter. The 273 mechanical pushrod is approximately 0.100 in. longer than the T/A.

Pushrod diameters for racing are 3/8 in. Most racing engines use mechanical cams, so the racing pushrod has one ball and one socket for the ends. For high rpm use, it's important that the length of the pushrod be selected (custom-made) so that the proper valve lash is set with one thread of the adjusting screw showing below the arm itself. This obviously requires that the pushrods be made after all other engine machining is completed. The larger diameter race pushrods should also be clearanced to the heads/block to make sure that there isn't any interference through the full lift of the cam.

ROCKER ARMS

The rocker arm translates the pushrod's motion to the valve. It redirects that motion from upward (as the pushrod comes up) to downward (to push down, and open the valve). The number one design characteristic of the rocker arm is the *ratio* at which it transmits this motion. Mechanical adjustment is also another major factor. For both the mechanical and hydraulic style rockers, the design ratio is 1.5-to-1. This means that the valve tip moves farther than the pushrod tip on the arm itself.

Ratios—For example, if the tappet and pushrod tip move upward 0.400 in., then the valve tip will go downward 0.600 in. (0.400 in. x 1.5). This is the ideal, or *design ratio*. Because of manufacturing tolerances, the actual rocker ratio usually isn't 1.5-to-1. It's more likely to range from 1.45- to 1.53-to-1 overall. Also, it isn't constant. The ratio right off the valve seat isn't the same as the ratio near maximum lift. Usually, the *overall ratio*

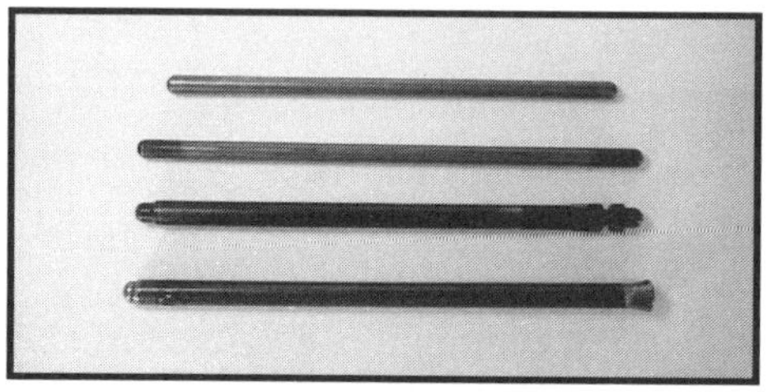

The various types of pushrods used in A engines. The top pushrod is the shortest and is used in a production hydraulic roller set-up. The second one down is a standard hydraulic pushrod. Next is the adjustable hydraulic pushrod. On the bottom is the typical mechanical pushrod. Note that the adjustable hydraulic pushrod has two male ends like the other two hydraulic pushrods while the mechanical pushrod his one male and one female end.

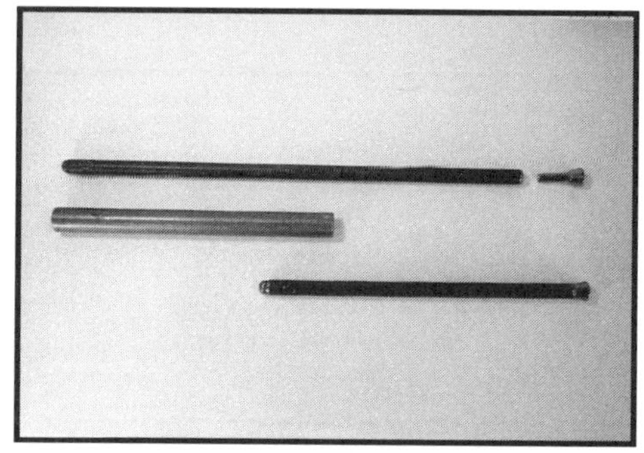

To make custom length pushrods (mechanical) you need a long tube with one end fixed and the other end (usually the top) loose. You also need an assembly tool as shown in the middle. Once cut to length and the end pressed in, the finished product is shown at the bottom.

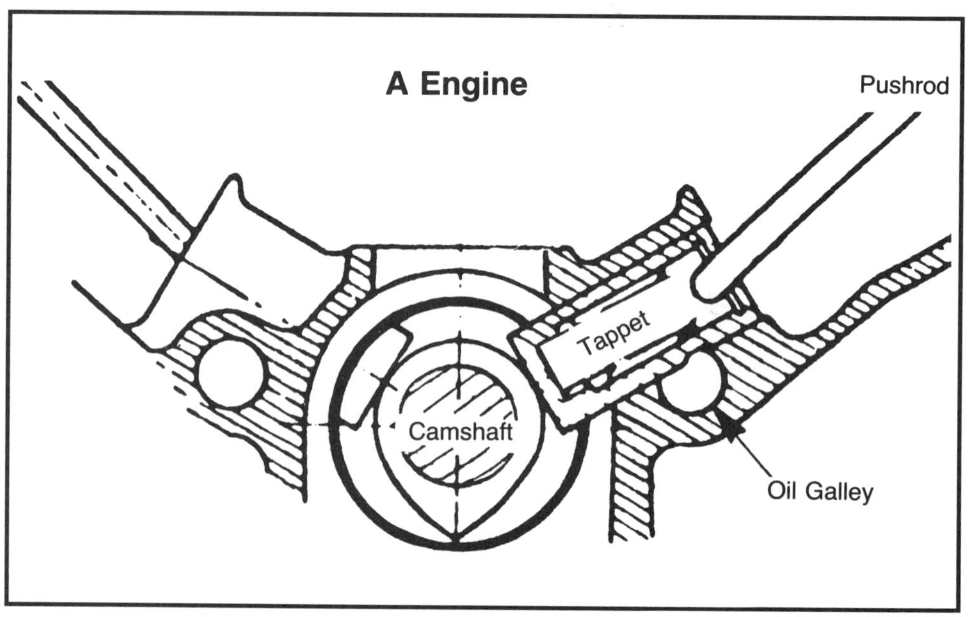

Camshaft, tappet and pushrod alignment. Note the location of the oil galley to the tappet and the angle of the pushrod. (Courtesy Mopar Performance).

(actual valve lift divided by actual cam/lobe lift) is the only thing that's considered.

For example, if the actual lift at the valve is measured at 0.624 in. and the cam lobe lift is 0.412 in., then the overall ratio is 1.515-to-1.

To check for ratios, use lighter springs than you intend to use. These are sometimes called *checker springs*. Racing valve springs have high loads, which can cause deflections and throw off the measurements. It's a good idea to know both light spring and race spring valve lift numbers.

High Ratio—There are high ratio rocker arms available for the A engine. Typically, they are aluminum with a roller bearing on the rocker shaft and a roller on the valve tip. Due to their construction, aluminum rockers tend to flex more than the steel rockers. The higher the ratio, the more the arms will flex. Most race cams are designed for the 1.5-to-1 ratio. Higher ratios in the 1.6- to 1.7-to-1 area are available. While these special rockers obviously increase the lift of a given cam, they will also add 2- to 3-deg. of duration at 0.050 in. for each 0.1 increase in the overall rocker ratio. As a general rule, don't change the rocker ratio without talking to the cam's manufacturer first.

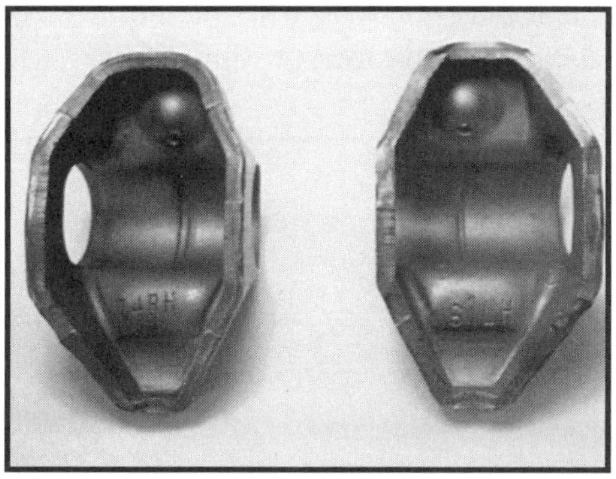

Typical hydraulic rocker arms. Note that there is a "left" and a "right" having to do with the offset or location of the pushrod ball in the arm itself.

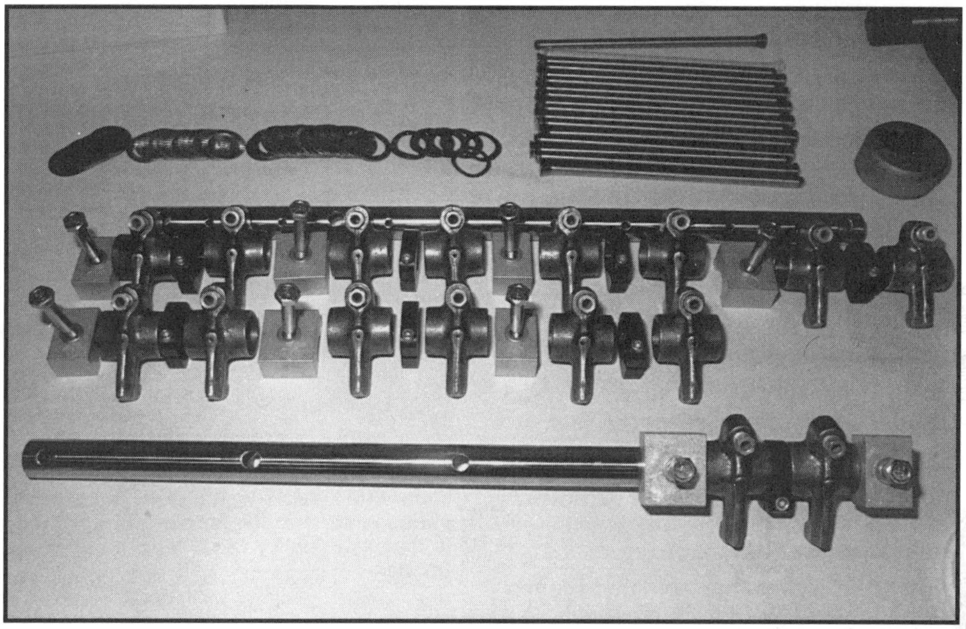

A typical rocker arm and shaft set-up ready for assembly.

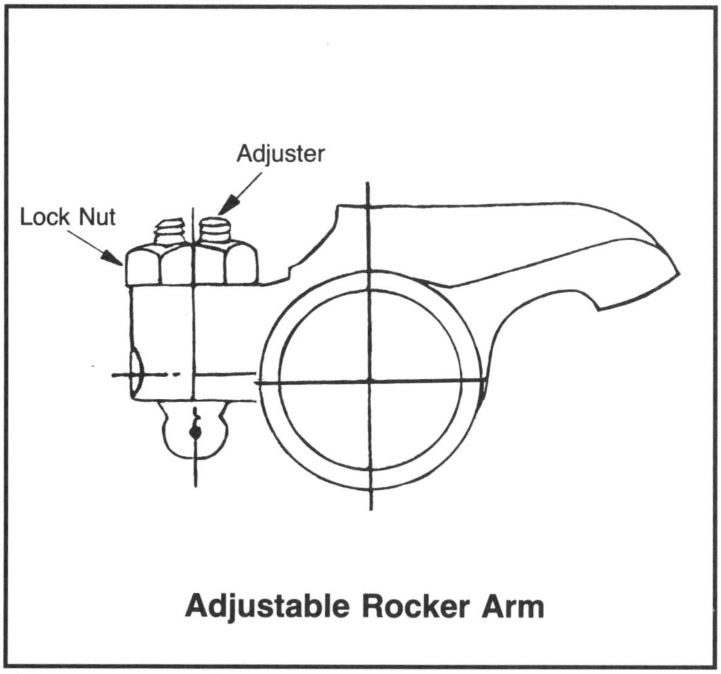

Adjustable rocker arms consist of three pieces. Rocker arm goes over a rocker shaft. Center pivot may be bushed or unbushed depending upon application.

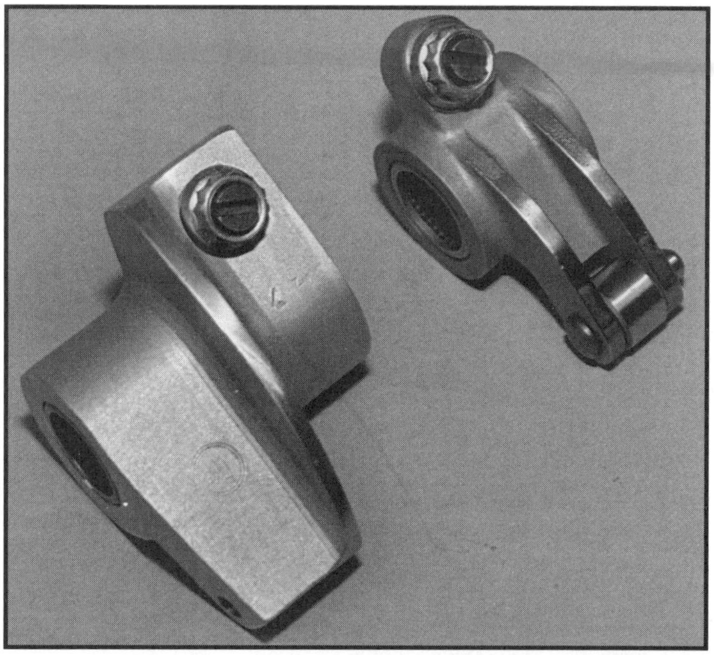

Dick Landy custom-made these aluminum rockers to adapt to the rearranged pushrod geometry of his 340 Pro Stock engine run in 1980. The rocker arms are no longer available, however.(Photo courtesy Hot Rod Magazine)

Unless they advise otherwise, always use 1.5-to-1 ratio rockers.

Installation—When setting up the valvetrain on the cylinder head, the alignment of the rocker tip on the tip of the valve stem should be checked. The center of the rocker arm tip should be directly over the valve's centerline. If it isn't, it should be corrected. There are special shims available to shim the rocker up the shaft (away from the stand). If the rocker needs to get closer to the stand to align properly, then the closest end of the rocker is ground down the proper amount. From a theoretical standpoint, the rocker arm geometry from a side view, should be such that the rocker arm's contact point at half valve lift should be in the center of the valve stem. Also this contact point should not fall off the edge of the valve anywhere in its travel. If this occurs, lash caps should be used to help solve the problem. This geometrical consideration is generally designed into the engine by the factory. Once built, it's very difficult to change. The W-2 head has the side view geometry specifically changed to help its high rpm characteristics.

Clearances—One of the biggest problems specifically relating to the racing, adjustable rocker arm is clearance to the valve spring and retainer on the underside of the arm itself. As the cams get bigger, the springs get larger in diameter. This extra diameter gets the spring/retainer very close to the underside of the rocker arm. With the bigger springs typically used with roller cams, this relationship will require a *radiused notch* to the rocker arm. This is true with either the forged steel rockers or the aluminum versions. The 273 mechanical rocker is a slightly thinner forging than the T/A or W-2 versions and requires less of a notch. Since this is a high-load, race problem, the T/A or W-2 rockers are still best. The aluminum rocker is thicker than the steel rocker and is more likely to need a notch. The race W-2 uses rocker stands instead of the cast-in pedestals, which moves the rocker shaft upward and away from the valve spring retainer to help relieve this situation.

Ideally a pushrod length should be selected that allows the adjustment screw in the mechanical rocker arm to have one thread showing below the rocker arm with the valve lash properly set. It shouldn't be over 1-1/2 threads. It can be less, but the pushrod socket may hit the rocker. With the large diameter pushrods used in racing, another clearance notch may not be required on the pushrod-side of the rocker, es-

The underside of the rocker on the valve end shows the valve spring relief which is required to clear the valve spring/retainer. Note the oiling hole in the middle of the relief. This is required to oil the valve tip.

If the stock relief won't clear the retainer/spring used, then it has to be enlarged. This is usually done by hand with a high-speed grinder.

Oiling holes in mechanical rocker arms are very important. They must extend through the rocker arm bushing to the rocker shaft. One hole must point at the valve tip and the other must point at the adjusting screw to properly oil the pushrod cup.

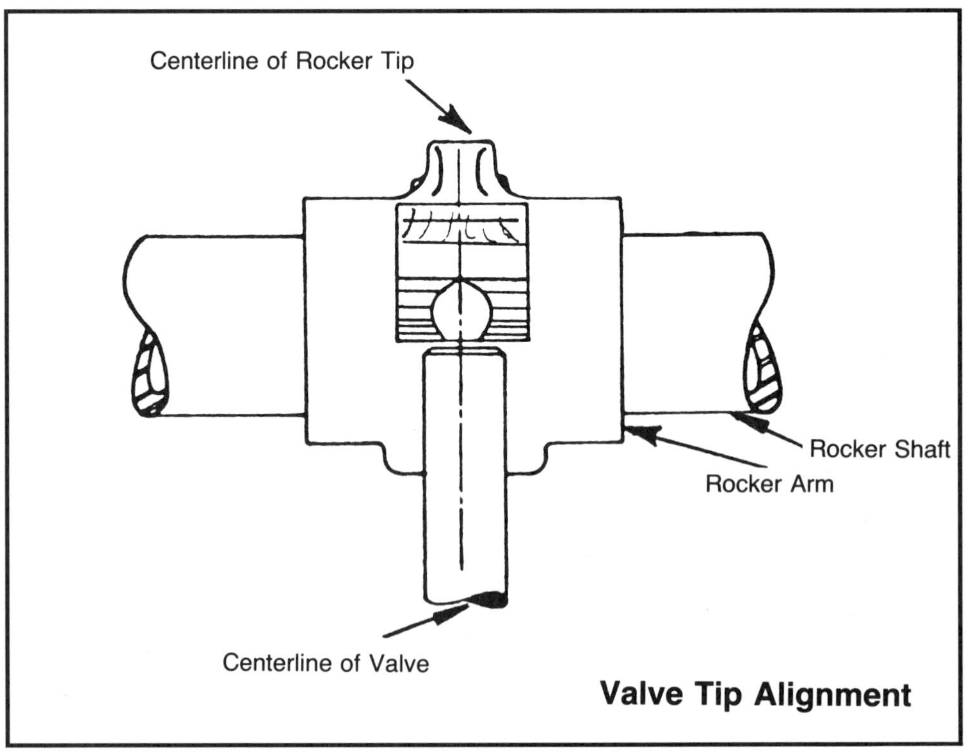

Valve tip alignment is critical in a race engine. The center of the valve stem should be directly below the center of the rocker arm pad. This is accomplished with rocker arm shims (spacers) and grinding, depending upon which direction (left or right) the rocker arm needs to go.

pecially with very high lifts. If the engine won't be turning high rpms, then more threads could show without causing problems.

Mechanical—Mechanical rockers (adjustable) can be used with hydraulic cams and are mandatory with mechanical cams. The pushrods must be changed if mechanical rockers are used with a hydraulic cam. Once the hardware is selected and is compatible, the rockers must be adjusted. They are set the same way as mechanicals, except that instead of setting a *positive* valve lash, they must be set for *negative* valve lash. This is done by adjusting the valve lash first to zero. Each valve is done separately (only four at any given crank position, same as with a mechanical cam) with the tappet on the base circle. From the zero lash position, the adjusting screw is turned down another 1-1/2 to 2 turns and the lock nut is then tightened.

ROCKER SHAFTS

In the 70's, the racing community generally felt that individual ball-studs were a better race set-up for valvetrain than a *rocker shaft* arrangement. But research and development has proved that there are advantages to using the rocker shaft over the ball-studs, and today, most top all-out race engines use rocker shafts.

Types—All Chrysler A engines use rocker shafts. Mechanical rockers need a special shaft for oiling purposes. The 340 shafts were the best of the hydraulic shafts, but they all should work equally well with hydraulic rockers. The 273 mechanical set-up

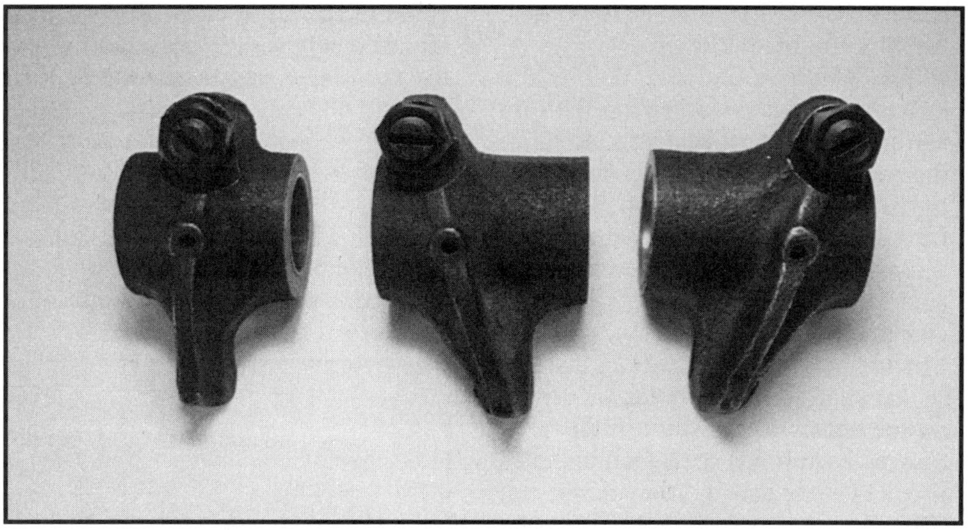

Standard W-2 mechanical rockers. The exhaust is on the left. The left and right intakes are shown to the right. You need 8 exhaust and 4 each of the two intakes to make a set.

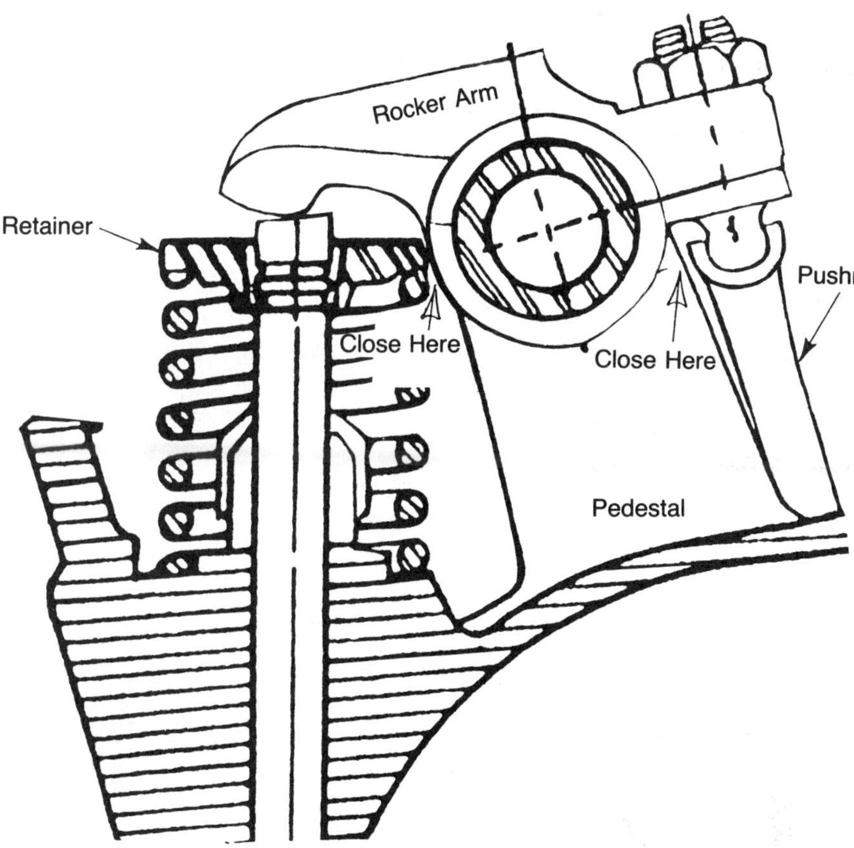

Valve Gear Clearance Problems

The A engine has two valvetrain clearance problems when used in all-out racing. First, the valve spring retainer gets very close to the underside of the rocker arm. This usually requires the underside of the rocker arm to be notched. The W-2 race rocker gear offsets the shaft away from the retainer which helps this situation. The second problem is the larger diameter mechanical pushrods that are used. The cup (top end) gets very close to the underside of the rocker next to the adjusting screw. Again this can be solved by grinding clearance on the rocker arm body itself. The valve side is the greater problem because as the notch gets larger it actually makes the rocker arm weaker and may cause it to fail. (Courtesy Mopar Performance).

Aluminum rocker arms for mechanical valve gear tend to look boxy or rectangler when compared to forged steel rockers. Note the width of the valve end of the rocker.

used a rocker shaft similar to the one used by the hydraulics.

The 340 T/A and the W-2 shafts have *banana grooves* on the bottom surface of the shaft to distribute oil to the rocker arm. This is needed to oil the rockers at high loads and high rpm. Be sure not to put them in upside-down. They go on the bottom or head-side of the shaft.

The difference between the 340 T/A and the W-2 rocker shafts is that the rocker shaft attaching holes are drilled on the center of the shaft in the T/A. The W-2 shaft has the attaching holes offset to move it away from the valves. The attaching bolts are in the same location in the cylinder head.

Advantages—One of the big advantages of the rocker shaft set-up over the ball-stud arrangement is that no stud girdle is required. This means big savings to the racer using a high-performance small-block. Another advantage is that the rocker shaft can accept the side loading of offset pushrods better than the ball-stud can. The offset pushrods come about because of the large intake ports used to make horsepower. The big ports move the pushrod over for clearance. Since the valve stays in the same place, an offset rocker is now required to connect the various pieces. Both the 340 T/A and the W-2 use offset intake rockers. The rocker shaft also holds the rocker alignment better once the alignment is blueprinted. Additionally, it helps decrease tip wiping and scuffing, which can be problems with ball-stud arrangements. Other than the elimination of the stud girdle, these advantages are only going to be found in high rpm, high output applications.

This shows the underside of the mechanical rocker on the pushrod end. For best operation (especially at high rpm), the adjusting screw should only show one thread below the arm.

ROCKER STANDS

A *rocker stand* is used in place of the cast-in pedestal. The pedestals are machined off in the Race W-2 cylinder heads so the rocker shafts can be moved. There are aluminum and steel versions of the rocker stands. The Race W-2 stands have the attaching hole offset relative to the centerline of the rocker shaft hole. The 305 Trans-Am race engines in 1970 also used rocker stands but the attaching holes were on-center. The Econo W-2 conversion puts the attaching hole back on-center using the cast-in pedestals. For more details on the conversion refer to the previous Chapter on *Cylinder Heads*.

VALVE SPRINGS

Valve springs can be a big problem in professional racing. For less de-

Rocker arms fit snugly on the chrome-moly rocker arm shafts with only 0.005 in. of play between rockers on this custom DLI set-up. Note the use of the peanut-style spark plugs to clear Hooker headers. (Photo courtesy DLI)

manding applications, it's a matter of picking the spring to go with the selected cam. In most cases, the cam manufacturer will suggest a spring to be used to get the best results. If no spring recommendation is available, the camshaft will provide the proper direction.

Selection—For general high-performance applications of the A engine, a different valve spring should be used for each of the following steps: up to 0.435-in. lift; 0.440-0.510-in. lift; 0.500-0.600-in. lift; 0.580-0.620-in. lift; 0.620-0.640-in. lift *special*. All of the springs used with these cam lift specs are installed at the stock A engine height of 1.69 in. The high-lift special cam (0.620-0.640-in. lift) uses the same valve spring but in conjunction with a special retainer that increases the installed height to 1.73-

The best performance features of valve springs are not really observable by eye. From right to left—318/2BBL spring, 360/2BBL, Mopar Performance performance spring and Mopar Performance high-performance spring. All are primarily for use with hydraulic cams. Note that the two performance springs (on the left) have dampers.

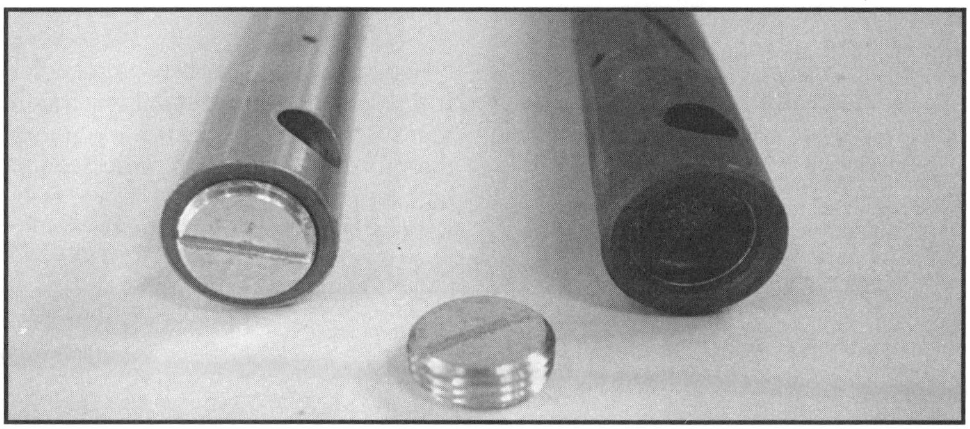

In addition to the offset attaching holes, another difference in the two mechanical rocker shafts is that the W-2 shaft (left) has a screw-in plug in the end while the T/A shaft has a press-in plug.

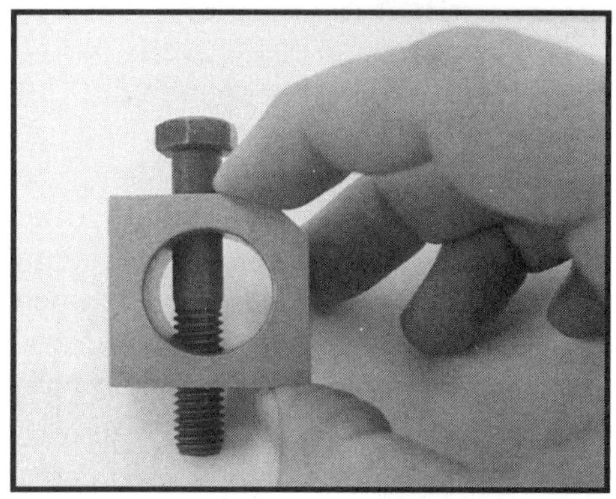

The W-2 stands have the attaching bolt offset to move the rocker shaft assembly away from the valve springs. Obviously, the hole in the shaft and brackets has to be offset also.

Rocker shaft stands for a W-2 cylinder head. Similar stands are also used on the 1970 340 Trans-Am race engines. They come in steel or aluminum versions.

1.75 in. For all-out performance, the installed height needs to be increased to 1.86 in. or more, which is the B engine, racing Hemi installed height. By using this height, you can also use the larger B engine valve springs. All of the small-blocks, including the A engine, are limited by room for springs. In order to use larger springs, longer valves can be used. This will move the keeper grooves up, increase the installed height and allow the big, high-load springs to be used. Once the installed height is changed to allow for the big springs, then the spring seat will need to be machined to a larger outside diameter (1.56 in. or more) because these bigger springs are also larger in diameter. It's a good idea to measure the outside diameter of the spring before sending everything off to the machine shop.

Preparation—To prepare valve springs for race use, the loads on the seat (closed) and at maximum lift (open) should be measured and recorded and the solid height should be checked. For all-out racing with roller cams, the new spring seat load should be between 169 to 180 lb. The open load should be near 600 lb. Once used, these same springs will drop to 150 to 160 on the seat and 550 open. If the seat load drops below 145 lbs., discard/replace the spring. These numbers apply only to all-out racing applications.

Dual-Springs—The inner, outer and damper of the racing dual-spring should be checked for *coil-bind* or *solid height*, separately and together.

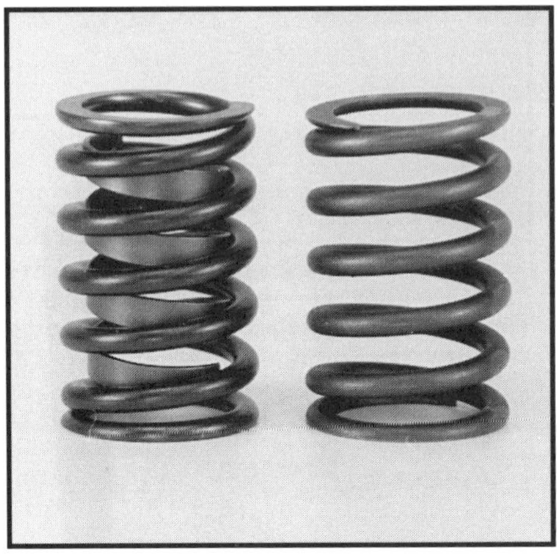

Close-up of two similar valve springs. Note that while both are similar in height and wire diameter, the spring on the left has a damper. Note that the damper is wound in the opposite direction from the spring itself.

If the spring is operated near this height it will break! If the head is set-up properly, then the outer spring is the limiting factor. Once the solid height and the installed height are known, the *durability factor* can be calculated. First, take the solid height (measured) and add 0.100 in. Then add the cam's lift at the valve and the answer should be less than the installed height. If the answer comes out greater than the installed height, then the spring's durability will suffer.

When prepping the spring, the *dampers*, (the counter-wound piece between the inner and outer springs) on a dual-spring should be cut (ground) square. Also, when measuring the solid height of the outer spring and damper assembly, it is sometimes found that the damper goes solid first. If this is the case, the damper can be ground shorter to provide the extra clearance required.

For bracket racing, the long-valve, all-out approach is not recommended. It's too costly for the small benefits in that type of racing. If for some reason the proper valve spring choice cannot be used, then the cam will have to be adjusted accordingly. At that point the

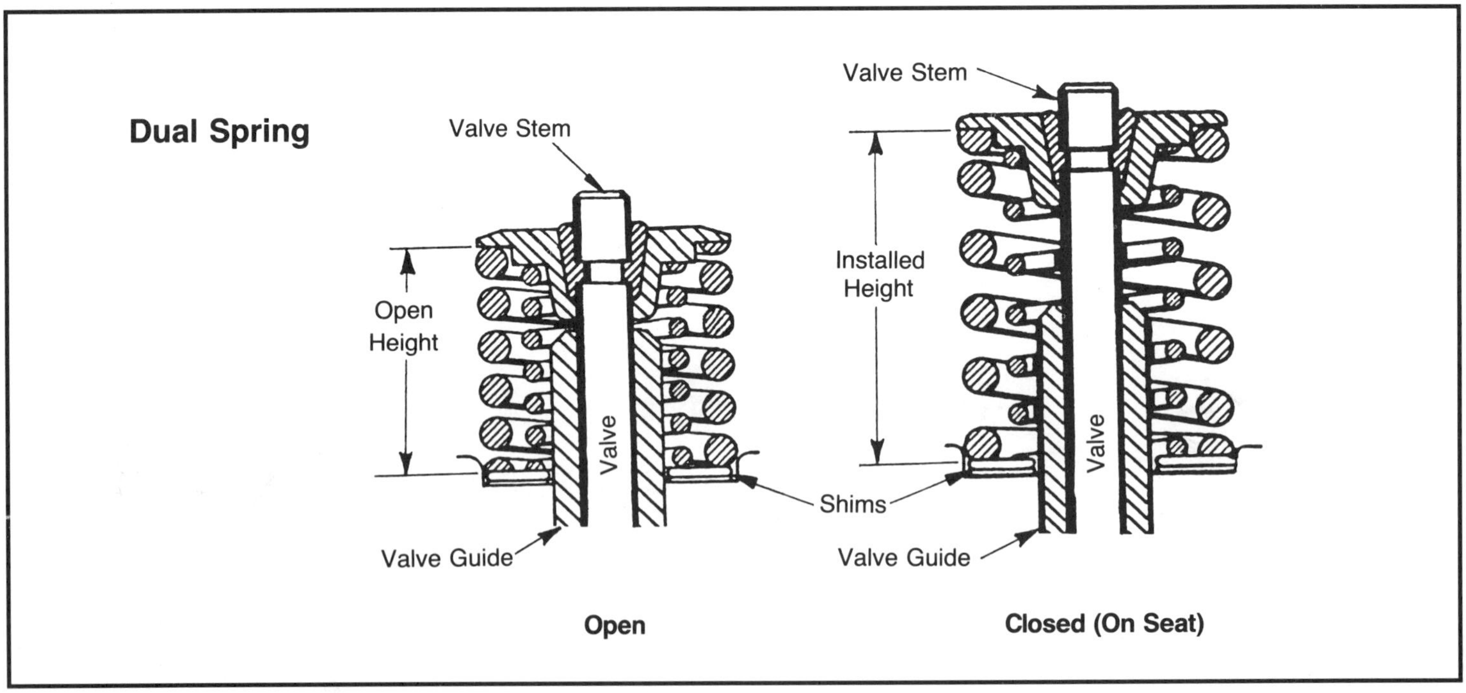

Valve spring loads and heights should always be checked before use. Measure the installed spring height or closed height—valve on the seat. The installed height is measured from the top of the spring seat or from the top of any shims used, to the underside of the valve spring retainer at its outside step. A dual spring also has an installed height for the inner spring. They are not the same. The open height is the height of the valve spring at max lift. This can be determined by subtracting the valve lift from the installed height. Double-check by actual measurement in the engine. The open and closed loads are checked at the heights determined above.

spring becomes the limiting factor for the cam and will dictate the cam profile that should be used.

Retainers—Valve spring *retainers* transfer the valve spring *loads* to the valve *stem*. The retainer itself stays the same with either a hydraulic or mechanical cam. All the production retainers are 8 deg. designs. Serious racers should use 10-deg. retainers. They don't "pull through" as easily. However, be forewarned that the two types of retainers are *not* interchangeable! Each retainer requires that the keeper match the retainer. They can be switched as a keeper and retainer unit.

There are four different types of material used for performance retainers: steel, chrome-moly, aluminum and titanium. The production retainers are steel. The chrome-moly and aluminum retainers are intermediate performance choices. The titanium is the all-out retainer. It's very strong and light but is also the most expensive.

Other than checking the fit of the retainer on the spring itself, the only clearance that directly affects the retainer is the *retainer-to-guide clearance*. This clearance should be greater than 0.125 in. at max. cam lift.

When starting with a stock A engine, check to see if the valves use *rotators*. They are special exhaust valve spring retainers used on trucks to increase durability. For performance

These are two different styles of titanium retainers. The two on the left are 10 deg. retainers. Note the larger center hole.

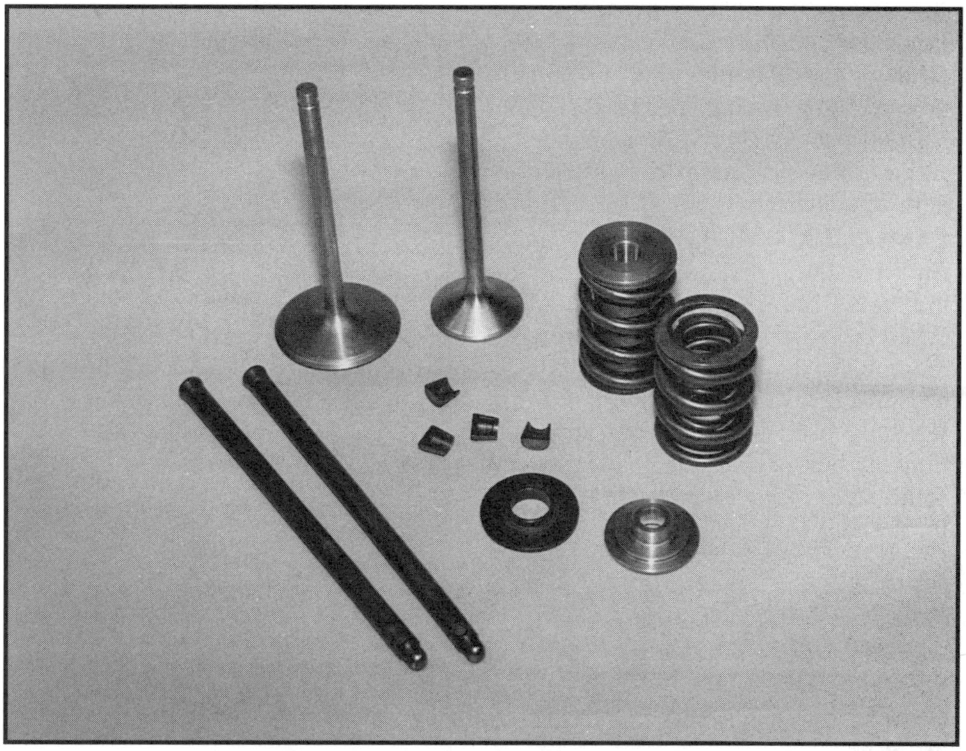

Selecting a combination of valve gear is dependent on so many factors. Dick Landy chose Manley 2.20 in. intake/1.65 in. exhaust titanium valves, Crower red stripe valve springs; Crane valve spring retainers, seals and chrome-moly pushrods. (Photo courtesy Hot Rod Magazine)

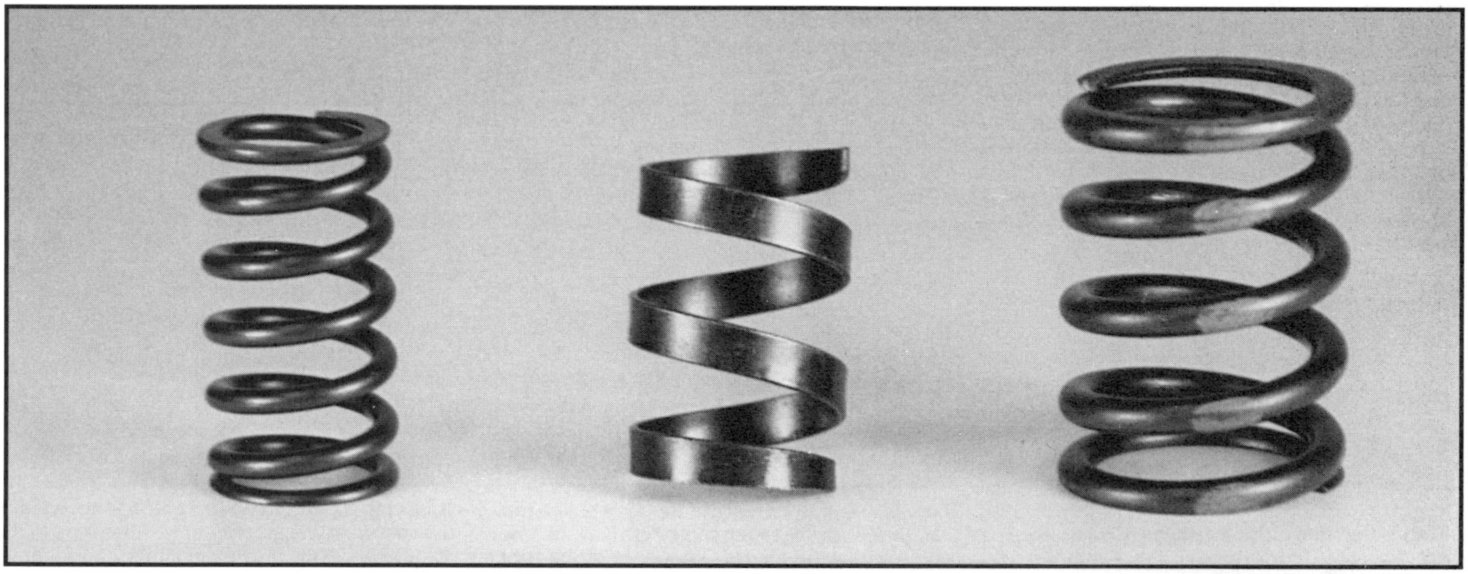

A single spring with damper is sometimes mistakenly called a dual spring, but this is a real dual spring. From left to right—inner spring, damper and outer spring. Note that the inner and outer springs are wound in the same direction while the damper is wound opposite.

applications, they should be discarded and the stock retainers used.

Keepers—The keeper or *valve lock* is used to attach the retainer to the valve stem. The two basic types are 8 deg. and 10 deg. All production engines use 8-deg. locks. There are three styles of 8-deg. locks—single, double and triple groove. The triple groove is used on the production exhaust valves. The single- and double-groove locks are used on the intake valves. Special race valves are only single groove. The 10-deg. locks come only in single groove. For racing purposes, only hardened locks should be used in the single-, double-, or triple-groove 8-deg. applications.

Lash Caps—The *lash cap* is a small cap (hardened) that goes on the top of the valve stem. Typically, it adds 0.080 in. to the tip height and increases the outside diameter. The total thickness of the lash cap is about 0.160 in. It's only used in the all-out applications. It's more common on the special small-stem race valves such as 11/32 in. and 5/16 in.

Valve Seals—It's hard to think of a valve seal in the valvetrain area but it's still part of the overall system. Basically there are two types of the valve seals—the *PC* and the *umbrella*. The umbrella seal is used in production. If using an umbrella seal for high-performance, then it should be made of *Viton* material because it's more heat-resistant. This style seal is the most durable and effective seal. However, it can be used only with single valve springs or single springs with dampers. It's too big to fit inside a dual-spring. The PC seal is used for dual-springs. It's smaller in diameter and attaches directly to the valve guide. It's not as durable as the umbrella seal. For street rods or street machines, the umbrella seal should be used. This will limit the choice of valve spring, which in turn will limit the cam profile. This is one of those areas where the valve seal dictates the cam. Without the seal, or if the seal wears out, oil usage will go up.

Retainers are pretty simple—but there are many styles and varieties. These are all 8 deg. retainers shown from the top and the bottom. From right to left—the production steel retainer, high-performance chrome-moly retainer, high-performance aluminum retainer and the race titanium retainer. Notice that the diameter of the retainer tends to get larger as it goes from production to titanium. This is because the valve springs used in the applications require the retainers to be bigger. In other words, the production spring is the smallest in diameter.

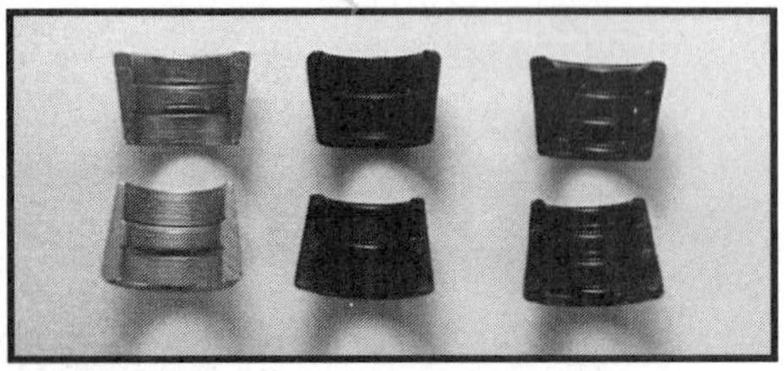

These are three different styles of basic keepers (valve locks). All are 8 deg. From left to right—single groove, double groove and triple groove.

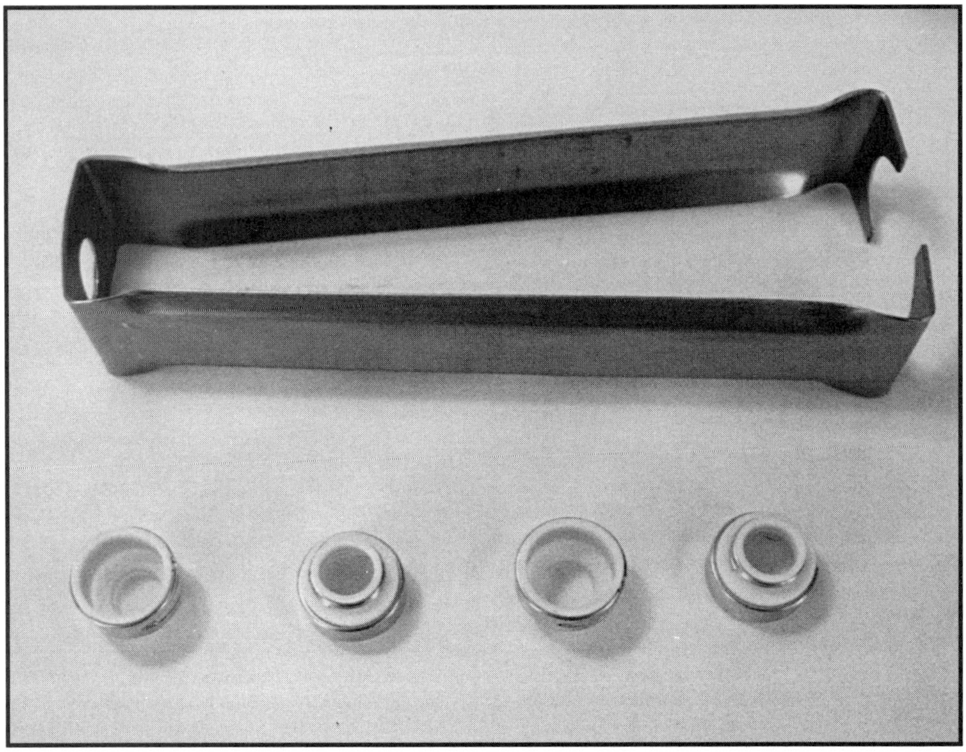

The teflon or "PC" seal attaches to the valve guide. Typically the production guide must be shortened and narrowed to accept this style seal. The tool shown at the top is required to install the seals onto the guide.

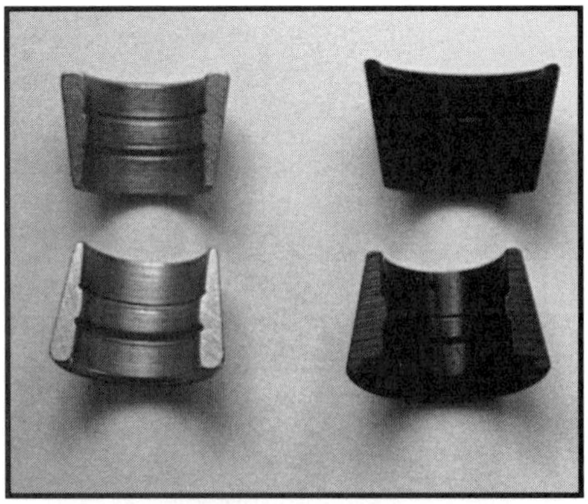

This photo compares the 8-deg. keeper (on left) to the 10-deg. keeper (on right). Note how much wider the 10 deg. keeper is (at its top) and how much more cross-sectional area it has.

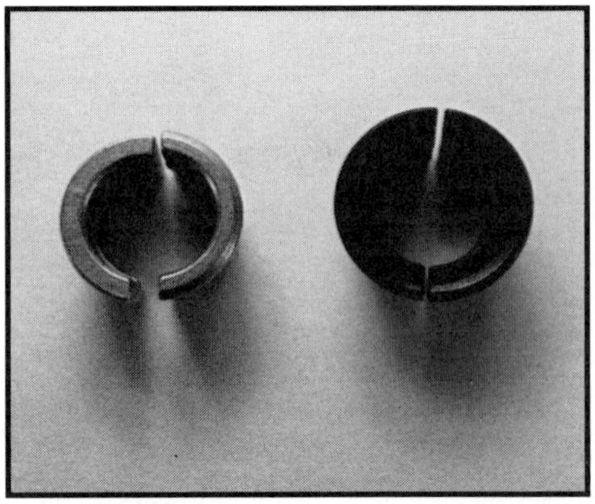

Looking down on the top of the two styles of keepers, the only noticeable difference is the diameter. Since they are not interchangeable, this becomes an important means of identification.

One of the unique features of the 10-deg. keepers is that they have a recess cut in the top so that a lash cap will fit inside. Here the lash cap is added to the top of the keepers as it would look on the end of the valve.

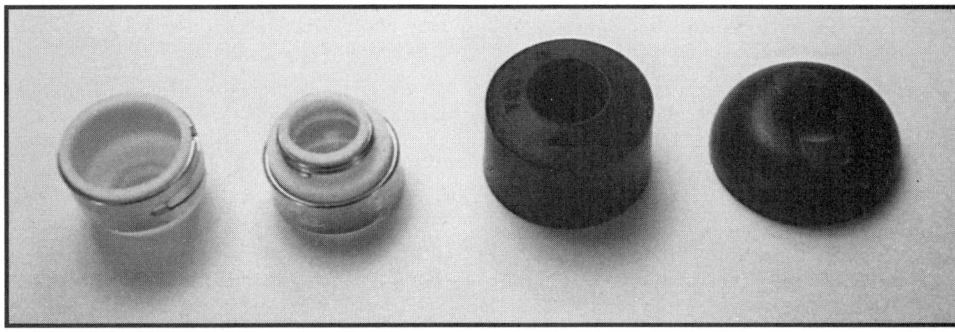

The two seals on the left are teflon while the two on the right are umbrella. The larger diameter of the umbrella seal prevents the use of dual valve springs.

CHRYSLER PARTS LIST

CAMSHAFTS

Hyd. .405 in./240 deg. P4452755 kit
Hyd. .425 in./256 deg. P4452757 kit
Hyd. .430 in./260 deg. P4452759 kit
Hyd. .465 in. 272 deg. P4452761 kit
Hyd. 0.410 in./248 deg. P4286667 kit
Hyd. 0.430 in./260 deg. P4286669 kit
Hyd. 0.455 in./272 deg. P4286671 kit
Hyd. 0.474 in./280 deg. P4286630
Hyd. 0.484 in./284 deg. P4120231 kit
Hyd. 0.508 in./292 deg. P4120233 kit
Mech. 0.528 in./284 deg. P4120653 kit
Mech. 0.557 in./296 deg. P4120655 kit
Mech. 0.590 in./312 deg. P4120657 kit
Mech. 0.620 in./324 deg. P4349266 kit
Mush. 0.630 in./320 deg. P3690589 kit

Stock Eliminator "Cheater" (340)
(Cam Only) P4349243
360 (Cam Only) P4349245
Roller Tappet (360 Super Stock)
(Cam Only) P4349564
Roller Tappet (340 Super Stock)
(Cam Only) P4120976
Roller Tappet (340-355 All-Out)
(Cam Only) P4120977
MuscleCar Repro Cam 340-360 HP
"2699206" P4452782

CAM ACCESSORIES

Cam Bearing Set P4120260
Roller Timing Chain &
Sprocket Set P4120262
Cloyes 5 Keyway Sprocket
& Chain Set P3690866
Camshaft Sprocket P3690695
Crank Sprocket P3690696
Roller Timing Chain P3690697
Cam Bolt & Washer Package
(with eccentric) P4007694
(without eccentric) P4007695
Offset Cam Key Package
1-2-3-4-5 P4286500

VALVE SPRINGS (Std. length valves)

Up to 0.440-in. lift P4286813
0.440-in. to 0.510-in. lift P4120249
0.500-in. to 0.600-in. lift P3614542
0.525-in. to 0.620-in. lift P3412068

VALVE SPRINGS (Long Stem Valves)

Up to 0.510-in. lift P3690933
0.500-in. to 0.620-in. lift P2806077
0.600-in. to 0.650-in. lift P3462887

RETAINERS

Chrome Moly P4452033
Aluminum P4286573
Aluminum HP P3412067
Titanium (8 deg.) P4007178
Titanium (10 deg.) P4120785
Titanium (10 deg. triple) P4120783

NOTE: The first 3 retainers are 8 deg. Only last two are 10 deg.

SHIMS (standard spring set)

0.060 in. P4007216
0.030 in. P4007217
0.015 in. P4007218
0.060-in. cup P4007227

SHIMS (inner spring seat removed)

0.060 in. P4007222
0.030 in. P4007223
0.015 in. P4007224
0.060 in. Cup P4007225

NOTE: Above shim packages come 50 pieces per package.

VALVE STEM SEALS

Viton H.D. P4120492
(use with single springs)
PC P3690963
(use with dual springs)

KEEPERS (3/8-in. stem, 8 deg.)

Single Groove P4120618
Double Groove P4120619
Triple Groove P4120620

KEEPERS (3/8-in. stem, 10 deg.)

Single Groove P4286612

VALVE LASH CAP PACKAGE

3/8-in. stem P4120626
11/32 stem P4120049

ROCKER ARMS

H.D. Hyd. Rocker Arm
(Right) P3690710
(Left) P3690711

ROCKER ARMS (Adj. W-2 Race)

Intake (right bushed) P4120200
Intake (left bushed) P4120201
Exhaust (bushed) P4120202

ROCKER (Adjustable W-2 Race)

Intake (right bushed) P4120200
Intake (left bushed) P4120201
Exhaust (bushed) P4120202

VALVETRAIN HARDWARE

Rocker Arm Shaft Supports
(W-2, T/A)
Alum. stands w/screws P4120098
Steel stands w/screws P4120102
(Both support pkg. use offset shaft).
W-2 Rocker Arm Shaft P4120589
(Attaching holes offset)
(Mech. rocker & offset stands)
T/A Rocker Arm Shaft P3577086
Rocker Arm Spacer Pkg.
(12 pc., 0.15-in.) P3690896
Rocker Arm Adj. Screw
& Nut Pkg. P4120636
(To service W-2 and T/A adj. Rockers)

PUSHRODS

Adj. Hydraulic Pushrods P4286808
Mechanical Pushrod Kit P4007284
(Cut to desired length-assembly required)
Pushrod (individual) P3690988
(service package for above kit)

TAPPETS

Hydraulic Tappets P3614321
Mechanical Tappet P2843177
Mushroom Tappet P3690137
Roller Tappet P4452919
Roller Tappet-Engle P4286591
Guide Bar for Engle Roller P4286592

7 BLUEPRINTING

Blueprinting is the process by which the part or the whole is optimized relative to a set of specifications. In the previous six chapters, I have covered the blueprinting of each *individual* part of the engine assembly. However, there are blueprinting procedures that depend on several parts. In a sense, blueprinting is the step that is performed *before* final assembly. Some call it *pre-assembly*. In the next few sections, I will discuss in detail the special operations that require more than one part. I will be relating several parts to each other or to the whole engine, on topics such as compression ratio and TDC (top-dead center). For example, the compression ratio is affected by the cylinder head, block and piston, and TDC is affected by the crank, block, rod and piston.

At this point in your engine build-up you should have several pages of specifications and measurements. I cannot stress how important record-keeping is to the overall engine building process. However, the most important measurements are coming up. Keep in mind that once the engine is final-assembled ("buttoned-up"), it would have to be disassembled to obtain this same information if you hadn't taken a few minutes to write it all down. It doesn't sound like a big deal now, but wait until you have a problem. Then, you may need to know the compression ratio, the cam centerline, individual combustion chamber cc's, etc. If all of the information is written down, you may be able to solve the problem without taking the engine "down" (disassembling). Without these records, you may not be able to remember the specific details, or worse yet, you may remember them incorrectly, which could send you off in the wrong direction looking for the solution. This can be twice as expensive and very time-consuming.

SHORT-BLOCK

In this section, you should be primarily interested in how the parts of the engine interrelate to each other, not the individual part itself.

By this time you should be well-acquainted with the various precision tools that are needed to build a top-quality high-performance engine. Micrometers, torque wrenches, dial-indicators, degree wheels, etc. are all specialized tools used commonly in precision engine building. The dial-bore gauge is such a special tool, and is not generally found in the average garage. It is used to measure the cylinder bore for taper and roundness. It can also be used to measure bearing bores in the block and rod's big-end. The dial-bore gauge is more accurate than the "snap gage" (telescoping gage). The extra accuracy is needed in many of the measurements taken on a racing engine's short-block. If a dial-bore gauge is unavailable, then a machine shop will have to do these operations. Be sure to get these measurements from the machine shop and write them down.

CRANKSHAFT INSTALLATION

I covered the preparation of the crankshaft in Chapter 2, but it should be cleaned again thoroughly. Don't forget the oiling passages I mentioned earlier! The crank now needs to be measured accurately for clearances. Be sure to check these measurements several times to be sure. Mistakes here could easily result in disaster.

The crank should be installed in the block to check for straightness. It's checked at the #3 journal and supported on main bearings in the #1 and #5. For this measurement try to get your dial-indicator to be as perpendicular as possible in the center of the journal. Note "360" cast into this crank to the right of photo.

Journal Diameters—Measure each of the rod and main journals (outside diameter) to check for size. The journal diameters will be used to measure the bearing clearances shortly. The crank journals can be checked for roundness by measuring each journal twice—the second measurement 90 deg. to the first. The *taper* of the crank's journal can also be checked by measuring once in the center of the journal and then once to either side next to the edge or "cheek" of the journal. Caution must be used not to get into the radius area on a full-radius crank. If the crank is used with a manual transmission, check the pilot bushing as mentioned earlier and machine as required for the proper pilot bushing to match the transmission. Now the crank can be installed in the block for further inspection and prep.

Crank Straightness—At this point, the first check is for *straightness*. This is usually checked twice, using two different methods. First, install the #1 and #5 main bearings (using a small amount of engine oil) in their proper place before the crank is set in the block. Then gently lower the crank in place and set a dial-indicator gauge on the #3 main bearing journal. With an eye on the dial, rotate the crank 360 deg. to check its straightness. It should fluctuate no more than 0.0020 in.

The second set-up is similar to the first. Remove the crank after its first check, then remove the #1 bearing and install the the #3 bearing with oil. Reinstall the crank and set-up the dial-indicator on the #1 main bearing journal. Again, rotate the crank 360 deg. The reading for both #1 and #5 should fluctuate no more than 0.003 in. If these measurements are off, then take the crank back to the machine shop and tell them of the findings. Go no further.

End Play—This is a good point to check the *end play* of the crank. Install the #1, #3 and #5 main bearings with oil and install the crank. Torque the three main caps down properly. Then push the crank all the way forward in the block. Set up the dial-indicator on a flat counterweight parallel to the centerline of the crank and zero it. Then push the crank all the way rearward, using long-handled screwdrivers. The indicator will read the amount of end play. With a new thrust bearing it should be less than 0.010 in., but can be as high as 0.018 in. on a used bearing. On an oval track or road race engine, the end play should be less than 0.008 in. in a new or freshly rebuilt engine. If the reading is higher than 0.018 in., the #3 bearing should be replaced. While the thrust bearing may last the entire race-cycle life of a brand-new engine when used with an automatic transmission, it is quite common to replace the thrust bearing several times between rebuilds if a manual transmission is used.

Bearing Clearances—Remove the crank from the block, install all five main bearings and caps, then torque them down to spec. Insert a dial-bore gauge inside all 5 main caps and jot down the measurement. Now compare the diameter of each main bearing to the diameter of each respective crank journal, which should have been measured and recorded earlier. The difference between these two numbers yields the exact main bearing clearance. For most high-performance race engines, the clearance should be very close to 0.0025 in. For example, if the 318's #1 crank journal is measured and found to be 2.4997 in. while the #1 main bearing I.D. is found to be 2.5018 in., then the actual bearing clearance is 0.0021 in. or 2.1 thousandths-of-an-inch. Each main bearing shell should now be marked as to location-number; 1,2,3 etc.

Main bearing clearance can also be checked with *Plastigage*. To do this, install the main bearings and the crank, clean off a section on the crank journal, lay a strip of Plastigage on the journal, then torque down the main cap. Remove the main cap and read the width of the crushed Plastigage on the scale provided. Repeat the procedure with each main cap. This method is faster and adequate for high-performance street use. However, for critical racing engine building, I recommend measuring with a dial-bore gauge because it is more consistent and more accurate.

The rod bearing clearances can be checked in a similar manner. Their clearances should also be close to 0.0025 in. in a race engine. The common method to check rod bearing clearances is to place the rod securely in a rod vise, install the bearing shells and torque the rod bolts to spec. The

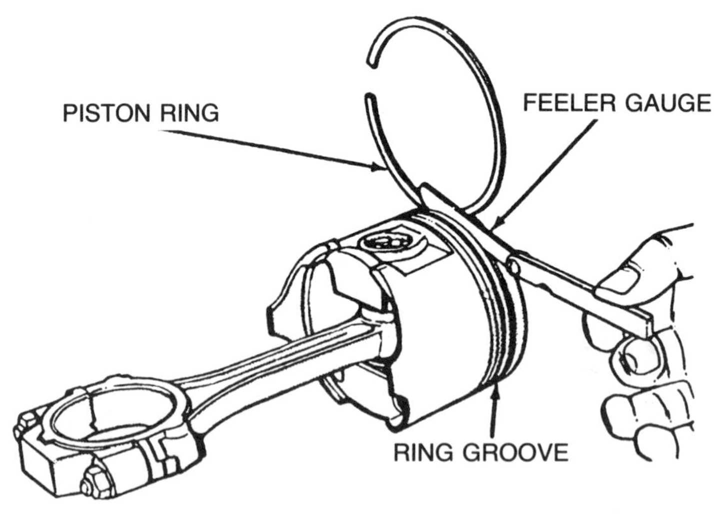

Ring Clearance In Groove

The clearance of the ring in the piston's ring groove should always be checked. For racing, it should be held to a minimum—0.001 or less *but not* zero or a press fit. Head-Land and Dykes rings can't be checked as shown but should still be checked once installed. Measure between the bottom of the ring and the bottom of the ring land. (Courtesy Mopar Performance).

dial-bore gauge can again be used to measure the inside diameter. Always measure 90 deg. to the bearing's parting line.

If any of these clearances are off, then the bearing sizes may be wrong, mislabeled by the manufacturer. Consult the machinist if you run into a clearance problem.

Stroke—Next, measure the exact stroke of the crank. Install all five main caps and torque down to spec. By using one piston-and-rod assembly and installing it in cylinders 1-3-5-7, each crank throw can be measured individually for its exact stroke length. Though this may seem redundant and time-consuming, it's better to be safe than sorry. No measurement in high-performance engine building should ever be assumed—always, always measure it to be sure.

If installing a billet crank or a full-radius crank, additional clearancing will often be necessary. These cranks are bigger and stronger but tend to "rub" or interfere with parts of the block, usually the bulkhead.

PISTONS & RODS

With the crank sub-assembled in the block, you can turn to pistons, rods, pins and rings. The pistons were prepared in Chapter 4. However, now each individual piston assembly must be installed in the block—carefully. It is important to install each piston into its proper cylinder bore. Usually, the block will be bored and honed by a machine shop. At that time, you should have also given the machinist the pistons to be used. The machinist then hones each cylinder to the proper clearance for each *individual* piston and marks the pistons accordingly. If this is true, line-up and install the numbered pistons, one through eight.

Rod side clearance is measured with feeler gauge between the two rods.

If they are not numbered, then measure the eight pistons with a micrometer and mark the diameters on top with a marker. Line up the pistons, largest diameter to smallest. Then, measure each cylinder with a dial-bore gauge and mark the bore size on the top of the block next to the measured bore. The largest piston will go into the largest cylinder bore, the second largest piston will go into the second largest bore and so on. The pistons are now marked with the number that corresponds to the proper cylinder.

By recording all the measured data, you can calculate each cylinder's *piston-to-wall* clearance. If you know which hole each piston goes in, then all you have to do is measure the piston with the micrometer and the bore with the dial-bore gauge and compare those two figures to get the piston-to-wall clearance for that cylinder. For example, if the #3 cylinder bore in a 340 measures 4.071 in. and the #3 piston has an actual diameter of 4.064 in., then the piston clearance in that cylinder is 0.007 in. (seven thousandths).

Mounting To Rods—The pistons can now be mounted onto the proper rods. The preparation of the rods is

To measure the crank end play, set dial-indicator on front end of crank, parallel to its centerline. Be sure to pick a machined surface. Use the counterweight and the block's bulkhead as leverage for the screwdriver to force the crank forward and rearward to get the total end play.

Deck height is measured from the top of the block to the flat of the piston. This is best done to the center of the piston if it's flat. However, this centered approach has to be adjusted if it's a domed/dished piston.

covered in Chapter 3. If pressed pins are used, be sure that the rod and piston are mounted properly the first time because they shouldn't be separated once assembled. With floating pins, they can be assembled and disassembled as many times as desired but be sure not to reuse the pin locks. With the pistons mounted on the rods, check the clearance between the side of the rod and the piston pin tower with the rod centered. There should be at least 0.100 in. on each side. Now put each piston and rod assembly, without rings, into the proper cylinder bore. Compare the valve notches in the pistons to the actual valve locations in the head. For example, starting at the 1-3-5-7 cylinder side and observing from the front of the engine, the first valve should be an exhaust. Therefore, the first notch on the #1 piston should be the smaller of the two notches (i.e. the exhaust notch). If this is true, then the second notch has to be an intake. The head indicates that the third notch should also be an intake (big one) and so on.

Domed pistons require special attention. Be sure that the dome is lined-up to stick up into the combustion chamber of the head. With a flat piston, be sure that the notches cut in the piston for the valves are actually below the valve itself, and not on the other side of the piston's centerline! Think about that one for a minute. It is not the same as what's discussed in the paragraph above.

Final Assembly—Remove the piston and rod assemblies and set aside. Gap the rings in the bore where they will be installed. Once gapped, mark and separate them according to their respective cylinders. Install the rod bearings, then check the bearing internal diameter (I.D.) with a dial-bore gauge and compare it to the rod journal's outside diameter (O.D.) to determine the rod's bearing clearance. Mount the rings on the piston with a ring expander, dip the piston, rings and all, in a can of motor oil (the same grade that will be used in the engine), put plastic sleeves or brass guides on the rod bolts and gently slide the pistons down into the bore. Use the wooden handle of a hammer to help slide or tap them down. Torque the rod caps to specification.

CAMSHAFT

The short-block itself is nearly complete, except for the camshaft. The cam should already be selected, so it needs to be installed. But before you do, let's quickly check several items.

With a hydraulic cam, check to make sure that the tappet bores are not bushed and that the tube isn't installed in the right side oil galley. With a mushroom lifter cam, check to make sure that the bottom of the tappet bore is spot-faced. If the cam is a roller, then make sure that the guide bar has clearance to the block. In all cases, the cam bearing should be centered in the journal and the oiling holes in the bearing should line-up with those in the block.

The first step in centerlining the cam is to line-up the "dots." Note the white circles on the top of the crank gear and the bottom of the cam sprocket. The important point here is that this is the first step not the final or last step.

Now, install the camshaft into the block. However, be sure to coat the camshaft generously with cam lube or molybdenum grease. Most cam manufacturers supply a tube of cam lube with purchase. This is critical. If the cam isn't lubed well enough, it will scour and quickly fail within a few moments of first ignition. One quick tip: If you are going to use a mushroom lifter cam, you should flip the engine over on the engine stand (crank up, decks down), drop in the lifters first, slide in the cam, then turn the engine back over. This saves a lot of time and hassle.

Sprocket—The relationship of the cam to the crank is very important. You can line up the dots on both sprockets like the Chrysler service manual says, which would define the position of the cam and the crank. That puts you close, but it isn't very accurate. There are approximately 25 teeth

Now we bring in the racer's tools—the degree wheel and the dial indicator and go for accuracy and maximum performance.

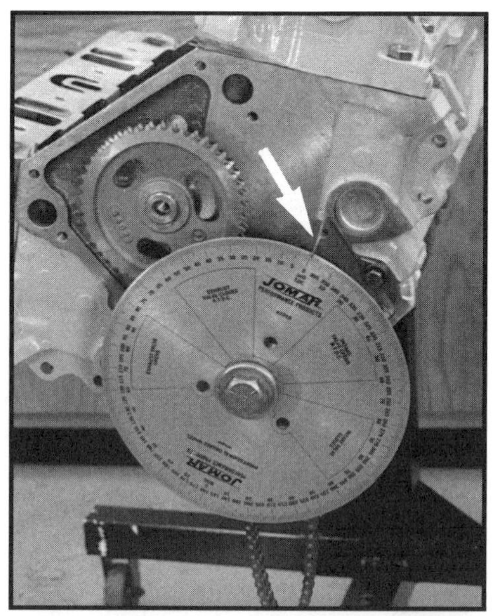

The next step is to zero the degree wheel by adding the pointer and zeroing to top dead center for the number one piston. Note the sharpened coat-hanger wire (held on by one of the front cover bolts) pointing at "0" on the degree wheel.

on the cam sprocket. Holding the crank in one position, this means that the cam sprocket and chain can be installed in any one of 25 different positions. Lining-up the dots cuts this down to one position, more or less! But the difference from one tooth on the sprocket to the next is approximately 15 deg. This means that accuracy is only plus or minus 15 deg. With the rest of the engine built to tolerances in thousandths-of-an-inch, this doesn't sound very good. It is bad for the engine's performance and output also. So lining-up the dots is the first step in cam installation but you need something else to get the cam-to-crank relationship a bit more precise.

Offset Keys—The cam sprocket is *keyed* to the camshaft itself. With a *straight* key, the relationship of the cam lobes to the sprocket teeth cannot be changed. However, with an *offset* key, you can divide the 15 deg. (representing one full tooth) down into smaller steps and therefore increase the accuracy. An offset key has the top of the key offset relative to the bottom. They come in 1 through 5 deg. offsets. By installing them with the offset to the left or right, virtually any cam-to-sprocket, and therefore crankshaft, relationship can be obtained.

Cam Set-Up—There are two methods commonly used to set up the camshaft. One is called *degreeing* and the other is *centerlining*. Both methods require a dial-indicator gauge and a degree wheel. The cam's centerline information is supplied with the cam from the manufacturer. For the first check, just line-up the dots and see how close you are to what is desired. If you're really lucky, you might be close enough, within 1 deg. either way. If not, then select one of the following two methods.

The centerline method defines the centerline of the #1 cylinder relative to TDC. This is the more accurate and more easily understood method of the two. "Degreeing the cam" refers to the degrees that the intake and exhaust valves open and close relative to TDC and BDC. With this second method, there are actually four numbers to check and adjust, which means there is a better chance you'll make a mistake. Therefore, I strongly recommend the centerline method. This is the method that I'll explain in detail.

If centerlining the cam yourself, you will need a TDC stop. This simple three-piece tool fits all A engines and is required to check TDC on the engine, especially if the engine is already assembled or installed in the car.

The first step in the centerlining procedure is to install the degree wheel on the front of the crank. You will need a pointer. A short piece of coat-hanger bent against one of the timing cover bolts can be used. Next, rotate the crank until the #1 piston is at the top of the block. Bend the pointer to point at zero—TDC on the degree wheel.

There are two ways to measure for the exact TDC—*positive stop* and with a dial-indicator gauge. With the dial-indicator installed perpendicular to the piston top and zeroed, rotate the crank to drop the piston to BDC (bottom dead center). Then rotate the crank to bring the piston up to 0.250 in. below TDC and stop it. Read the degree wheel. Reverse the direction of rotation and take the piston back down to BDC and continue on up again to 0.250 in. below TDC. Read the degree wheel again. The exact TDC is halfway between the two readings. The

Next the dial indicator is set-up on the #1 intake valve. If the heads aren't assembled or installed at this point, you can set the dial-indicator up on the #1 tappet. Note that the intake valve is the second valve to the right not the first valve. Try to get the indicator parallel to the valve stem.

TDC stop installed in number one spark plug hole. Be sure to adjust down far enough so that the piston is stopped before it gets to the top of its travel. If you miss on the first try, just lower the piston back to the bottom and lower the TDC stop adjustment.

pointer is bent to reflect this. More on this bending procedure in a moment.

With the positive-stop method, a threaded metal stop is attached securely to a head bolt hole or through the spark plug hole. If the cylinder head is on the engine this is the only reasonable method to use. There is a special positive-stop tool for use through the plug holes. In either case, adjust the metal stop with the piston at BDC such that it protrudes into the piston path and will stop it from reaching TDC. Then secure the stop. Rotate the crank clockwise until the piston comes up gently against the stop. Read the degree wheel. Rotate the crank counterclockwise back down to BDC then back up against the stop. Read the degree wheel again. The exact TDC is halfway between the two readings.

For example, rotating in the clockwise direction, the first stop reads 21 deg. Rotating in the counterclockwise direction up to the stop, the degree wheel reads 27 deg. The pointer is obviously 3 deg. off (27-21 divided by 2). If you bend the pointer 3 deg., everything should be zeroed.

Bending the pointer from 27 to 24 deg. is the proper direction. To double-check, rotate the crank clockwise again and it should stop at 24 deg. on the degree wheel. Note: If you'd bent the pointer 3 deg. in the wrong direction, the second set of numbers would have been 12 deg. apart—30 and 18. The two identical 24 deg. readings mean that the zero on the degree wheel is now the true TDC. You can remove the stop or dial-indicator and rotate the crank to zero on the degree wheel, TDC. Note that there are two TDC positions that are not the same but are exactly 360 deg. apart. The degree wheel cannot tell them apart because it's attached to the crank. The 2-to-1 cam drive creates the double position. At one position, the exhaust valve is closing and the intake valve is opening. At the next position 360 deg. away, but also TDC, both valves are on their seats. This second TDC position, where both valves are closed, is the one you want to use. It is considered the TDC #1 firing position. All references to TDC from now on will refer only to this position. Also, from now on the crankshaft must always be rotated *clockwise*. It's never backed-up. Now that the degree wheel is zeroed, care must be taken not to hit or bump it or the pointer or the "zeroing" procedure will have to be repeated.

You are now ready to centerline the camshaft. Switch the dial-indicator to the #1 cylinder's intake tappet. If the head is on, then use the #1 intake valve's retainer. Install the intake tappet on the camshaft. Lubricate the tappet with a very light oil. Install the dial-indicator parallel to the tappet's centerline.

Rotate the crankshaft clockwise until the tappet reaches its maximum lift. Zero the indicator at this point. Rotate the engine two full revolutions, which brings the tappet back up to max. lift. Check the zero reading. Re-zero, if necessary. Rotate the crank one-and-a-half revolutions clockwise. Continue rotating slowly until the indicator reads 0.050 in. below max. lift (zero). Stop and record the degree wheel reading. Continue on clockwise through zero (max. lift) and down to 0.050 in. indicator reading again. Stop and record this reading. The centerline of the camshaft is halfway between the two readings, so add the two readings together and divide by two. The answer is the cam's centerline.

For example, after zeroing the indicator, rotate the crank clockwise up to the first 0.050 in. below the max. lift position. The degree wheel reads 72 deg. Continuing on clockwise yields the second 0.050 in. reading of 158 deg. This means that the cam centerline is at 115 deg. (72 + 158 = 230, 230 divided by 2 = 115). If you want the cam to be at 110 deg., then install a 5-deg. offset key. This would make the two new readings 67 and 153 deg. respectively. This changes the center-

CRANE CAMS DATA SHEET

PART NO: 694571 SERIAL ____________

GRIND NUMBER: CCH-312-2-NC HYDRAULIC COMMANDER

ENGINE IDENT: 1964—UP PLYMOUTH-DODGE-CHRYSLER "LA" V-8 273, 340, 360 & 67-UP 318 CU. IN. RGH. IDLE PERF. USAGE, GOOD UPPER R.P.M. H.P.

VALVE SETTING: INTAKE: .000 EXHAUST: .000 → HOT

LIFT:			ROCKER ARM RATIO
	INTAKE @ CAM 3520	@ VALVE 528	
	EXHAUST @ CAM 3680	@ VALVE 552	1.5

ALL LIFTS ARE BASED ON ZERO LASH AND THEORETICAL ROCKER ARM RATIOS

CAM TIMING		OPENS	CLOSES	ADVERTISED DURATION
@ .0044	INTAKE	48° BTDC	84° ABDC	312°
TAPPET LIFT	EXHAUST	89° BBDC	53° ATDC	322°

SPRING REQUIREMENTS

	TRIPLE	DUAL	OUTER	INNER	RECOMMENDED RPM RANGE WITH MATCHING COMPONENTS	
PART NUMBER		99838			MINIMUM RPM	3500
LOADS	CLOSED	110 LBS.	@ 1.688	OR 1-11/16	MAXIMUM RPM	7000
	OPEN	344 LBS.	@ 1.161		VALVE FLOAT	7000

CAM TIMING		OPENS	CLOSES	MAX. LIFT	DURATION
@ .050	INTAKE	18° BTDC	44° ABDC	103° ATDC	242°
TAPPET LIFT	EXHAUST	59° BBDC	13° ATDC	113° BTDC	252°

Crane supplies this data sheet with every cam like most aftermarket manufacturers. The problem is, there is too much information, so it is easy to make a mistake. There are 14 duration numbers to pick from! Which one will you use? Setting up a camshaft to 4 numbers—intake open and closed, exhaust open and closed—is 4 times as hard as setting up to *one* centerline.

line to 110 deg. (67 + 153 = 220, 220 divided by 2 = 110). If you had installed the key backwards (offset in the wrong direction), then the readings would have been 77 and 163 deg. This would mean that the centerline was now at 120 deg., which would indicate that the key was backwards.

COMPRESSION RATIO

The short-block is now complete. However, before installing the heads, think about *compression ratio*. The engine's compression ratio is one of the most important factors. Unfortunately, many engine builders estimate or assume what the actual compression ratio is. This can be fatal. The best time to check for final compression ratio is after the short-block assembly and before the heads are finally installed.

Definition—The definition of compression ratio is the volume above the piston at BDC (bottom dead center) divided by the volume above the piston at TDC. Obviously, the volume above the piston at BDC is equal to the volume above the piston at TDC plus the cylinder's displacement (this is for one cylinder only, not the whole engine). Although the compression ratio itself is dimensionless, the volumes used to calculate it are given in cubic-centimeters (cc's). This means that the displacement of one cylinder (calculated from the exact bore and stroke of the engine) is converted from cubic-inches into cc's. The bore and stroke of all A engines are given in inches so the volume calculated from these numbers will be in cubic-inches. Since all the rest of the volumes in the compression ratio calculation are in cc's (metric measurements), then the cylinder volume has to be converted to cc's. A 340 has a single cylinder displacement of 42.5 cubic-inches, which translates to 696.5 cc's (the conversion factor is 16.387). Now, what other volumes are needed? Obviously, the cylinder head's combustion chamber volume. You know this from Chapter 5 when you cc'd the head and recorded the answer. Next is the head gasket thickness. Finally, there is the piston itself. The first part of the piston consideration is for a flat piston. The volume comes from the relationship of the flat of the piston (top) to the cylinder block's deck surface at TDC. This height is called the *deck height*. If the piston is below the deck surface then the volume is added to the head and gasket volumes because it makes the overall chamber bigger. On the other hand, if the flat piston top is above the deck surface, then the volume is subtracted from the head and gasket volume. Additionally, the piston may not be flat. It may have a dome on top, or it may have a *dish*. It may also have valve notches and chamfers. These are not as easy to measure as the deck height itself. It requires a cc'ing technique similar to that used on the cylinder head.

Calculation—First, set the piston at TDC. Place the dial-indicator on top of the piston parallel to the centerline of the cylinder. Then drop the piston 0.500 in.

Seal the piston to the cylinder wall with a light grease, just above the top ring. Wipe any excess grease away. Take the same clear plate used to cc the head and seal it to the deck. With a burette and cc'ing fluid (parts solvent can be used) measure the volume above the piston and below the deck. This volume figure includes any piston dish, dome, notches, etc. The problem is, that the volume just measured isn't what you want. It's the comparison of this volume to the volume of the perfect 0.500-in. thick disc that you really want. It's easier to

The next step is to zero the dial-indicator at maximum valve lift. Note the second valve (intake) is at high lift and indicator is reading zero. In this step be sure that the tip of the indicator doesn't come-off the retainer. If it does just reset the indicator for more travel or better alignment while keeping it parallel with the valve stem and start zeroing procedure over.

If using high-lift race cams, valve spring loads may be very high. Using a checking spring (shown on left) can allow you to check clearance and centerlining measurements more easily. However, they must be removed before starting the engine.

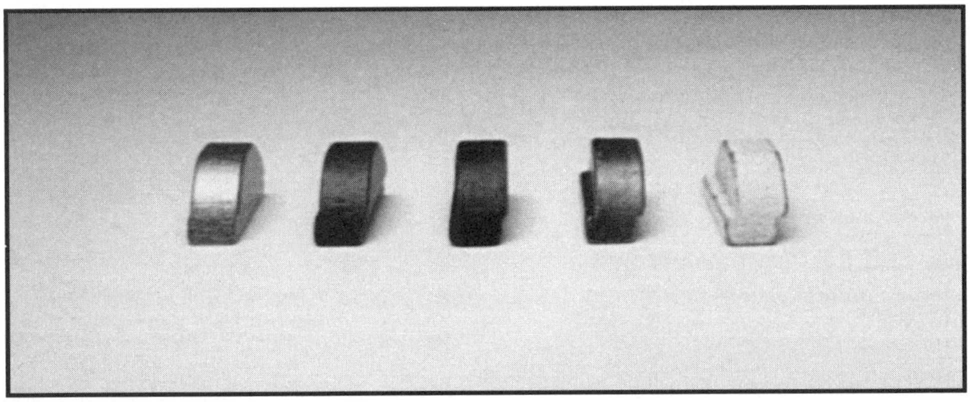

Offset keys allow you to adjust cam centerline and get more accurate to match a manufacturer's recommendation or to experiment. Multi-keyed sprockets are one method, but these offset keys are less expensive. Notice that the key offset increases as it goes from left to right (1 deg.to 5 deg.).

explain by example.

Using a 340 c.i.d. engine, assume that the piston is flat and has a deck height of zero. Now, if you lower the piston 0.500-in., the perfect disc volume is 105.2 cc (6.42 cubic-inches) by calculation. Since it's a flat piston with a zero deck, the measured volume should also be 105.2 cc. This means that the piston isn't helping or hurting the compression ratio. Stated another way, the measurement shows the piston has no notches, chamfers or dish. The combustion chamber is the sum of the head cc's and the gasket volume with a zero deck and flat piston. Getting back to the example, if the deck height is decreased to 0.050 in. below the deck with the same flat piston, the chamber gets bigger. Now the actual measurement will be 115.7 cc's. If you compare the number to the perfect disc (115.7 vs. 105.2) you'll find that the chamber got bigger by 10.5 cc's. This is added to the sum of the head and gasket volume to determine the total chamber volume, which is the volume above the piston at TDC. Now let's assume that the piston has a 10 cc dish in it and it's still at 0.050 in. below the deck (lowered 0.500 in. for cc'ing). This would yield a measured volume of 125.7 cc's (115.7 + 10). If this same dished piston had a 5 cc intake notch and a 3 cc exhaust notch then the 0.500-in. fill volume would be 133.7 cc (125.7 + 5 + 3) . Each time the total fill volume is compared to the perfect disc (105.2 cc in this case) and the result is added to the chamber size. That covers the conditions that increase the chamber volume.

Now, look at domes and see how they fit into this 0.500-in. down technique. Domed pistons are known to increase the compression ratio by decreasing the chamber volume. So, if the dome on top of the piston is 15 cc, how will this affect your 0.500-in. disc? If the flat of the piston is at zero deck, then the measured volume will be 90.2 cc's (105.2 − 15). Since it is smaller in volume than the perfect 0.500-in. disc, the chamber volume is decreased. Subtract the difference between the ideal and the measured (105.2 − 90.2) from the sum of the head and gasket volumes. Why do we so this? In the real world, you'll measure the 90.2 cc's but you won't know that you have a 15 cc dome until after you measure the 90.2 and subtract it from the perfect disc, which is 105.2 cc in this case. Now if the *flat* of the domed piston has a deck height of 0.050 in. below the deck, i.e. we've "dropped" the piston 0.050 in. down the cylinder, the measured volume would be 100.7 cc's. This is still smaller than the perfect 0.500 in. disc, so you still subtract 4.5 cc's (105.2 − 100.7) from the head and gasket volume. Now if you add the 5 and 3 cc notches to the piston, it will measure 108.7 cc's. This is larger than the 105.2, so the 3.5 cc's (108.7 − 105.2) are added to the chamber volume.

This example is relatively easy to understand. In reality, you don't know how big the dome is in cc's and you won't know the size of the piston notches. The technique will give you the right answer for compression ratio purposes in spite of this lack of specific knowledge. You measure the 0.500-in. downfill volume and compare to the perfect disc. If the fill volume is larger than the perfect disc then the difference is added to the chamber volume. If the fill volume is smaller, then the difference is subtracted from the chamber volume. Obviously, the perfect disc changes for every exact bore size. For example, it's not the same number for a 0.030-in. overbored (4.070 in. actual) 340 as it was for the stock bore (4.040 in. actual) that I used in the above example. Each bore is calculated separately. This method won't tell you the exact size of the dome by itself. It gives you the result of dome, notches, chamfers and deck height together. Using this method, you do not really need to know the deck height itself. However, you should still measure the deck height because it can affect the piston-to-head clearance. If you wanted to know the dome's true volume you would have to make a second measurement. Once the the total fill measurement was completed, you could fill in the valve notches with modeling clay and cc the 0.500-in. down measurement again. From these two measurements and the exact deck height (to the flat of the piston, not to the top of the dome), the dome volume could be calculated. It's not needed and can be confusing.

After centerlining, install front cover and gasket. Note where the timing tab is on the cover. It's best to leave the engine at TDC after cam centerlining, since it will make final assembly easier.

To calculate the engine's actual compression ratio, the combustion chamber size, which is the volume above the piston at TDC, must be determined first. The total chamber size is the result of three volumes—head, gasket and piston. The head and gasket are always positive and are added together. The piston could be positive or negative. Using deck heights, the volume is added if it's below the deck surface. The volume is subtracted if the piston top (flat) is above the deck surface. With the 0.500-in. downfill volume method, the calculated volume is added if it's larger than the calculated perfect disc and subtracted if the fill volume is smaller than the perfect disc. At this point, you have the total chamber volume. To this number, add the cc volume of one cylinder of the engine. This sum is the volume above the piston at BDC. The actual compression ratio of the engine is this sum divided by the total chamber volume. The 0.500-in. downfill method is more accurate and is recommended for building racing engines.

CYLINDER HEADS

With the short-block completed, the heads can now be assembled. I covered cylinder head preparation in

Chapter 5, which also included a discussion on valve work. Each valve should be marked to correspond with the cylinder it was prepped to.

ASSEMBLY

The first step to assembling the heads is to gather all of the valves, springs, retainers, keepers and shims. Next, make sure you have a good engine build sheet to record the installed heights and spring loads as you measure them. For details on these areas, go back to Chapter 6 on *Camshafts & Valvetrain*.

Installed Height—You'll need to check the *installed spring heights* of all springs. The typical A engine installed height is 1.69 in., but special retainers are available that will increase this to 1.73 to 1.75 in. Additionally, there are special long valves for the A engine that increase the installed height to 1.86 in. or more. The particular height needed should have been determined in Chapter 6. I will assume for this section that it is 1.69 in. In this example, keep the installed heights the same. There are spring shims available in 0.015-in., 0.030-in. and 0.060-in. thicknesses that are used to adjust each valve's installed height separately. Let's assume that the following measurements were made on one head—1.690 in., 1.695 in., 1.705 in., 1.705 in., 1.720 in., 1.710 in., 1.725 in., 1.695 in. The one 1.690 in. and the two 1.695 in. heights are left alone. The two 1.705 in. and the one 1.710 in. heights should have one 0.015 in. shim added. The one 1.720 in. and the one 1.725 in. should have a 0.030 in. shim added. Each spring set should be marked and organized in such a way that it will not get mixed up. Each retainer and set of keepers must stay with the valve spring it was checked on. A spring set is really the spring marked for position, its required shim, the retainer and keepers.

Load—Next, check all 16 valve springs for load. The biggest value for the actual load numbers is for comparison after the springs have been used. However, many engine builders like to match the springs. This generally requires two or three sets of the same spring to start out with. This gets expensive when building just one engine. In either case, all springs are load-tested at the selected installed height and then arranged from lightest to heaviest load at the height. If you have several sets of springs, then pick the 16 springs with the highest loads. If using only one set of springs, then it is best to put the 8 with the highest loads on the exhaust valves. The most important thing is to record the loads, installed heights and spring part number or identification.

Clearance—The last item to measure on the cylinder head itself before it is installed is the valve-to-valve clearance on the seat. This should be a minimum of 0.030 in. It can be measured

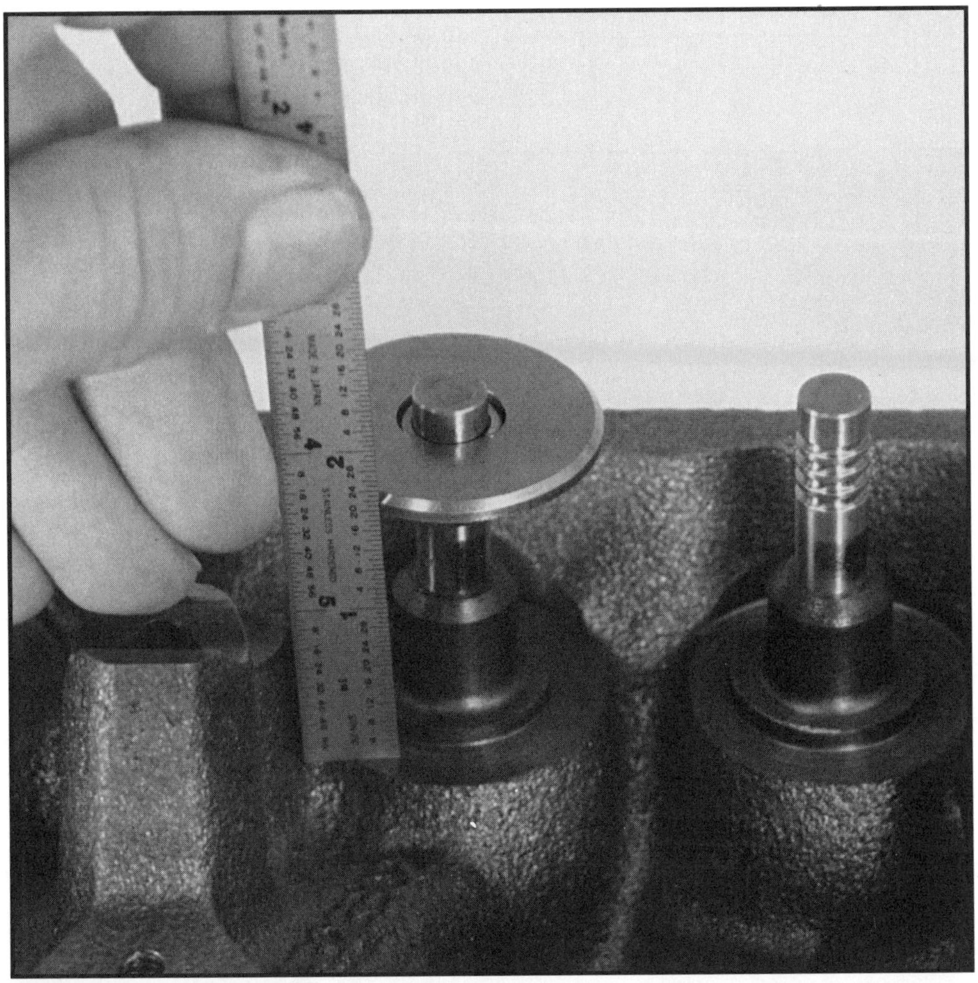

The most important specification (number) relating to valve springs is installed height. This is the distance from the spring seat on the head to the underside of the retainer and can be measured with a steel scale.

Another, more accurate method is to use a snap gauge then it is a scale when measuring installed height. This method is easier and more accurate, but not as quick.

with feeler gauges similar to those used for gapping spark plugs.

Gaskets—Before bolting on the head, check the head gasket to be used for fit and alignment. If 0-rings are used, check the compression ring flange on the gasket itself to be sure that it's going to seal on the 0-ring. Obviously, this compression ring must extend past the 0-ring in order to seal. Not all gaskets meet this criterion. It's also recommended that the actual gasket be measured by a micrometer for thickness. For the next step, which is putting the heads onto the short-block, it's best to use a clean, used gasket or none at all. Remember, we're still "practicing." This isn't the final assembly.

INSTALLING HEADS

The next step is to install the cylinder heads onto the short-block. Usually they are bolted down with a bolt on either end of the head and one on both sides of the cylinder that is going to be checked. The bolts are not torqued but they are tightened with a wrench.

Valvetrain—The first thing to do once the heads are on is to install the valvetrain—rocker arms, shafts and pushrods. The first check is for pushrod clearance. Rotate the engine through two full revolutions, stopping every 45 deg., to check the pushrod clearance to the cylinder head, gasket, block, etc. If there is a close spot then go back and check more closely and at finer intervals.

Valve Tip—With the valvetrain working, check the wipe pattern on the valve tip. By putting *Dykem Blue* (also called *Machinist's Blue*) on the valve tip and setting the lash correctly (set to zero if it's a hydraulic cam) you can observe the resulting pattern. Rotate the engine at least two full revolutions (4-6 may give a better pattern) then inspect the valve tip. The Dykem will be removed wherever the rocker rubbed. Ideally, the rocker should be centered on the stem and not approach the outside edge.

Rocker Ratio—Check the rocker arm ratio by installing a dial-indicator parallel to the valve stem on the retainer. Once the engine is rotated through the maximum lift of the cam, the actual rocker arm ratio can be calculated. The lift of the cam was measured at the cam in Chapter 6. Now, measure the lift at the *valve*. Dividing the at-the-valve measurement by the lift at the cam, will yield the actual rocker ratio. For example, take a 0.630-in. lift roller cam with the lift at the cam measured at 0.420 in. Now measure the lift at the valve at 0.600 in. (you should do this at zero value lash setting also.) This calculates to a ratio of 1.43-in. (instead of the design ratio of 1.50-in.).

CHRYSLER PARTS LIST

BLUEPRINTING TOOLS

Degree Wheel	P4286552
TDC Indicator Tool	P4349737
Plastigage Assortment	P4286819
1 blue, 2 red, 2 green	

Valve-to-Piston—The next step is to check the valve-to-piston clearance. There are two methods for checking this; clay and dial-indicator. Using modeling clay is the easiest especially for beginners. That method is detailed on page 41. Using checking springs with a dial-indicator is more accurate. If no head gasket is used for checking, be sure to add the actual gasket's thickness to the valve clearance number actually measured. If no notches are cut in the piston yet, expect the valve to hit the piston, especially in higher compression ratio engines and engines with bigger cams. Nothing can be accomplished if the valves hit the pistons in the checking stage. To guard against this, cut out two cardboard shims 0.050 in. to 0.100 in. thick to fit between two of the bolts around the cylinder—one shim to either side of the chamber. The installed thickness of this cardboard shim can be checked by inserting feeler gauges between the head and the block. This should provide adequate clearance for the engine to rotate without the valve hitting or

Once the snap gauge is set, then the height is read on a micrometer.

getting bent. Compare the measured, installed shim thickness to the gasket thickness. It's assumed that the shim will be thicker than the gasket, so subtract the gasket thickness from the shim's thickness and subtract the answer from the measured valve clearance to find out what it would be.

For example, suppose the actual gasket is 0.040-in. thick and the cardboard shim measures installed at 0.080 in. The measured valve-to-piston clearance with the 0.080-in. shim in place is 0.050 in. on the intake and 0.030 in. on the exhaust. This indicates an actual 0.010-in. clearance on the intake but a 0.010-in. interference on the exhaust. These are the numbers used to calculate the depth of the valve notch. However, if the valve notches have already been cut, giving measured clearances of 0.090-in. intake and 0.100-in. exhaust, then this set-up would not hit the valves, even without the head gasket. Now in this case the valve-to-piston clearance *without* the head gasket would measure 0.050 in. (0.090—0.040 in.) on the intake and 0.060 in. (0.100—0.040-in.) on the exhaust side.

Valve-To-Block—While checking valve-to-piston clearance, check also the valve-to-block clearance. It should be 0.030 in. minimum. This is usually only a problem or consideration on the 273 and engines with large, oversize valves on small bores.

Piston-To-Head—The final clearance check is the piston-to-head clearance. The best way to generate this number is to rotate the engine to TDC firing, meaning both intake/exhaust valves are closed. Remove the rod cap and nuts and install the dial-indicator on the rod, parallel to the centerline of the cylinder. Zero the indicator. Then push the piston straight up until it contacts the cylinder head. The indicator will read the piston-to-head clearance. It should be no less than 0.055 in.

Conclusion—The preliminary checks are now complete. The engine can now be disassembled and final precision adjustments should be made before final assembly. Since the engine has been rotated several times during the checking procedure, check each bearing, rod and main, as you remove them. Check closely for scratches on the bearing shells. If you find a scratch, check the crank until you find the cause and polish it out before final assembly.

If you want to decrease the installed height, then a spring shim can be added between the bottom of the spring and the cylinder head. Shims come in different thicknesses and different styles—outer only, inner and outer, at sizes varying from 0.030 in., 0.060 in., etc.

A valve spring tester is needed to check spring load. Springs should be tested upon installation and when they are removed. To be useful, both the load and the height at which that load occurred need to be known. So even though the spring tester reads 120 pounds, you need to know at what height the reading was taken.

8 ENGINE ASSEMBLY

All measurements and modifications are now complete. Everything has been inspected, checked and inspected again. Chapters 1 through 6 prepped all the individual parts and Chapter 7 tied all these individual parts together. Never try to final-assemble the engine with work still left to be done. All of the machining, polishing, grinding, etc. should now be complete and every part should be cleaned and cleaned again. Now, and only now, are you ready to put everything out on a *very* clean workbench in a *very* clean environment and do the final assembly of the engine. At this point it's a good idea to remove any unnecessary tools, parts and equipment. Then lay out the tools to be used and position the parts accordingly. I've devised a checklist of the engine assembly that outlines the steps in order.

SHORT BLOCK

☐ Remove the block itself from its plastic bag and clean again as required. Mount block onto engine stand.

☐ Install the oil galley plugs using sealer.

☐ Check all water-jacket core plugs, especially those in the bellhousing area.

☐ Flip the block over so that the main bearing area is upward and remove the main caps.

☐ Insert the bearing shells into the block as marked earlier (ones with holes go in block). Install the other half of the shell into the caps as marked. Be sure that shells are radiused if a full-radius race crank is to be used.

☐ Oil the bearings in the block. Spread over the full bearing surface using your finger. Be sure to cover the thrust surfaces on the #3 main—front and rear.

☐ Check crank flange for pilot bushing if it is to be used with a manual transmission. Not required with automatic transmission.

1. Before final assembly, gather all parts, organize them and be sure you have everything that you need. This is how Dick Landy and his crew at DLI set up before assembling an engine. (Photo by Michael Lutfy)

☐ Install the lower half of the rear main seal. Be sure the scraper section is pointed in the proper direction—sharp end pointing toward #3. The older "rope" seal doesn't have a direction.

☐ Carefully lower the crankshaft into the block.

☐ Oil the shells in the caps and install the caps onto the block in the proper order and orientation. They cannot be rotated 180 deg. (side-to-side) even though they look almost symmetrical.

☐ On #5 cap install the other half of the rear main seal. Consider putting small amount of RTV behind the seal to help hold it in place. Rope seal doesn't need RTV. For racing, a *neoprene* (rubber) seal is recommended.

☐ Lightly tap the caps into place (fully seated against block).

☐ Oil the main cap screws and insert into caps. They should tighten down by hand easily.

☐ Torque the main cap bolts to 100 ft-lbs in three equal steps. After torquing, rotate crank. It should spin freely—no tight spots.

☐ Assemble the connecting rods onto the pistons. Each piston and rod is marked to go together in sets. If you are using pressed pins then the piston and rod assemblies will already be together. With floating pins, oil the pin and rod small-end upon assembly. Install the lock rings and be sure that they are fully seated.

☐ Install the piston rings onto the proper piston. Each set should be marked as to which piston they go on. Install the oil ring first. Then install the second ring. The top ring goes on last. Check manufacturer's instructions to find "dot" or other reference to indicate which surface of the ring goes up (toward piston top) on installation. Be sure to use a ring expander here. Also use expander carefully—rings do break quite easily. With the oil ring, try to get the gaps in the top and bottom rail to be approximately 3-in. apart from each other and to get the expander's gap to the front of the engine as installed while the gaps in the two rails are toward either side.

☐ Remove the rod caps and install the rod bearings.

☐ Oil the rod shell in the rod and install crank journal shields over the bolts. The fancy shields are around nine inches long and made of brass threaded and countersunk to screw on over the bolt. A cheaper solution is 1/2-in. OD fuel line. The best arrangement is one long one at 9-10 in. and one short one at 2 to 3 in. The long one goes on the high side bolt as it's being installed and serves as a guide as the bolt and rod slides down around the crank journal.

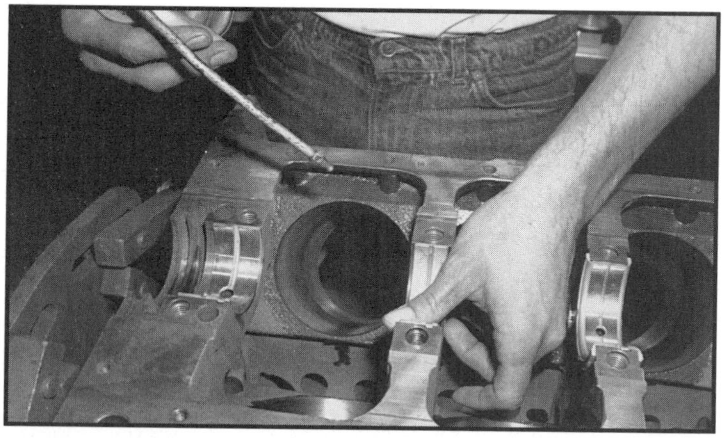

2. With the block on the engine stand and upside down, install the main bearings. Coat them with engine oil after they are in place. (Photo by Michael Lutfy)

5. The bearing shells are installed in the caps and the caps and bolts are installed on the proper bulkheads. Using a brass or plastic hammer, tap caps into location. Once the best alignment is achieved, the bolts are tightened. (Photo by Michael Lutfy)

9. Be sure floating pin pistons have one pin lock installed before putting on rod (the 2nd lock is added later). On a pressed pin rod, double-check the press (interference fit) before assembly. With pressed pins, the piston-rod assembly is usually done separately by the machinist.

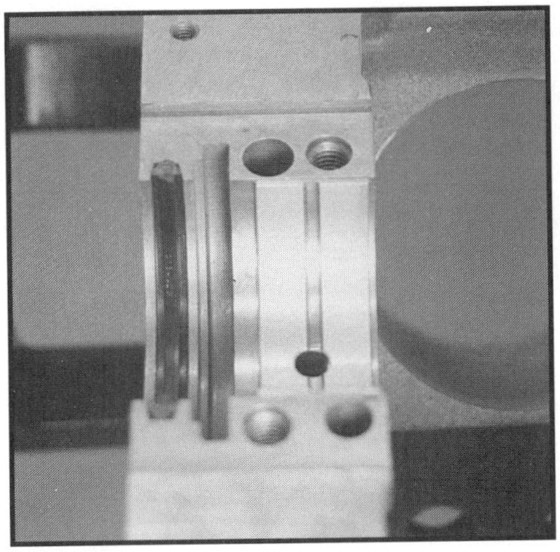

3. The rear main seal goes in with the "sharp" edge toward the front of the engine.

4. Carefully lower the crankshaft into place. (Photo by Michael Lutfy)

6. With all caps snug, the crank is pressed forward and the #3 (thrust) cap is torqued. This pressed better aligns the top and bottom half of the #3 bearing shells.

7. Always double-check crank end play with the actual pieces being used and at the time of final assembly.

8. Once all 5 caps are torqued properly, the crank's turn-over torque can be measured using an inch-ounce or inch-pound torque wrench. The result should be written down on the engine's build-up sheet.

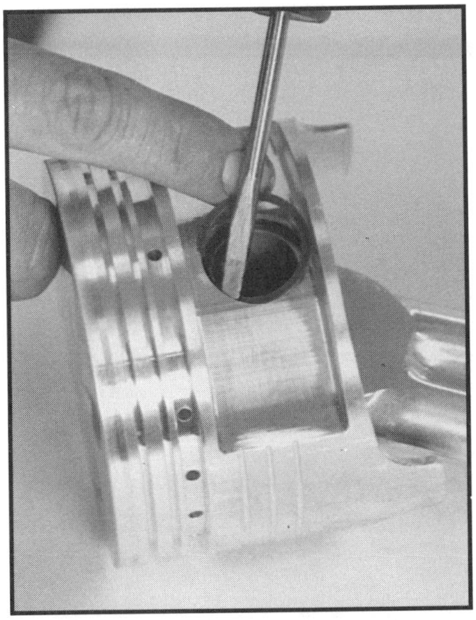

10. Installing the second pin lock is always much more difficult than installing the first. With the low pin end clearance desired in racing, it's hard to get the whole assembly together. Patience is required here.

11. Once the piston and rod are assembled, check the rods' side clearance in the piston using feeler gauges. Each piston should be done separately.

12. Once the rod has been attached to the piston, the rod cap must be removed and the rod bearing shells inserted. Be sure to keep the proper caps with their rods and do not accidentally reverse the cap on the rod.

☐ Rotate the crankshaft so that the #1 journal is down.

☐ Oil the piston skirts on the #1 piston.

☐ Align the ring end gaps on the piston such that the top ring gap is to the rear, second ring gap is 90 deg. away and the expander gap is to the front.

☐ Install a ring compressor and compress the rings tightly. Check piston alignment for the cylinder that you are installing. Check dome location, spark plug relief, valve notches and notch orientation—intake under intake valve, etc. Guide the piston and rod assembly into the cylinder bore. Don't force it or you may break a ring. Once all the rings are below the deck and the ring compressor is free, set it aside.

☐ Align the long rod bolt guide—brass or fuel line hose onto the crank journal. Push the piston assembly down carefully. As the two rod bolt shields approach the journal be sure that the piston doesn't rotate in the bore causing a misalignment of the rod on the journal. Once aligned, push into place.

☐ Install the rod bearing shell into the cap and oil.

☐ Install rod cap and shell assembly onto the rod. Check alignment of rod and cap numbers and bearing tangs.

☐ Oil the rod bolt threads and tighten nuts until snug.

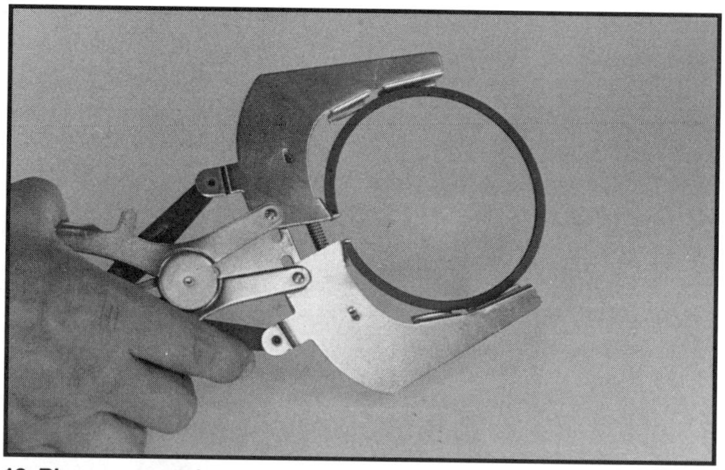

13. Rings are next. Install oil ring, expander and its two rails first. Then the two compression rings—the second ring goes on first and the top ring goes on last. These two compression rings will require a ring expander to install properly.

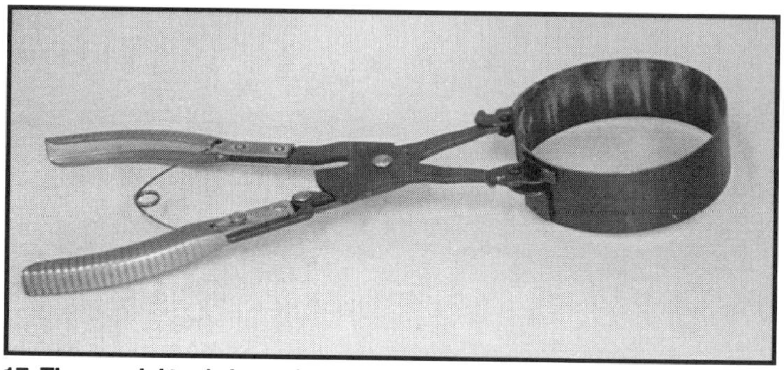

17. The special tool shown is very handy and is definitely the easier ring compressor to use. Note the one way lock on the lower handle.

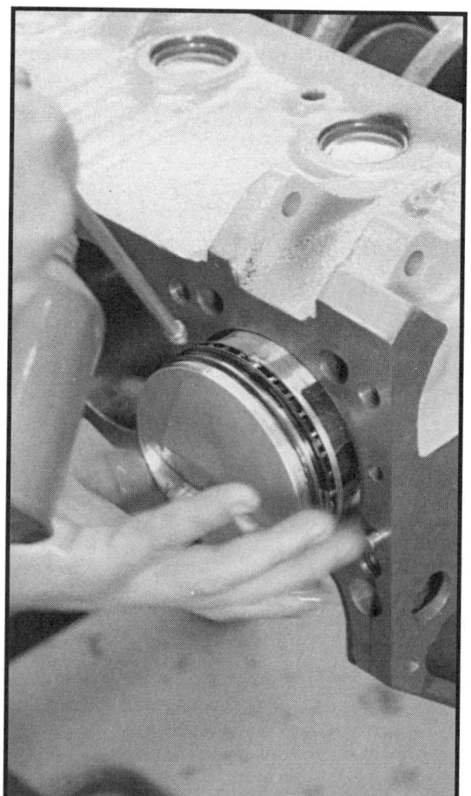

20. Oil all rings and the rod journal well.

21. Attach ring compressor. At this point, it's best to have the oil ring about a half inch above the deck surface. Double-check notch alignment at this point. Don't install piston with the valves notches upside down.

22. Using a plastic or wooden hammer handle tap the piston quickly and firmly into place. As soon as the top of the piston passes below the top of the block—stop.

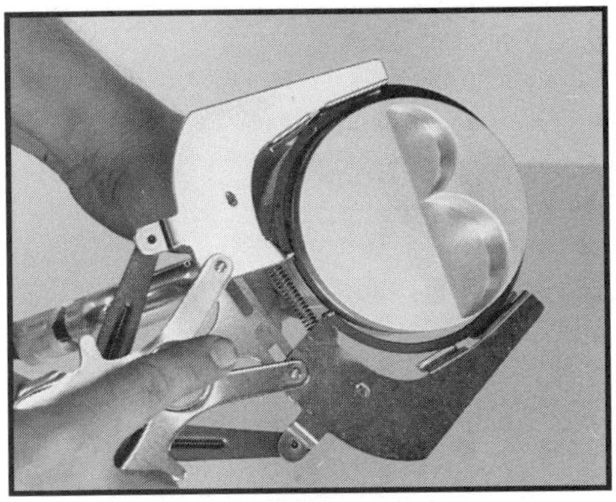

14. With the expander the ring is carefully placed on the proper piston (#1 cylinder ring on #1 piston, etc.). Be sure that the top of the ring is to the top of the piston. The code for the top of the two compression rings is included with the ring package from the manufacturer.

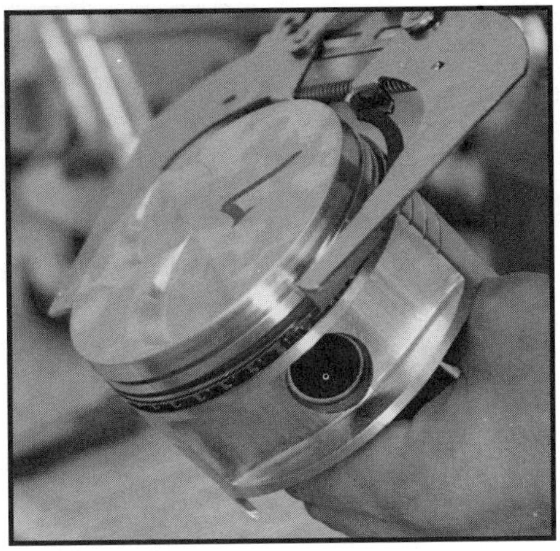

15. Be sure to install second ring in the second groove. Also don't try to install the second ring if the top ring is already on the piston.

16. For installation into block, a ring compressor is necessary. The most common is the adjustable band. It's inexpensive but more diffcult to use. You need to use more caution and have more patience.

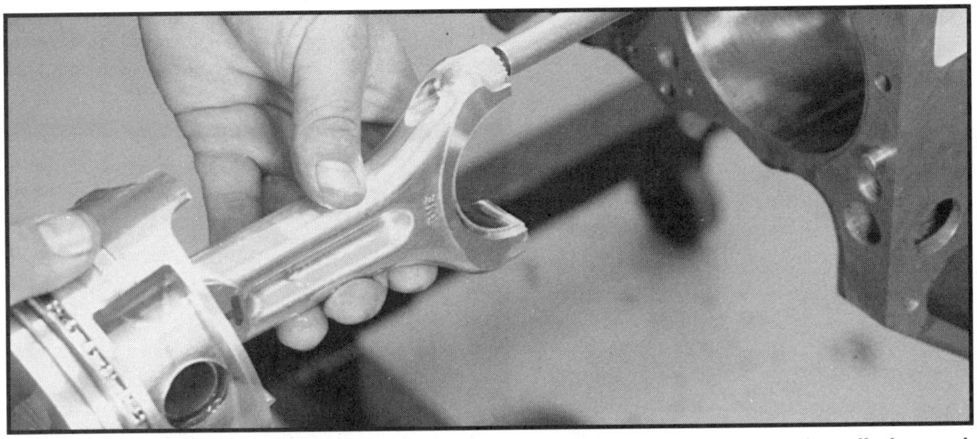

18. Install a brass guide or piece of plastic fuel line over the rod bolts to protect the cylinders and crank. Tip engine over on its side, and set crank throw at BDC. Make sure the right piston is going in the right bore, and located properly in relation with front and back of block.

19. The piston can be inserted about half-way without the ring compressor. To the top right you can see the rod guide protruding.

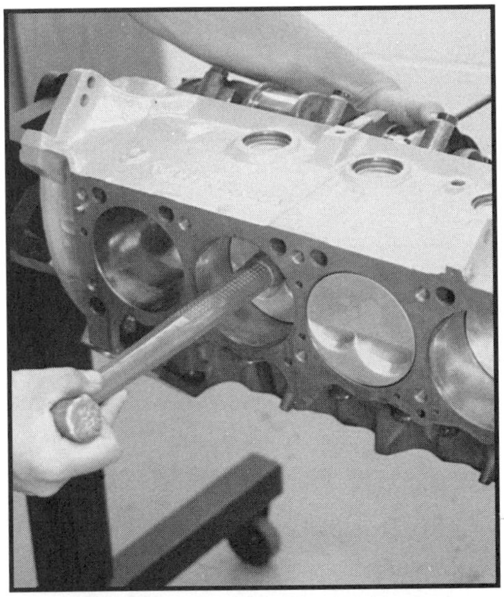

23. Check the rod and rod guide for proper alignment with the rod journal. Keep to the rod bolts straight up-and-down relative crank's centerline. Hold rod end to guide it over crank journal and proceed with tapping piston until rod seats against journal.

24. Oil rod caps with bearings and install. Torque each nut to proper spec. Once all eight rods are installed, retorque all rod bolts.

25. Installed pistons are not all the same. Some have the small valve notch to the right as #6 does in the photo and some have it to the left as #4 does. This is as it should be because the small notch must line up with the exhaust valves. On an A engine head the exhaust valves are the two centermost valves and the two end valves. Always use this to double-check your piston installation.

- [ ] Torque the nuts to specification—45 ft-lbs for production or 50-55 ft-lbs for high-performance. For special bolts use the manufacturer's recommendations.

- [ ] Install the vibration damper bolt at front of crank to enable you to use a large wrench to rotate the crank. The damper itself is not installed at this point.

- [ ] Rotate crank with #1 piston attached, checking for binding, tight spots, etc.

- [ ] Stop crank with #3 rod journal down and repeat above steps. Usually, it's easier to do cylinders 1-3-5-7 in order and then switch over to the other bank and do 2-4-6-8.

- [ ] Complete the installation of all eight pistons. Go back and check each rod bolt torque.

- [ ] Install crank sprocket if not already on crank. Rotate crank so that dot on crank gear is up.

- [ ] Oil the five cam journals, lube cam and install into block.

- [ ] Install cam sprocket and chain such that dot on cam sprocket lines up with dot on crank sprocket.

- [ ] Check cam's centerline using same method as in Chapter 7. This requires the dial-indicator and degree wheel and starts by checking the TDC location.

- [ ] With centerline set, install camshaft bolt using Loctite. Torque to 40-45 ft-lbs.

CYLINDER HEADS

- [ ] Support a cylinder head on wooden blocks tall enough to hold head securely. Valve guides should be parallel to the work bench or the valve spring side should be somewhat higher.

- [ ] Line-up valves, springs, retainers, shims, etc. in order, next to the valve and cylinder of the same number. Do cylinders 1-3-5-7 first then cylinders 2-4-6-8 next.

- [ ] Oil the #1 exhaust valve stem and install into its proper location in the head.

- [ ] Install the valve seal and push the seal down on top of the guide. If the PC-style guide seal is used it is installed onto the guide just before the valve is inserted.

- [ ] Install the marked shims onto the spring seat. Place the spring on top of the shim(s). Align the retainer on top of the spring and directly over the valve stem. Use a valve spring-compressor to compress the retainer/spring assembly. Install the two keepers onto the valve stem and carefully release the spring compressor.

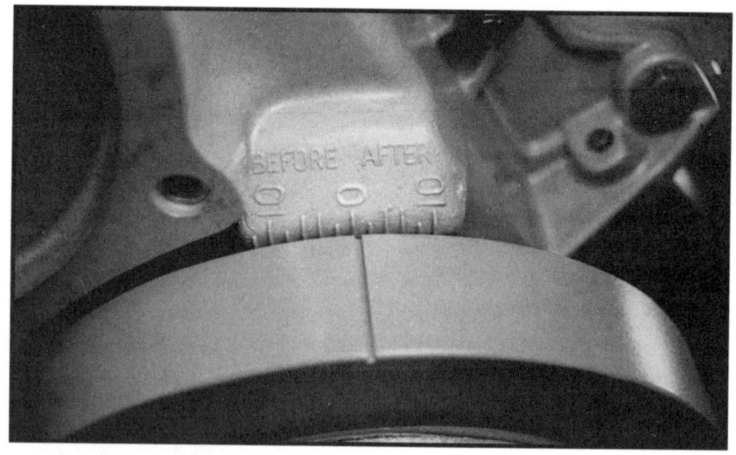

26. Once the short-block has been assembled, rotate the engine until the #1 piston is at TDC and install the vibration damper. The timing marks should line up.

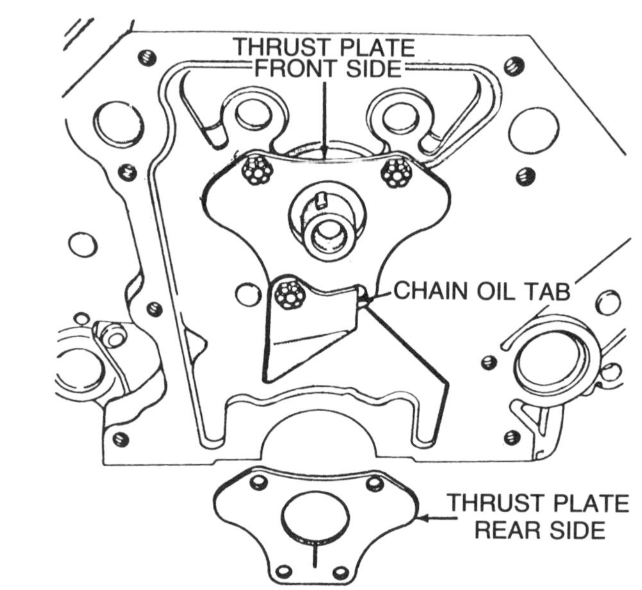

Camshaft Thrust Plate Installation

29. The thrust plate has a front and rear side and must be installed with the rear side next to the block itself. (Courtesy Mopar Performance).

33. Once all the pistons are installed and the heads are assembled, then each head can be installed on the engine. Install all the head bolts and tighten slightly. Then torque in sequence in steps of approximately 15 ft-lbs.

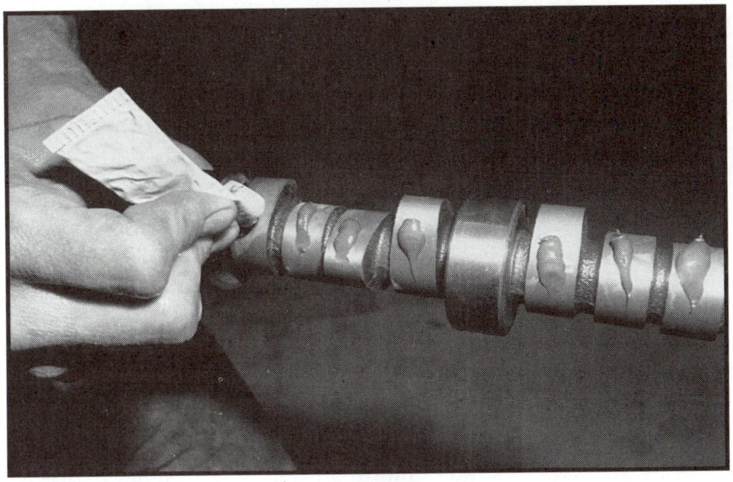

27. Oil the five cam journals with engine oil, then coat the cam itself with cam lube. Coat each lobe generously. (Photo by Michael Lutfy)

28. Gently guide cam into block. (Photo by Michael Lutfy)

30. When installing the cam for the final time always centerline it (degree it in) using the dial-indicator and degree wheel. Never assume that the same pieces will automatically go together the same way twice. Make sure!

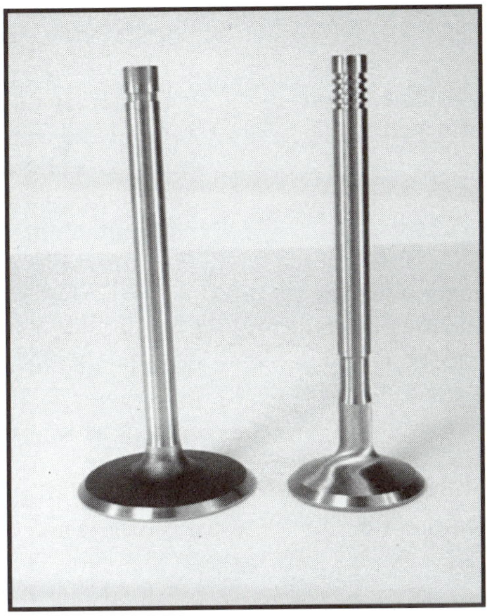

31. When assembling heads, note the number of valve lock grooves (one on the left, three on the right) and be sure you have the right pieces to work with the valves. If you are using 11/32 or 5/16 stem valves, be sure that the rest of the hardware (keepers, retainers, seals, etc.) is compatible.

32. Check valve spring loads and installed heights before final assembly and write down this information on the engine build-up sheet.

34. Observe that the center bolt on the upper row of head bolts gets very close to the valve springs and retainers. Tightening will require the use of a 3/8 drive socket and extension.

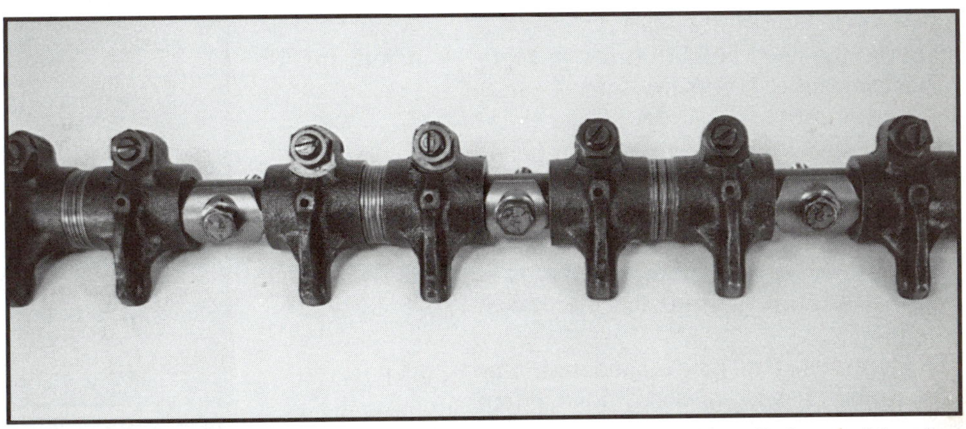

35. The rocker shaft assembly should be put together separately and installed on each head as an assembly.

☐ Repeat procedure and install the #1 intake valve.

☐ Complete installation of the rest of the valves in this head following the same procedure for each individual valve.

☐ Place a large socket on the retainer over the valve tip (but not touching the tip or the keepers) and strike the top of the socket with a hammer. This will seat the keepers. A loose keeper may pop-out and will need to be re-installed. Double-check the height in the retainer of the two keepers. They should both be at the same height.

☐ Once one head is completed, use the same procedures on the second head.

MOUNTING HEADS

☐ Throughly clean block's deck surface of excess oil from piston installation.

☐ Double-check dowel-pin height by putting both heads on the block without gaskets. They should sit flat against the deck surface.

☐ Remove heads and re-clean both deck surfaces.

☐ Double-check valve notch alignment—intake under intake, etc. Also check dome location relative to combustion chamber location. Be sure head Numbers 1-3-5-7 are going on the left side of the engine on cylinders 1-3-5-7.

☐ Spray or coat head gaskets with head gasket sealer unless gasket manufacturer specifies dry installation, i.e. no sealer.

☐ Once dried, place head gasket over dowel pins. Check that top side of gasket is facing up.

☐ Install cylinder heads over dowels.

☐ Oil head bolt threads and install bolts. Lightly tighten all the head bolts upon installation.

☐ Torque the head bolts in order in 25 ft-lbs increments. Final torque is 100 ft-lbs.

☐ Double-check final torque on all head bolts.

VALVETRAIN

☐ If not already done, rotate engine on stand to be in its normal position—heads up, crank down.

☐ Dip tappets in cam lube oil and insert into tappet holes. If hydraulic tappets are used, don't pump-up.

☐ Install the sixteen pushrods onto the tappets.

☐ Set-up each rocker shaft. Install exhaust rocker first, then intake. (Note left and right rockers on hydraulic valve gear.) Offset intakes are both right- and left-handed. Select proper offset for each position that will allow the pushrod tip and valve tip to align with that rocker arm. Use the rocker shaft attaching bolts to hold rocker arms in sets of two. If you are using mechanical valve gear and rocker shafts with banana grooves, be sure they are located on the bottom. The long-stamped steel retainers go on the #2 and #4 attaching bolts.

☐ Lubicate valve tips and pushrod tips.

☐ Install rocker arm and shaft assembly onto the cylinder head. Insert all 5 shaft-attaching bolts into their appropriate pedestals and start on threads. Note the "notch" on end of hydraulic rocker shaft. It should be pointing to the centerline of the engine and toward the front of the engine on the left bank and to the rear on the right bank.

☐ Align each pushrod with its rocker arm. Tighten the second and fourth rocker shaft bolts. Double-check each pushrod's alignment. Tighten the #1, #3 and #5 rocker shaft bolts. Once shaft is snug against the pedestals, the shaft attaching bolts should be torqued to 30 ft-lbs using the following order: 3-1-5-2-4.

☐ Double-check cam centerline as installed using degree wheel and dial-indicator and same procedure as outlined in Chapter 7. If the centerline is not what was desired, correct as required. Once correct, install cam bolt with Loctite and torque to 40 ft-lbs. If checked at the tappet upon cam installation you may omit this step. However, it is often easier to check centerline at the valve than it is at the cam. The valve springs can provide a tappet pre-load that helps the accuracy.

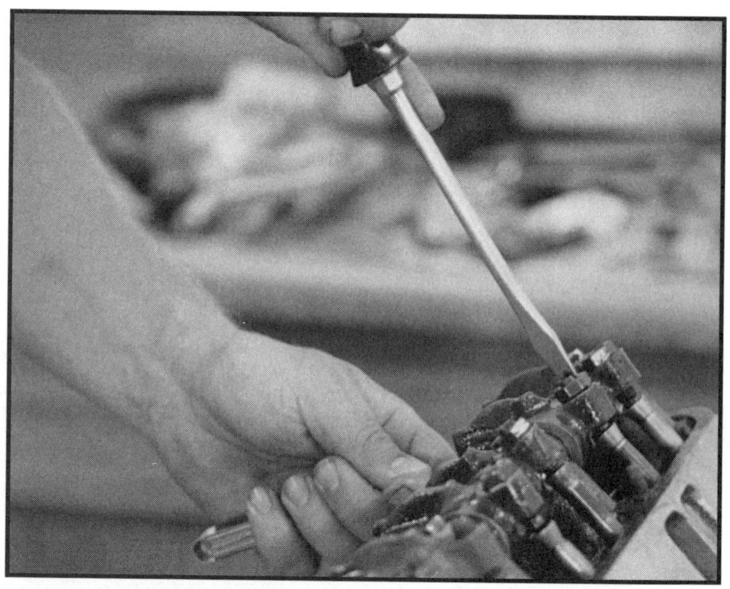

36. Once the pushrods and rocker shaft assemblies are installed, valve lash must be set. Usually four valves are set at each position—TDC, BDC, etc.

37. Always check the assembled valve tip alignment. Note that the rocker arm pad is directly over the valve tip. Observe the limited clearance between the rocker arm and the retainer. This clearance should also be double-checked upon assembly.

38. RIGHT—The short-block, cylinder heads and valvetrain are now together. Note that the engine has been rotated back to TDC (white dots line up).

39. The A engine intake manifold gasket comes in four pieces—two sides and two ends.

40. On engines that have been milled, place a small amount of RTV in each corner where a side gasket meets an end gasket.

41. LEFT—Once the gaskets are in place, lower the intake manifold carefully into position, being extra cautious not to move the gaskets.

EXTERNAL ACCESSORIES

☐ If you have one, install the oil shield (thin, flat disc) on the front of the crank next to the crank sprocket.

☐ Check the crank seal in the timing cover and replace, if required.

☐ Install the front cover and torque bolts to 30 ft-lbs.

☐ Rotate the engine on the stand and put the crank side up.

☐ Install the oil pump. Torque attaching bolts to 35 ft-lbs.

☐ Install windage tray or scrapers if they are being used. The standard windage tray attaches to the #2 and #4 main cap bolts.

☐ Install the oil pan. Torque bolts to 15 ft-lbs. Be sure oil pan drain plug bolt is installed.

☐ Rotate engine on stand back to its normal position, i.e. crank down.

☐ Pour cam lube oil onto all lobes of the camshaft itself. Rotate the crankshaft one revolution and pour the remainder of the lube oil over the cam.

☐ Install vibration damper on front of crankshaft. Torque damper bolt to 135 ft-lbs. Check damper timing mark against timing mark on front cover and compare to the known TDC as checked earlier in Chapter 7. Crank will rotate during torquing procedure so rotate it back to TDC #1 firing position.

☐ Add sealer to the intake gaskets—sides only.

☐ Install intake gaskets—sides first then the ends.

☐ Carefully drop intake manifold down into place on the gaskets.

☐ Install intake manifold bolts and torque to 35 ft-lbs in the order shown in accompanying photos.

☐ Lubricate drive gear, teeth and shaft.

☐ Install the distributor drive gear noting alignment of the slot relative to the centerline of the crankshaft/camshaft. It should be pointing to the first intake manifold bolt on the left side of the engine. (Left side is 1-3-5-7 and first bolt is at front of engine.) Remember, crank is in TDC #1 firing position for this step.

☐ Install the distributor with the rotor pointing to the #1 cylinder.

☐ Install valve covers and torque bolts to 40 in-lbs.

☐ Install dummy or used plugs into the spark plug holes to keep dirt and dust out of engine during storage and/or installation.

☐ Tape over the exhaust ports in the head and the carburetor flange on the intake manifold.

☐ Also tape over the fuel pump flange.

☐ Install the oil filter, a used oil filter or tape over the holes.

☐ Cover the complete engine assembly so that it doesn't get dirty until it is installed.

☐ SPECIAL NOTE: Remember to remove tape and old plugs after engine is installed!

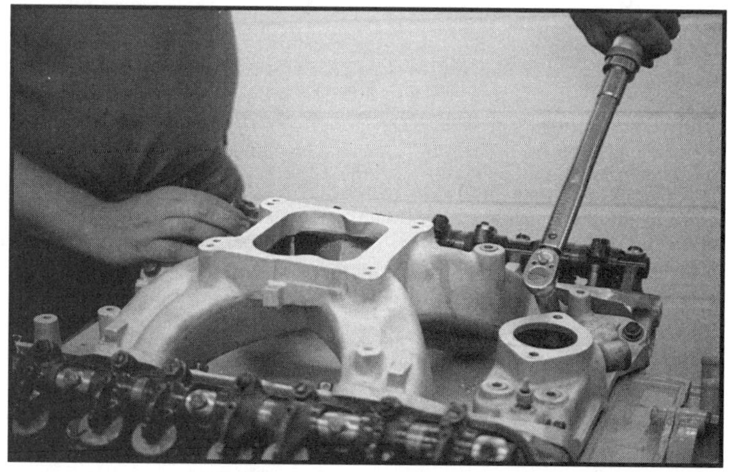

42. All the manifold bolts should be started before any are torqued down fully. The two bolts on either side between the front and rear runners are difficult to get a big wrench on.

46. The front cover is ready for the water pump and fan. Note the radiused water passages going into each side of the block

CHRYSLER PARTS LIST

ENGINE ACCESSORIES

Chrome Valve Cover Set	P4120608
Valve Cover Accessory Set	P4120272
Chrome Breather Cap	P4120446
Chrome Water Inlet Neck	
'64 to '78	P4286759
'79-'87	P4452025
Chrome Fuel Pump	
Block-off Plate	P4286567

GASKETS

Valve Cover Set	P4120095
Race Valve Cover Set	P4452089
Teardown Set	P4120692
Race Teardown Set (except 360)	P4452110
Race 360 Teardown Set	P4452111

PAINT

Black ('83 to '87)	P4120753
Red ('64 to '69)	P4349218
Blue ('72 to '83)	P4349217
Street Hemi Orange ('70 to '72)	P4349216

43. Manifold bolts should be torqued using the sequence shown in fig. 47. This single-plane, 4-barrel intake is now ready for a carburetor. At this point, if the engine is to be stored, shipped or installed without the carburetor, then be sure to tape-over the carb opening.

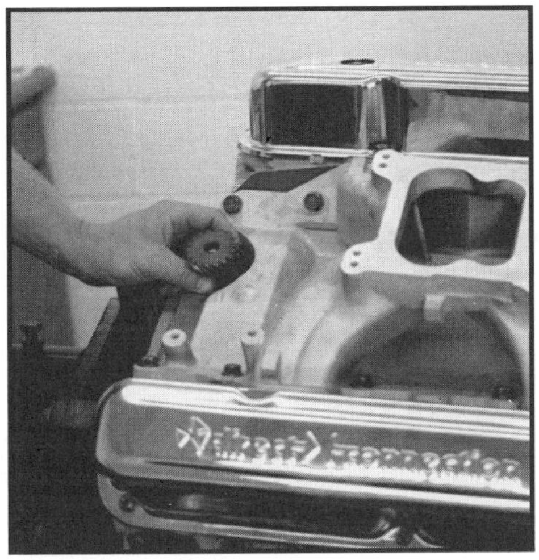

44. Installing the intermediate shaft gear is done late in the assembly process. The camshaft and oil pump should be on. Make sure that the slot is aligned properly with the engine at TDC.

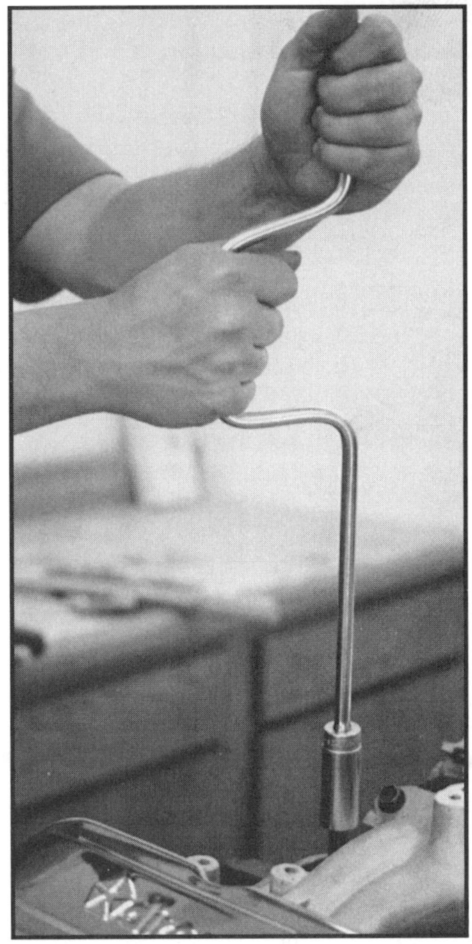

45. Once the engine is together, oil pan and valve covers installed, prelube the engine by hand with a speed handle or by using an electric drill. This is done through the distributor hole at the rear of the intake manifold. The intermediate shaft is removed, and a priming shaft inserted directly into the pump and turned. This is best done as you are closer to installing the engine.

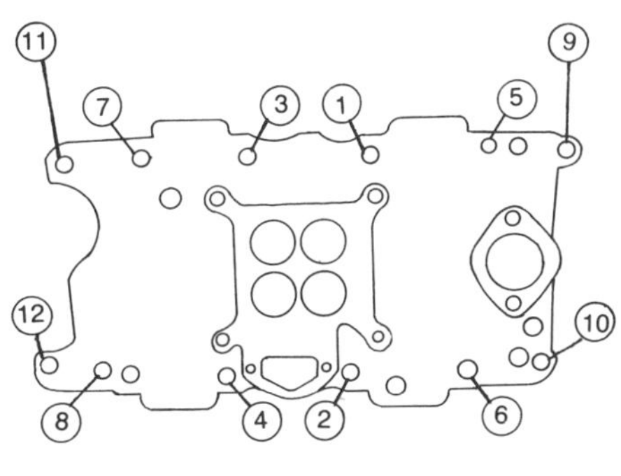

Intake Manifold Torque Sequence

47. The A engine intake manifold torque sequence is shown. Note that the thermostat housing on the right of the figure indicates the front of the manifold abd engine.

48. Since we now have a complete engine, we must consider its overall function in the vehicle. You must consider external balance. Refer to the text for details. The small weights on the left are for the cast crank 340 while the large weights on the right are for the cast crank 360.

49. LEFT—Once the engine is completed, you can run it on a dynomometer for testing and further development.

9 OILING SYSTEMS

The oiling system consists of everything that provides oil to the engine. Its primary function is to lubricate all moving parts. Its secondary job is related to engine performance and horsepower. The sophistication of the oiling system is dependent on the type of racing that the engine is going to be used for. It is comprised of oil pump, oil pan, windage tray, oil filter, oil pump drive, breathers and PCV valve. All this equipment bolts onto the assembled engine. In fact, the majority of it attaches to the bottom of the block. Obviously, the engine is all assembled and ready to make power. With that in mind, most people don't give the oiling system too much thought by itself. Many just think they can install the stock parts and go racing. In reality, this couldn't be further from the truth. Strange as it may seem, this "stock parts work on race engines" theory relating to the oiling system was held by most racers up to the early '70's. The truth of the matter is, the physical parts, like the oil pump and oil pan, are only part of the overall oiling system. There are oiling passages in the block, heads and crankshaft. There are oiling holes in the rocker arms, rocker shaft, cam, rods and pistons. There seems to be a lot more to this "simple" oiling system than meets the eye.

In the last ten years or so the horsepower/performance aspect has gotten a large amount of press and a great deal of attention from the racers. With all this visibility, you would think that this performance aspect was the only thing to consider in an oiling system and that it obviously must be top priority. Well, to this end I feel that the #1 function is to keep oil in the bearings. There are oil systems that can deliver 30 hp gains, but if this system didn't keep oil in the bearings where would you be? You'd be without an engine pretty soon, that's where. You would have all this horsepower potential and a pile of broken parts. All of this can happen in one lap or one run.

HISTORY

Chrysler has been involved in racing since the early 1960's. Through research and development, Chrysler engineers soon discovered windage losses in an engine's oiling system. They were the first major auto company to make this discovery. Factory engineers soon reduced these windage losses by using a *windage tray*, which soon found its way into thousands of production vehicles. Most of the early work was done on Hemis and 426 Max Wedges. However, the 340 had a windage tray in production when it

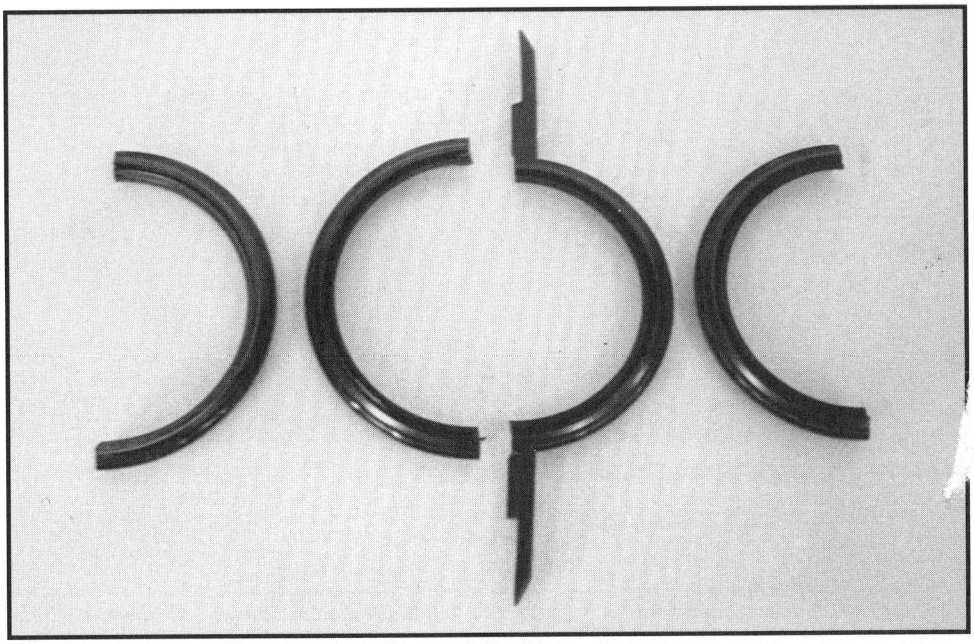

Although the rear main seals are similar on all A engines, the 360 uses a unique diameter. Note that the two seals on left have a larger diameter than ones on the right.

was introduced in 1968. The high-performance A engines have carried on this tradition ever since.

It's interesting to follow the modernization of the oiling system. The Chryslers were all playing with big, special oil pans and oiling systems in the early 60's. In those days, Chryslers were popular among the drag racing set. However, Chryslers began dominating too much and were soon handicapped by various sanctioning bodies to slow them down. Chrysler continued to support their racers, both factory pros and amateurs alike. There were so many Chryslers racers, that Chrysler soon began offering factory racing seminars around the country. One of the main topics frequently discussed was the oiling system. Chrysler told their racers how to make special oiling systems, and explained how they worked. Because the seminars were open to the general public, Chrysler's oiling system theories soon spread to other people, including the competition. By the mid-'70s, several aftermarket manufacturers produced oiling systems for general availability in racing and advertised than heavily, based on Chrysler's theories. Now, everyone uses a similar oiling system, though most have forgotten its origin.

BASIC OILING SYSTEM

The block has many oil passages drilled into it. The oil pump bolts to the block and the block distributes the oil through these passages. This distribution is obviously a very important function. From the oil pump, the oil goes through the block, out the right side to the oil filter. On the A engine the filter attaches directly to the side of the block. However, if headers are used, a right-angle oil filter adapter is required. This adapter has several machined holes that intersect with each other, leaving sharp edges and sharp corners. By grinding the sharp edges away and rounding the corners, oil flow will be improved. If the oil pump doesn't have to work as hard to get the oil to flow, then it doesn't take as much power out of the engine to drive the pump.

CIRCULATION

The oil returning from the filter enters directly above where it left and goes straight up into the right side oil galley. At the right rear of the block, near the #5 main, is the access hole for the vertical oil passage. It's drilled from the main cap parting area straight-up to the rear of the right-hand oil galley. The trick here isn't the drilled hole itself but the plug that is pushed in part way up. On a stress-relieved block, it has to be replaced.

The A engine's oil filter mounting pad before anything is added.

On any new engine build-up check to be sure it's installed. Without it, the engine ends up with very low oil pressure, which means it can't be raced.

Pressure—As I mentioned earlier, Chrysler was the first to discover power losses in the oil system's windage, losses that could be reduced by modifying the oiling system. Cutting the losses means that the engine will make more power. This performance gain can be as high as 50 hp, but realistically it's more like 30 hp.

The oil pump supplies the oil to all moving parts in the engine by *pressure*. The basic output of the pump is volume and pressure. For racing applications, pressure is monitored closely. As a general rule of thumb, 10 psi of oil pressure is required for each 1,000 rpm that the engine turns. This means that 30 psi is needed to go 3,000 rpm, 50 psi to go 5,000 rpm, etc. These numbers are based on a hot engine with the oil up to operating temperature. The A engine doesn't need as much oil pressure as some other engines, so it can be slightly under these numbers and still be okay.

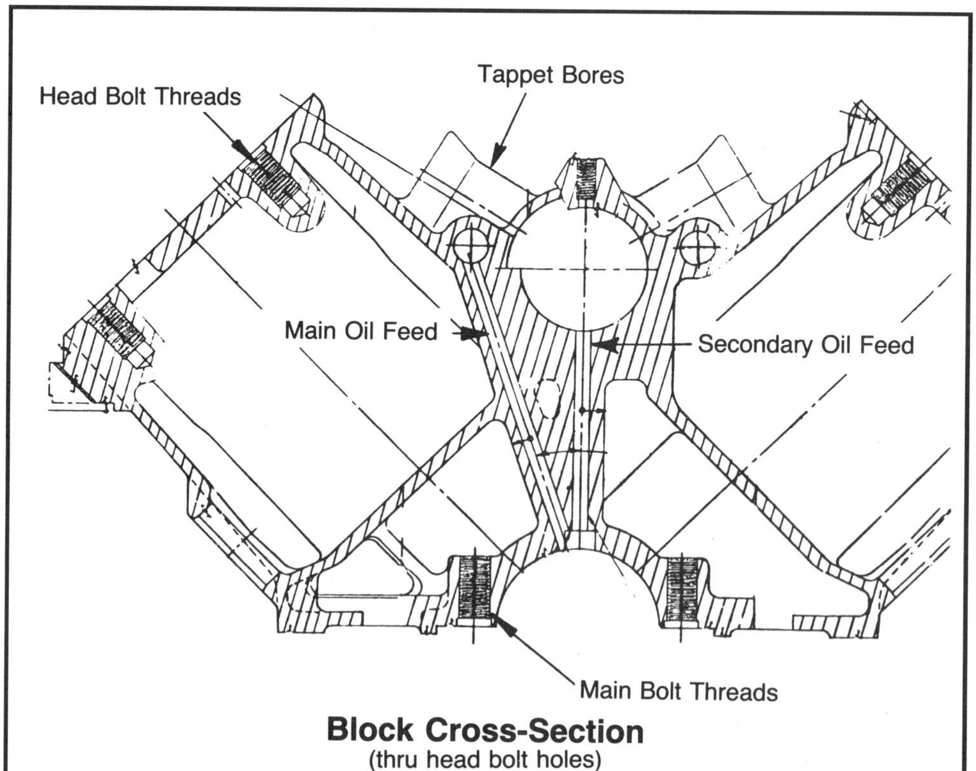

General block cross-section through center of main bolt holes and head bolt holes. Note the main oil feed passage from block right-side (left of figure) oil galley down to main bearing, and the secondary oil feed from top of main bearing to bottom of cam bearing

SHORT-BLOCK

In Chapter 1 on the *Cylinder Block*, I discussed the boring out of the right oil galley and the installation of tubes. This is obviously not done when hydraulic cams are used. The tube keeps the oil from leaking out around the tappets. Sixteen leaks can amount to a lot of oil and pressure loss. By plugging them (actually tubing would be a more accurate term), more oil is available for the crankshaft at the same pressure and pump power requirement. This is important in high rpm, high-output and long-distance (durability) engines. The original oil galley is reamed out so that the I.D. (inside diameter) of the tube is the same as the original I.D. of the galley itself. This keeps the flow characteristics the same.

The second step of the right oil galley tubing process is to drill upward from the mains into the tube. Otherwise, the oil would never get down to the crank. The drill that is used is the same size as the original drilled hole at the main bearing bore. But in some cases, the blocks are machined in production with a *stepped* drill. This means that the top part of the passage is smaller than the bottom. Therefore, when the hole is drilled into the tube, the upper passage is also enlarged so it is the same size from top to bottom. This small change would be good to do to any block, even ones without the tube. But the drill required is extremely long for its diameter and very hard to find unless you get a "tube" kit. By using a tube kit and drilling the main feeds, more oil is supplied to the crank and the resistance in the line is decreased, which again helps the oil pump to function with less power.

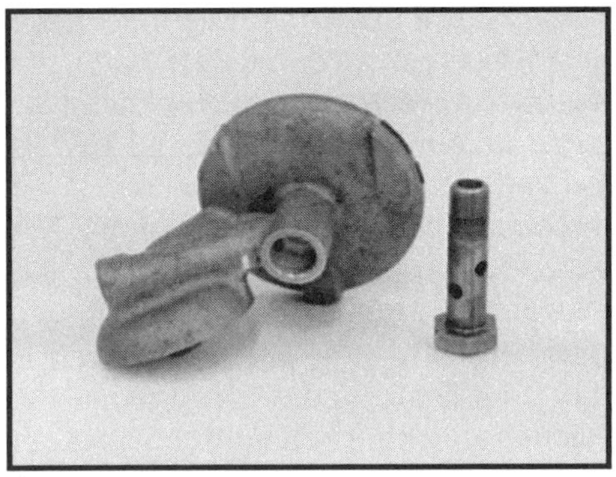

This photo shows the right-angle filter adapter and its unique attaching bolt. Note that the bolt is hollow with holes in its side.

Bearings—The bearings were discussed in Chapter 2 on the *Crankshaft*. Keep in mind that high clearances increase the engine's oil requirement. High rod side clearances also increase the oil demand. And a large clearance for the crankshaft end play will increase oil demand. If the crank end play and rod side clearance are minimized, then the high race clearances in the rods and mains themselves will have much less effect on the oil requirement.

In some cases, the rear main seal can give engine builders a problem. The seal itself doesn't cause the problem but the rear main leakage does. This is more common on street engines than race engines. In many cases, replacing the rear main seal doesn't solve the problem or doesn't fix it for long. The solution is a special #5 bearing. It has two grooves cut in the surface so that the crank's rotation helps pump the oil away from the rear main seal instead of directly at it. This will usually solve any rear main seal problems.

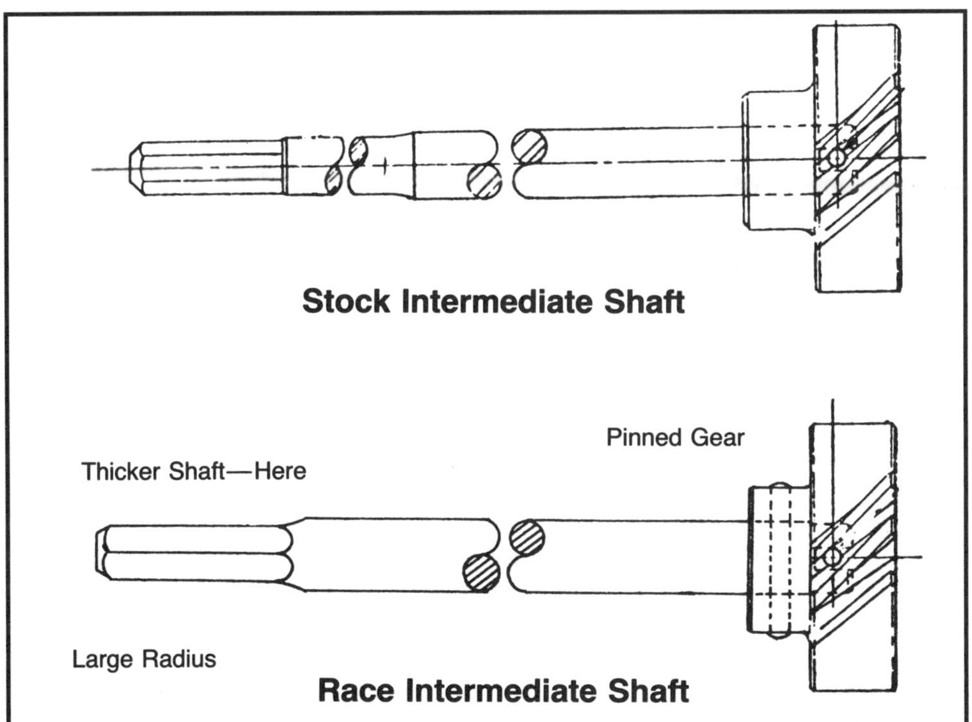

The intermediate shaft is used to drive both the distributor and the oil pump. With a high volume and high pressure pump you can sheer off the shaft. The race shaft has a thick shaft all the way to the hex and a big radius going from the machined hex to the round shaft. This is much stronger than the stock shaft. Note that the gear is pinned to the race shaft to keep the gear from turning on the shaft.

The right-angle oil filter adapter is installed onto the block.

Camshaft—The #2 and #4 journals on the cam have a hole drilled through them. The oil pressure comes from the top of the #2 and #4 main bearing straight up to the cam bearing. The next step is to go to the valvetrain in the cylinder head. This passage is in the top part of the cam journal. Therefore, the pressure has to come in the bottom, around the sides of the cam and out the top (actually slightly to one side). Obviously, there won't be much pressure left to go up to the valvetrain. To help with the pressure loss, most A engine cams have a hole drilled through the #2 and #4 journals so that the pressure that's on the bottom gets to the top. Another method is to groove the cam journals that oil the heads and valvetrain. A small 360-deg. groove will also help get oil to the valvetrain.

Pistons—The pistons keep the pins supplied with oil in several ways. They may have a drilled hole from the back of the oil ring; they may have a slot cut in the pin bore parallel to the centerline of the pin; or they may have a vertical hole or two drilled from the bottom of the pin tower into the pin bore and sometimes all the way out the top of the pin tower. Different piston designs use one of these three styles of pin lubrication. It's only important to check to see if they are there and that they are drilled through and clear of obstructions.

Rods—The pin oiling holes in the small end of the rod were discussed in Chapter 3. There is another oiling hole that is found in some production rods. It's a very small hole located toward the top (beam area) of the big end of the rod, toward one edge, not centered. The rod bearing shell (top) also has a small hole that lines up with the one in the rod. The purpose of this hole is to spray pressurized oil onto the cylinder bores. This is desirable in some truck engines and some low-performance passenger car engines. It is not best for a high-performance engine. It will put so much oil on the cylinder walls at high speed that the rings won't be able to control it. Then, the oil will seep into the combustion chambers and the engine's power will fall off, oil consumption will go up and detonation will be increased. If the *rod* doesn't have the hole, then it won't spray onto the cylinder walls, whether the bearing has the hole or not. On the other hand, if the rod has the hole and you want to block it, simply change the rod bearings to bearing shells without holes—that will stop the rod from spraying onto the cylinder walls.

VALVETRAIN

The valvetrain is oiled from the #2 and #4 cam journals. There's a drilled hole in the block that brings the oil up to the block's deck surface. The head gasket has a matching hole in it. Each head has a drilled hole up from the deck surface to the rocker stand or pedestal. Only the #2 pedestal has the oil hole in each head. This makes it the second stand from the front on the left side (bank), and the second from the rear on the right side.

Rocker Arms—The rocker shaft is hollow. The oil comes in around the second attaching bolt hole and goes out by a hole drilled where each rocker arm is. Hydraulic rocker arms have a small hole to oil the pushrod end but the valve tip is oiled by oil flowing over the end of the rocker. The mechanical rocker arms are more scientific. The mechanical rocker arm has a hole drilled from the pushrod socket area or the adjusting screw hole into the inside of the rocker. Since there is oil pressure in the journal, it will force oil out the hole, down the threads and onto the pushrod tip. The valve tip side is usually oiled by another small hole under the arm itself but pointing directly at the rocker pad area. For high rpm and/or high-output engines, the rocker shaft should have banana grooves. This spreads the oil pressure along the rocker and allows it to oil better and last longer. One thing to remember: Don't restrict the oil going to the valvetrain. There are a lot of press articles that recommend the oil at the block's deck surface be restricted. My experience indicates that if you restrict the oil and have a high rpm, high-output engine, then the rockers or valve tips will get scuffed or the pushrods will be smoked out. If the oiling system is prepared properly, there should be no need to cut back on the valvetrain oil.

OIL PAN

The type of oil pan that should be used on the engine should be determined by the type of racing/performance activities that the engine

The A engine uses three types of basic oil pans. Vans use a front sump, passenger cars use a middle sump and trucks use a rear sump. This is a rear sump pan that has been deepened for added capacity.

is going to be used in. The deepest part of the pan is called the *sump*. Production pans come with front sumps in vans, rear sumps in trucks and center sumps in most passenger cars. The 360 c.i.d. pans are different from the 273/318/340 c.i.d. pans. The 360's use a smaller radius (half-moon) in the ends of the pan. The main consideration in the pan (wet) is capacity. The stock pan at 4-5 quarts is not big enough for racing purposes. However, it is acceptable for street applications.

Modifications—The first step is to drop the sump 2 in. This increases the pan's capacity to 7-8 quarts. The next important items are the *baffles* in the sump area. There should be an *acceleration baffle* across the back of the pan (sump) and as high up on the back of the sump as possible. There should also be a *deceleration baffle* at the front of the sump, but it doesn't need to be as high as the one in the rear and it only needs to be half as big.

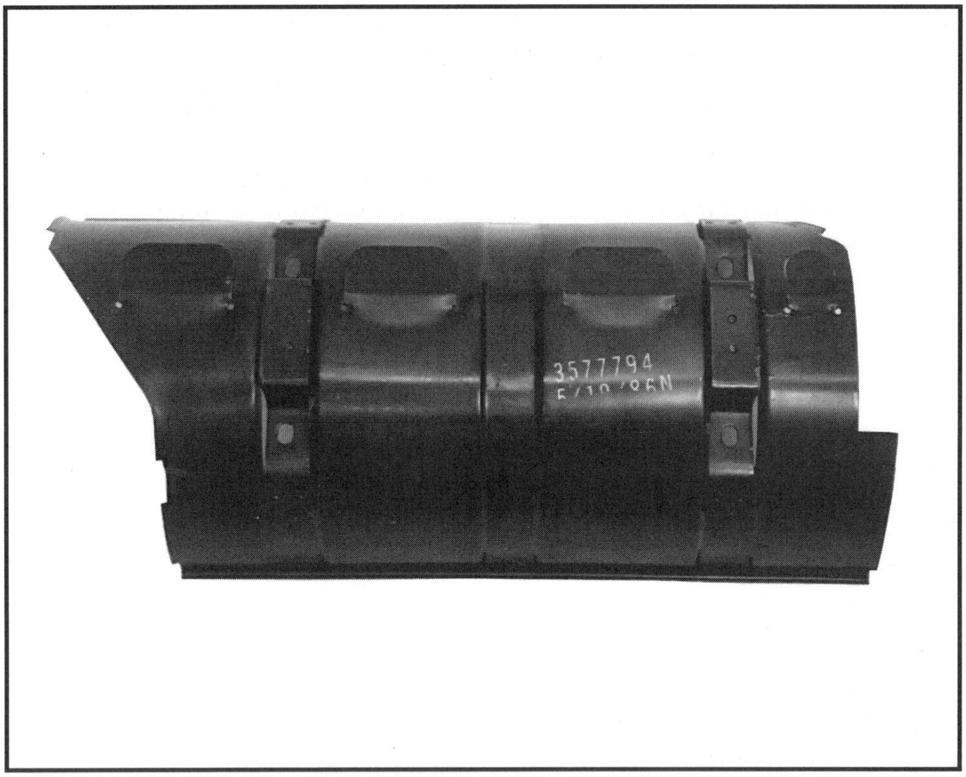

A typical A engine windage tray. It is designed to "fix" production oil pans.

The deepened pan requires a deepened oil pickup.

High capacity oil pumps use thicker rotors. Note the rotors on the left are 1/4-in. thicker.

The final consideration on the pan (wet) is ground clearance. As the pan is made bigger and deeper, it gets closer to the ground. Drag cars can get pretty low, but be careful with street machines.

There may not be enough ground clearance on a road race or circle track race car. A sump-style pan, lowered 2 in., is about the maximum that the street car can go, and even then the car has to be driven with extra care. On most drag race cars 9- to 10-in. deep (total depth—pan rail to bottom) pans are acceptable, but you have to watch the jounce and rebound bumpers in the front suspension if the car does wheelstands, especially on burnouts. Otherwise, the pan will be shortened the hard way—quickly.

With wet sump pans, the idea is to get the sump as far below the crankshaft as possible; to make the sump hold around 10 quarts; and to get the pan as large as possible. The best designed, large wet sump pans can net as much as 20-30 hp. With the ground clearance problem, the solution to the increased size without decreasing ground clearance is to make the pan wider at the bottom than the bottom of the block. These are called *side-bucket* pans. I will discuss them in more detail in a later section.

Windage Tray—The windage tray goes inside the oil pan. It bolts to four special main cap screws. It's a semicircular piece of sheet metal with slots located next to the crankshaft. Its functions are to "scrap" the flying oil particles out of the crankshaft's path and back into the pan, and to keep the foamy oil in the bottom of the pan away from the rotating rods and crank. Originally the 318/340 c.i.d. and 360 c.i.d. had two different windage trays because of the different spacing of the main bolt caps used on the 360 c.i.d. engine. Recently, production engineers have slotted the attaching holes in the tray so that the same tray can be used on both the 318 and 360 c.i.d. engines.

Oil Pump—The stock oil pump is also located inside the oil pan. It bolts to the block in the area of the #5 main. When capacity is increased, and if the pickup is good, then the stock pump is adequate. The big rotor (actually thicker), high-volume pump is a very popular swap for high-performance uses. If using a stock oil pump in racing/high-performance applications, check to see that the rotor is pinned to it's driveshaft. This only applies to the stock, production pump. If the engine is to be used for high rpm's, then the oil pressure relief spring should be changed. There are two types.

The stock relief spring tends to run around 50 psi while the high-performance one runs close to 70 psi (both numbers are "hot"). This relief spring

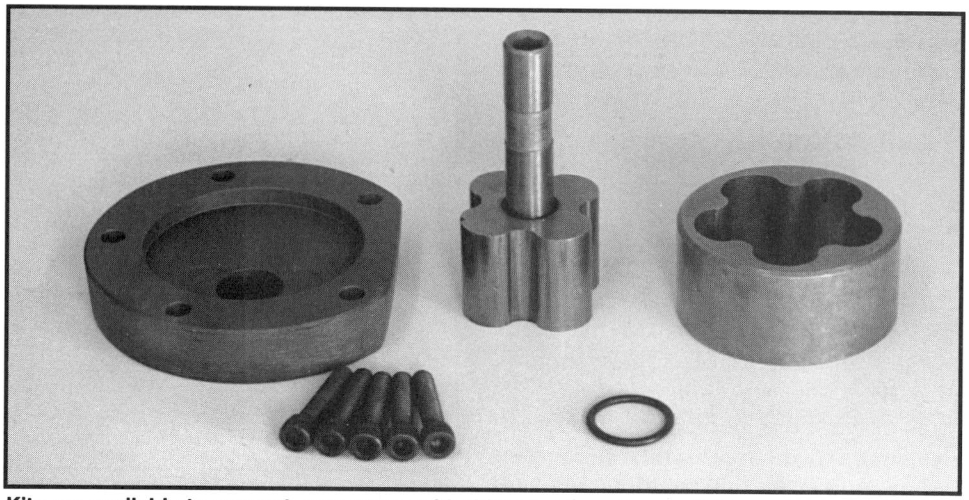

Kits are available to convert your pump with thicker rotors. To use thicker rotors either get a complete pump assembly or a kit to use the pump you have. The kit also includes the special, thicker cover and longer screws.

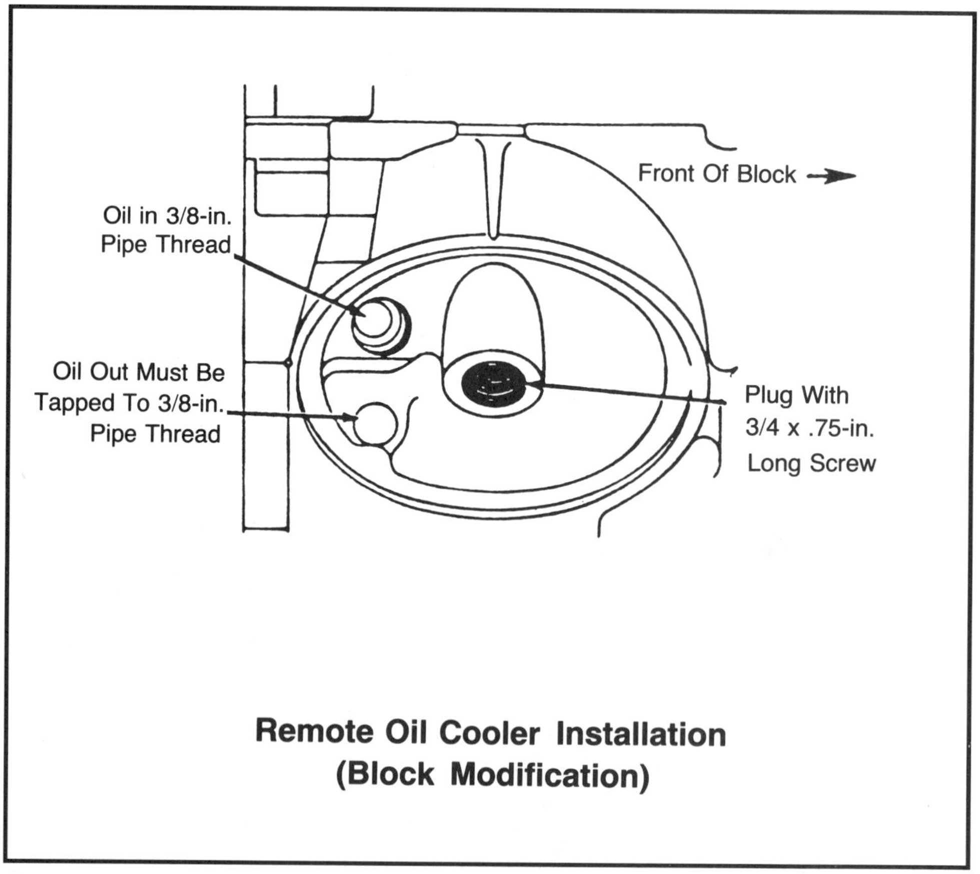

Detailed view of the oil filter pad for the remote oil cooler installation. Note that the large center hole that normally has a fitting to hold filter is plugged.

swap could hold true on all the stock pumps and most of the high-volume pumps. Keep in mind that high volume does not mean high pressure unless the relief spring is changed as well. The stock-style pickup used with the stock pump can be improved upon. For racing, a dual pickup set-up or a larger, single pickup should be used. It should be located close to the bottom of the sump and near the right rear corner in a drag car. It should also be able to swing, i.e. move back and forth with the oil. Don't hand-fabricate these pieces. Purchase them from one of the several special suppliers that are available. By the way, all of the pumps discussed above are for wet sump oil systems.

Oil Pump Drive—The oil pump drive is sometimes called the intermediate shaft. It's driven at the back of the cam. It's geared to the cam but also has a slot in its upper end that drives the distributor. For racing purposes, there are special intermediate shafts. The horsepower required to drive the oil pump at high pressure, high volume and high rpm can shear a stock oil pump drive. To solve this problem for racing, there are special heavy-duty shafts with larger cross-sections and large radii that are much stronger. These shafts also have the gear pinned to the shaft instead of just pressed on. Also, if the camshaft is a roller, then you must have an aluminum bronze gear on the shaft. All of these special parts are readily available through aftermarket suppliers.

Wet vs. Dry—A *wet* sump oil system is one that collects the oil by gravity, directly under the engine. These pans reached the peak of their development in the mid-70's in NHRA's Pro Stock class. They were 9- to 10-in. deep and held a minimum of 10 quarts of oil. They had a large side bucket on the right side of the pan, and in some cases, a smaller side bucket on the left. They all had swinging pickups in the bottom. The Pro Stock cars had been converted to front rack-and-pinion steering, so there was no center link and therefore no center link hole in the pan itself. These pans were good for an increase of 30 to 40 hp over a production (stock) pan.

The dry sump is not legal in many racing classes and in many sanctioning bodies. It was developed in road and boat racing, primarily because of the low ground clearance or limited space requirements. When Pro Stock drag racing cars started getting smaller and lower, the big oil pans began to be a problem. Originally, all Pro Stock cars were about the same size at 108-in. wheelbase. Then, NHRA began to allow lighter, sub-compact cars with wheelbases under 100-in. They were obviously faster than the bigger cars, so everyone started building them. A Dodge Colt or Plymouth Arrow fitted with a V-8 didn't have room for the big wet sump-style oil pan. Enter the first, serious drag racing dry sump. It allowed the cars to be built low to the ground. Unfortunately, it cost 15 to 25 hp in engine output. However, the smaller cars were still faster in most cases than the bigger cars because of their size advantage.

Chrysler began an extensive development program to improve the dry sump system at that time. After extensive testing, Chrysler developed a newly shaped oil pan and redesigned the overall system that made the dry sump as good as the original deep-pan wet sump. Then Chrysler applied that knowledge to the wet sump and produced a much-improved wet sump pan. This is how the *side-bucket wet sump pan* came to be. This type of pan is what all the top competitors use in racing today.

So now the wet sump was better, but what everyone really wanted was a

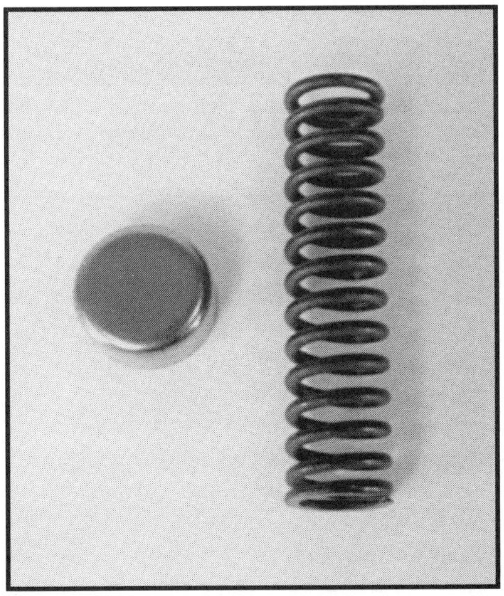

To change the oil pump's pressure, you need a stronger oil pump relief spring. Additionally, you will need the press-in plug in case the original is damaged in use or upon removal.

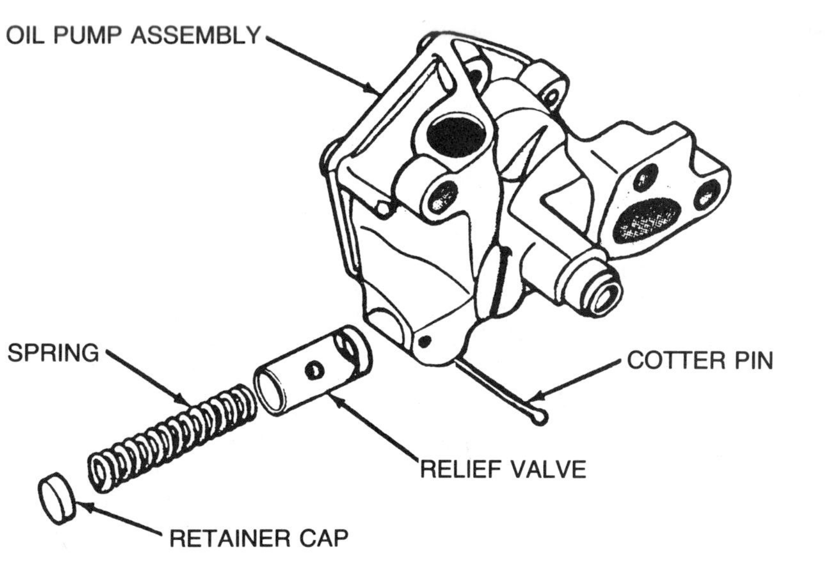

A Engine Oil Pump

A engine oil pump is located inside oil pan. Spring controls pressure output. Standard springs in stock pumps usually make 40 to 55 psi. High pressure springs are available to increase pressure to 60-80 psi. (Courtesy Mopar Performance)

better dry sump pan. In time, that is exactly what Chrysler did develop. Soon, Mopar racers started going FAST! Then everyone had dry sumps. This same basic attitude has carried over to present day. Most racers think that the dry sump is faster than a wet sump. In reality, the biggest advantage of the dry sump is that it allows the car to be lowered without a power loss.

Toward the end of this dry sump development came the 340 Pro Stock motor. Running a dry sump and squeezed into a Plymouth Arrow, the car won almost every race in NHRA Pro Stock in 1978, including the World Championship. The dry sump system that was used on the car has changed very little since. In fact, oiling development for Pro Stock really peaked during 1978. Once the dry sump was working, everyone went back to cams and heads to hunt for more power. Dry sump systems are more readily available today, but the basic science is derived from that 1978 340 Pro Stock engine.

The heart of the dry sump system is an external, Weaver 3-stage oil pump that has two scavenge stages and one pressure stage. It is driven by a Gilmer belt off the front of the crank. The pump is driven at 55-60 % of the crank speed. The 55 % is used for the higher rpm engines.

The drag race pan has two pickups. One is located in the right rear of the rear bay of four cylinders (numbers 5, 6, 7, 8) and the second one is located in the right rear of the front bay of four cylinders (number 1, 2, 3, 4). In an oval track engine, the pump uses another stage (i.e. 4-stage pump) and the extra stage is used to scavenge the right rear of the right-hand valve cover.

The *dry sump* of a dry sump oiling system is a remote reservoir which is really a holding tank. In a drag car, it's usually located ahead of the engine, near the radiator. In an oval track car they should be located on the left side of the car. The "drag" and "oval track" tanks are not the same. The drag dry sump tank is approximately 14-in. tall and 6-in. in diameter. It has baffles inside and holds 6 to 8 quarts. Other than holding the oil, the secondary function of the tank is to separate the air from the oil inside the tank. To do this, the oil must enter the tank tangent to the outer wall. The oil flow is designed to go around the outer wall. Because of the excess air that tends to accumulate in the tank and it's resulting pressure increase, the tank must be vented to the engine's valve cover. This allows the pressure in the tank to equalize with the pressure in the crank case.

The oval track dry sump tank is larger and should hold 3 gal. This will cause the tank's diameter to get larger. The tank vent should also be attached to the left side valve cover.

OIL

The oil itself should be a quality, high-performance grade. It must meet Mil Spec MS-DG. Anti-scuff additives such as *lubrizol* and cam lube oil should be drained from the engine after the engine is broken in, but before it is

CHRYSLER PARTS LIST

OIL SYSTEM PARTS	
Hi-Po Oil Pump Assembly	P4286589
Oil Pump Hi-Pressure Relief Spring	P3690944
Intermediate Shaft Bushing	P1737725
Oil Pump Priming Shaft	P4286800
Right Angle Oil Filter Adapter Package	P3690884
H.P. Oil Filter (Black)	P4349746
H.P. Oil Filter (Chrome)	P4349747
Oil Pan Bolt Set—Nylon Insert	P4120613
Oil Pump & Dist. Drive (Std.)	P3690715
Oil Pump & Dist. Drive (Alum./Bronze Gear)	P3690874
340 Pan Gasket Set	P4007572
360 Pan Gasket Set	P4007573
'64-'69 Race Pan Gasket	P4452095
'70-'86 318/340 Race Pan Gasket	P4452097
360 Race Pan Gasket	P4452093

Outside and bottom view of dry sump pan. Note the base pan was cast (rather than stamped) before it was modified. This is desired in oval track racing for durability and added rigidity.

This is an inside view of an oval track dry sump pan. Note the dual pickups.

raced. Usually the performance engines should use 20W-50W racing oil for general race applications. You can use high-quality 30W in summer and 40W in the winter for high-performance street machines as an alternative if racing oil is not available, or just too expensive. The only synthetic oil that I have had experience with in racing is *Mobil 1*. It seems to work well for turbochargers and superchargers. However, it's expensive and the effective viscosity in the engine seems to be low—in the 5 to 10W category. It performs much like a low viscosity oil. It's a favorite trick of the Pro Stockers, but if the engine's clearances are high, then the oil pressure will be low.

Temperature—It's best to keep the oil temperature below 200 deg. This usually isn't a problem in a drag car, but street cars might need an oil cooler. In a road race/oval track car, oil coolers are necessary, but even then the temperatures are likely to get somewhat higher than 200 deg. F.

PRIMER

A race engine should have the oiling system *primed* before it is started. This assures the builder that everything is working correctly and also provides a good supply of oil to all the moving parts of the engine before they actually move. To prime the pump, (wet sump style) remove the distributor and intermediate shaft (engine should be in the car and ready to go at this point). Insert the long hex-shaped shaft, (specially made to fit the hex on the oil pump) into the hole where the intermediate shaft was and engage with the oil pump hex. The shaft will stick-up above the block at the rear. Attach a 1/2-in. electric drill to the shaft. Use the drill to turn the engine's oil pump and create oil pressure (it should be as high as 40-50 psi). When the engine is started, there will be oil in the galleys.

The #5 main cap modified for use with a dry sump system. Note plug in middle of cap.

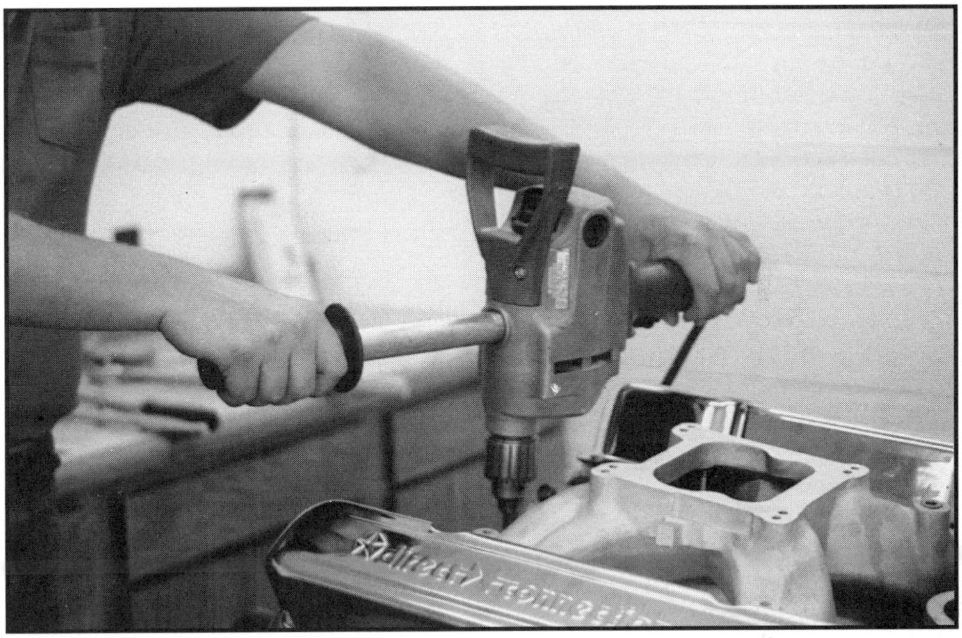

RIGHT—Once the engine is complete, it should be primed and prelubed using an electric drill and an oiling pump priming tool.

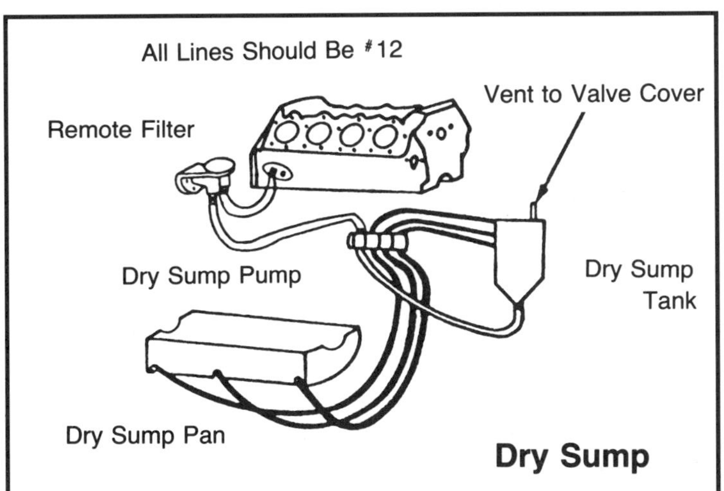

Basic dry sump set-up is the same for most applications. However, some of the details change from drag racing to oval racing (Courtesy Mopar Performance).

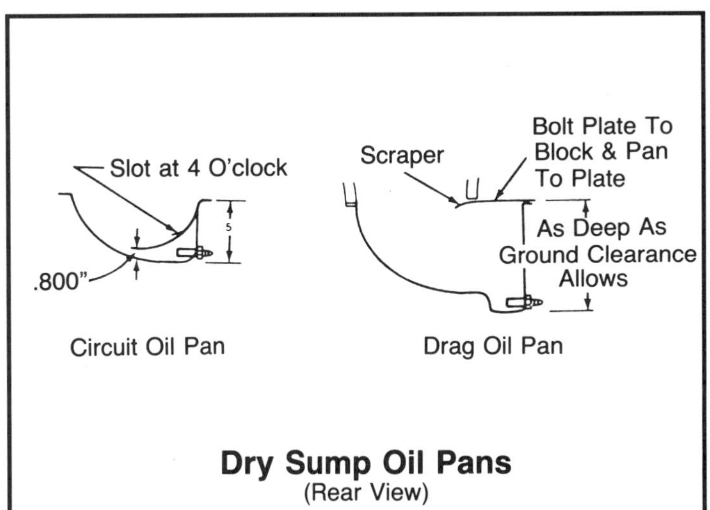

The drag racing dry sump pan tends to be deeper and has the oil scraper located differently (Courtesy Mopar Performance).

10 IGNITION SYSTEMS

There are not many physical parts to the ignition system. It is basically comprised of a distributor, spark plug, plug wires, coil, ballast resistor, control box and wiring. Many racers forget that the spark plug is part of the whole ignition system and they tend to take the plug for granted. A plug is just a plug—right? The plug itself may not be very trick and may not get much attention, but think what happens if it doesn't do its job? Generally, the engine builder doesn't worry about the ignition system. It's the last thing that is put on the engine. In most cases, it's installed after the engine is in the car. They are usually selected and installed by the customer. Well, when this customer and the engine builder are the same person, namely you, then you need to pay close attention to selection and installation of the proper ignition system. If it doesn't work perfectly, then all the work put into the engine goes right down the drain.

Much of the ignition system's hardware can't be built. The block can be honed, the heads ported, but the ignition tends to come under the category of "if it's broken, buy a new one." This can be very frustrating for the do-it-yourselfer. With this in mind, I'll cover specific details so you'll understand what you need for the ignition system. Keep in mind that, like so many other parts of the engine, it's important to consider what the engine is going to be used for. Equally important is the condition of the rest of the engine—how much output, how many rpm, etc. You will need to keep these questions and their answers fresh in your mind as I go through the basics of the ignition system.

HISTORY

The early ohv V-8's and the early A engines in the 50's had very little use for a high-output, high rpm ignition system. The production-based ignition worked fine. Maybe the coil was changed and the distributor recurved and more timing set. But those changes were really just adjustments rather than new systems. It wasn't until the '60's, with the advent of big-blocks and the exploding popularity of racing, that ignitions were developed further. The first big breakthrough was the *dual-point ignition.*

This was a fine performance ignition and was used on the 1968 340

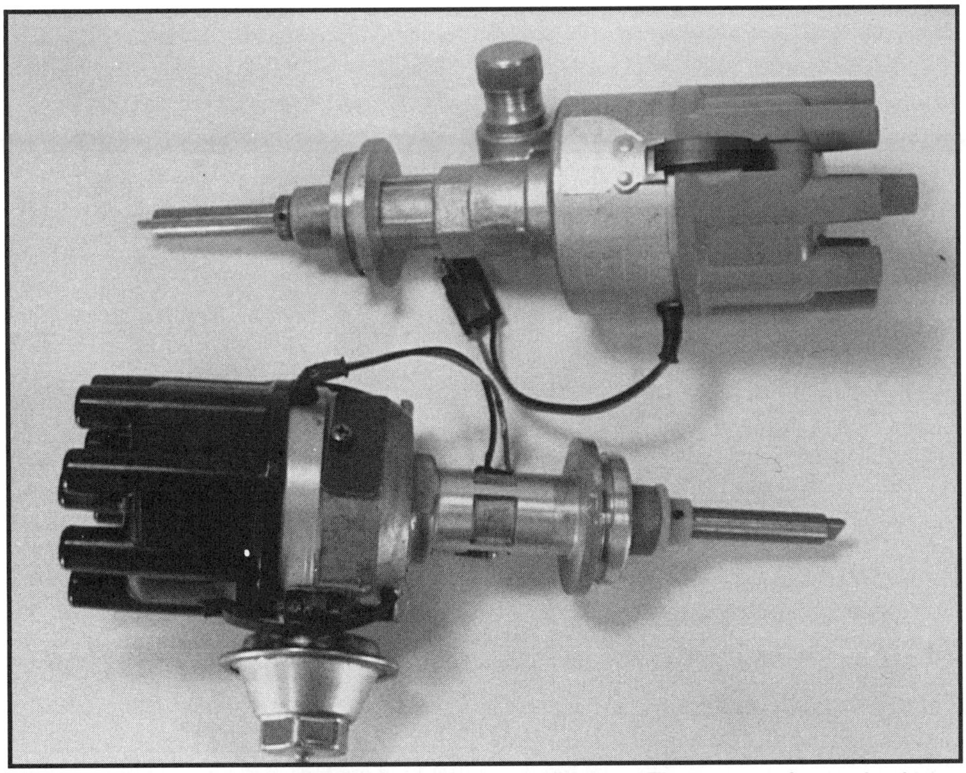

This photo shows the two types of electronic distributors. The top one is an aluminum, mechanical tach drive unit. The lower one is a vacuum advance unit.

c.i.d, the '70 T/A 340 c.i.d./6-barrel engines and some of the 273 c.i.d./4-barrel engines. These were all production engines. The real difference between the GM/Ford engines and the Chrysler engines in the 60's could be seen in racing. Almost everyone used dual-point ignitions in production on their high-performance vehicles in this era. The GM/Ford engines used them in racing also. They changed the coil and recurved the distributor but they were still dual-point ignitions.

Mini-Point—The first minor breakthrough on performance ignitions occurred in the early 60's, but no one outside the Chrysler racing family really noticed. It was the introduction of the *mini-point*. One-third the size of the stock point, it was much smaller and therefore lighter than the stock point. The lighter weight allowed it to function better at high rpm's. At the time, the stock-style dual-point was beginning to show its limitations. The Chrysler racing engineers on the scene began looking for something better. First, they developed the *Prestolite Transignitor* ignition with the black heat-sink, and later on, an improved version with a blue heat-sink. In the early 60's, the transistor was still pretty rare. In simple terms, the Transignitor was a *transistorized ignition*. It was pretty high tech for its day and was used as a performance and race ignition for many years. It still used the point-type distributor as a triggering mechanism, so it was used by the GM/Ford racers also. Many of these racers used this ignition up to the mid to late 70's with great success.

Electronic Ignition—At Chrysler, the race engineers weren't about to stop there. They had high-output engines that ran high rpm's, so they obviously ran into problems. They were aware of the inherent shortcomings of the point distributor (single or dual). Whereas the A engine was a late-comer in the area of race oil system development, it was first in the area of performance ignition development. The A engine was the first one that used an electronic ignition in 1971. In '72-'73 the 340 c.i.d. received a full electronic ignition system, and since then, virtually everyone else has converted, one way or another. These original electronic ignitions were used in production but worked even better in racing. The ignition was no longer dependent on the mechanical opening and closing of a set of points to trigger the ignition. Now it could be done electrically, without parts touching physically.

The bottom of each distributor shaft has a tab. This mates to a slot in the top of the intermediate shaft. This means that the distributor can be installed 180 deg. out of phase. When installing the distributor, it's always best to mark the direction that the rotor is pointing upon removal or go back to TDC and start from scratch.

The A engine shared in the next production "breakthrough" in ignitions. In the mid-70's, Chrysler introduced the Lean-Burn system on cars with both A and B engines. It was the first of the computer-controlled ignitions. It was an analog-based system, but had no real use in racing. The new digital ignition systems evolved from the original analog system. Though it is quite good, it hasn't yet been applied to V-8 racing (other than on experimental cars). It has been used successfully in the 2.2-liter race program. It is the future for production but racing seems to be staying with the "old" stand-by, electronic ignitions.

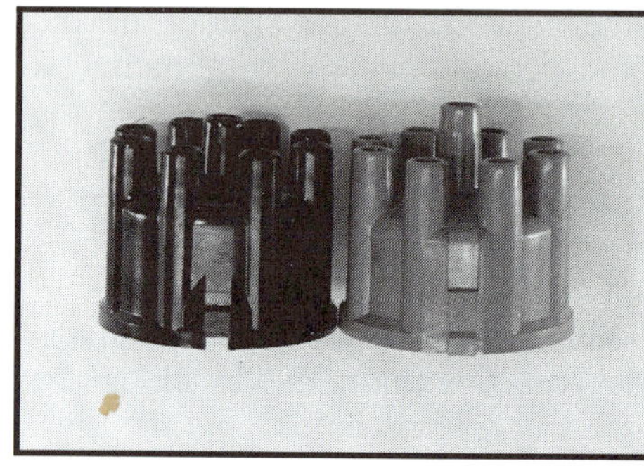

Two styles of electronic distributor caps—vacuum advance style (left) and mechanical tach. They are not interchangeable. Notice notch in one on left is a tab on the right.

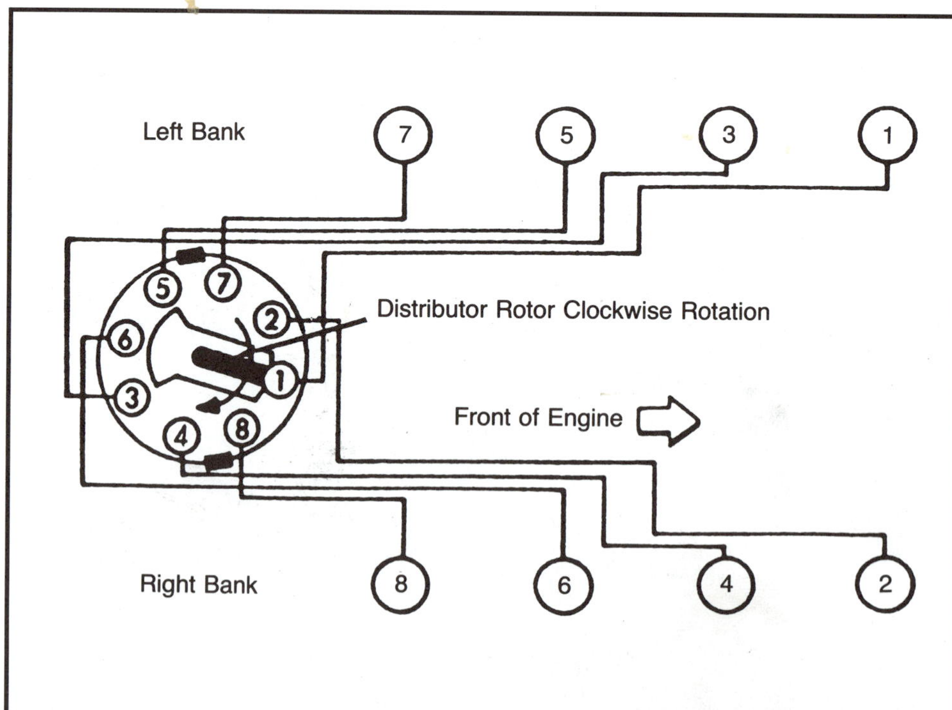

Basic A engine firing order. Note that the distributor is at the rear of the engine and that the cylinders are numbered so that all the odd numbers are located on the same side (left bank).

DISTRIBUTOR

In a point-type ignition, the distributor is the heart of the system. In the electronic systems, the distributor itself may not be the main element, but it still plays a key part overall.

The distributor on the A engine is located at the rear of the engine and rotates clockwise. There are four basic different distributors—single-point, dual-point, electronic and computer. They are all interchangeable in the engine as long as the rest of the system is changed and the wiring corrected. The distributor consists of the housing, shaft, advance mechanism, cap and rotor. If the shaft or housing are bent, then it's best to replace the whole distributor. Be careful with the service parts used for the cap and rotor. There are some counterfeit parts on the market, and they don't fit as well on the housing or shaft. A poor fit causes a loss in engine performance, reliability and durability. Also, be sure to know the type of distributor. The racing electronic distributor (called the *mechanical-tach*) uses a different cap than the vacuum advance models. They look almost exactly the same, but the cap fit is poor if the caps are switched. Chrysler may be the only reliable source for the mechanical-tach distributor service parts.

Two styles of electronic distributor. Notice the cut-down reluctor used on the mechanical tach drive distributor on the left.

Firing Order—The most obvious function of the distributor is to tell the firing order of the cylinders. This is determined by the firing order of the engine: 1-8-4-3-6-5-7-2. This is selected as soon as the plug wires are put into the cap. Be careful on the 5 and 7 wires, they're easy to cross. With a quality distributor cap, all distributors perform this function well.

Timing—The second function of the distributor is the "when." The "when-the-plug-fires" aspect is really determined by three separate items; the *initial advance*, *total advance* and the *curve*. The initial advance is set by rotating the distributor in the block.

This is the triggering mechanism for the vacuum advance distributor. Notice the height of the reluctor.

This is the triggering mechanism for the mechanical tach drive distributor. Compare this reluctor height and shape with that in the previous photo. This reluctor is machined down for increased accuracy and stronger magnetic signal. It's all part of the "Race" distributor.

The total advance is the sum of the distributor advance plus the initial advance. The distributor advance is built into each distributor so it tends to vary from one unit to another. The curve is how the advance gets from the initial setting advance to the total advance or from zero distributor advance to full distributor advance. This curve is controlled by the distributor's *advance mechanism*. This also varies from one unit to another. However, by changing the springs in the advance mechanism, the curve can be changed quite easily. This method is called *recurving the distributor*.

Triggering—The next aspect is how the distributer *triggers* the timing. There are two basic mechanisms—*points* and *magnetic*. Points are used in both the single- and dual-point distributors. Triggering occurs when the point physically opens/closes. Points require a lot of adjustment to maintain good operation. They also have a physical limit on their rpm capability because they must physically move.

The magnetic trigger is used in the electronic distributors. With the magnetic trigger, there is no physical opening or closing and no physical contact between the magnet and the reluctor. Therefore, they require little or no maintenance and have no real rpm limit by themselves. For racing, the electronic distributor is by far the best and is the only one that should be considered. The solid shaft, computer ignition distributors don't offer the adjustment capability required in performance and racing applications. Therefore, they are not recommended for those applications.

Vacuum Advance—The final function of the distributor is the *supplemental advance*. This is generally done by vacuum and is therefore called *vacuum advance*. It consists of a vacuum pot added to the side of the distributor that alters the advance mechanism to give more advance at a given rpm than the advance mechanism would normally yield because of its centrifugal aspect. The centrifugal advance is defined by the springs, distributor advance and rpm and could be considered a somewhat more

Top view of the typical distributor's advance mechanism. Note the small, short spring to the left and heavy spring to the right with the loop in the end. These two springs are changed to recurve the distributor.

MSD's Blaster 2 coil is oil-filled with specially designed components that make it perfect for use with the MSD 5, 6 and 7 series ignitions. (Photo courtesy MSD)

The Pro Power 3 coil from MSD uses a finned housing to dissipate heat. It is best for endurance racing and designed to work only with MSD 7 and 8 series ignitions. (Photo courtesy MSD)

scientific term for advance curve. For racing, the vacuum advance isn't considered because it is designed in such a way that the supplemental advance goes to zero at W.O.T. (wide-open throttle). The Chrysler vacuum advance distributors are adjustable to a certain extent since their "vacuum" effect can be decreased or dialed-out down to zero, if desired. The vacuum advance is desired for better fuel economy and driveability and is best for dual-purpose, street engines.

Selection—Selecting a proper distributor is dependent on what the engine will be used for. For stock, a single-point is fine, but for any performance application, it should be replaced with either the dual-point or the electronic distributor.

Though the dual-point is acceptable for performance purposes, the electronic offers many performance advantages, especially for racing. If that is the goal, then you must use an electronic. Likewise, if you have a solid shaft/computer distributor, this should be replaced by the electronic for any performance use as well.

Electronic—Since the electronic is most likely to be used, let's examine it in detail. There are actually two types of electronic distributors commonly used—the vacuum advance and the mechanical-tach. The vacuum advance unit has an aluminum housing and a vacuum-pot on the side, and has been used in production A engines since 1971. It is the unit that should be used in any engine for the street. In recurved or high-performance form, it can also be used in milder forms of racing and dual-purpose applications.

The mechanical-tach, on the other hand, should only be used for racing. There are two different types of mechanical-tach units—cast iron and aluminum. The aluminum units are newer and 2-3 lb lighter than the cast iron units. The only ones sold new today are the aluminum units.

Once the distributor is selected, check the advance curve. For performance purposes, a more aggressive curve is best, even if the unit is a production one. For the street, the curve should be fully advanced by 2,000 rpm. For racing only, the curve is fully advanced by 1,300-1,500 rpm.

COIL

The coil's primary responsibility is to supply power to the ignition system. The coil can't be looked at by itself. The *ballast resistor* and the type of system used must also be considered. If a coil is randomly selected, there will probably be problems. There are three common types of coil, and one highly specialized type that can only be used with the MSD 7AL-2 ignition system. They must be carefully selected to match the application. The three types are production, high-performance and racing.

The production coil is obviously a stock one. The high-performance coil can usually be used in place of the stock coil. In some cases, a different ballast resistor may be needed. Check with the manufacturer on this point. The race coil should only be used on a race engine. It requires a different ballast resistor. The special coil is used with the MSD 7AL-2 and is unique to that ignition system. That type coil is available from Autotronic Controls (MSD), Mallory and Accel. If you're using an MSD 7AL-2, then you must use this special coil because any of the

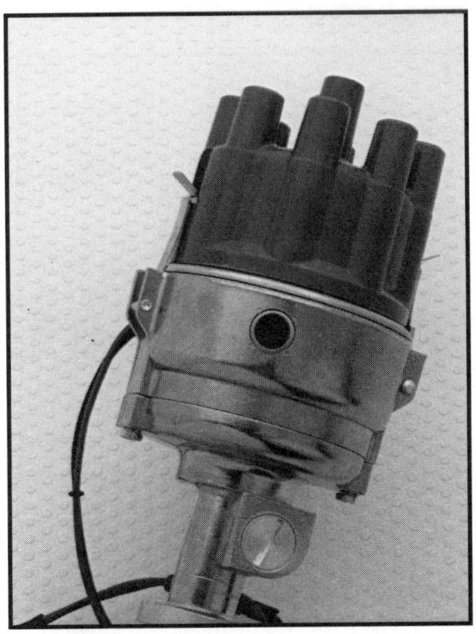

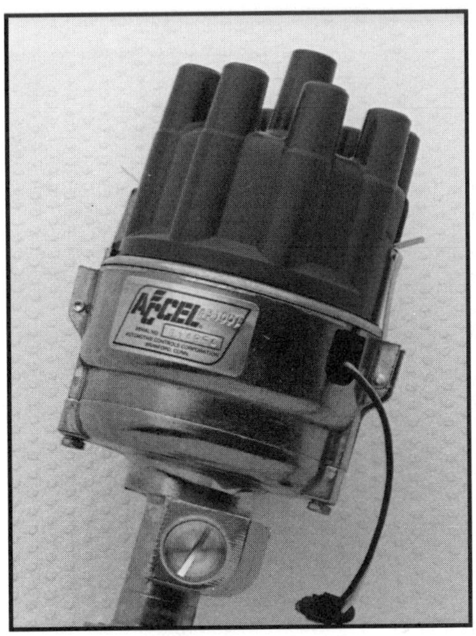

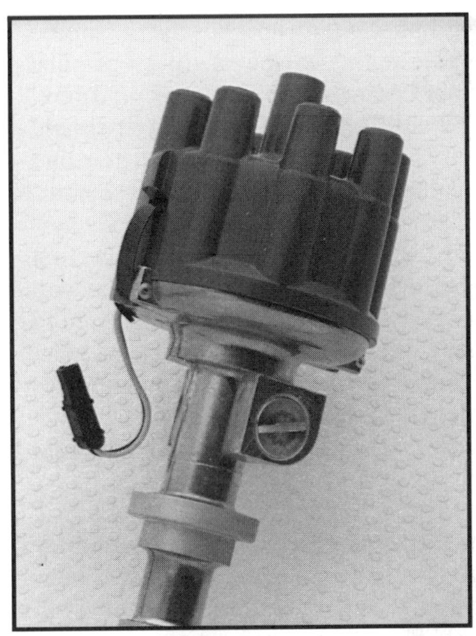

These three distributors represent the top from Accel. The 37 series (left) is a dual-point; the 43 series (middle) is a magnetic; and the 44 series (right) is a direct-drive unit. (Photos courtesy Accel)

other three types will not work with it. However, the MSD 7AL-2 is for very high rpm, high-output racing only. For example, most of the NASCAR racing engines find the MSD 6A or 6T to be sufficient.

CONTROL UNIT

The control unit, generally called an *ECU*, is the heart of any electronic ignition. Obviously, it is not used with a point-type ignition system. Basically, the ECU controls the ignition system's characteristics. It controls rpm capability, durability, driveability and to some degree, cost. Chrysler has four different levels plus an older race unit that are color-coded. Then there are the MSD ignitions and other styles of aftermarket electronic systems.

Types—The production ECU unit is black. It should only be used up to an rpm limit of 5,000-5,500 rpm. The next step up is the Chrysler orange box. It's a single ballast resistor unit and good to 6,000 rpm. It's a good choice for a high-performance street machine. The next level is the Chrysler chrome box. It's also a single ballast resistor unit and is good to 7,500 rpm. It makes a good choice for street machines, bracket racers and entry-level oval track racers. There also was an older unit (blue) that was a dual-ballast resistor unit. It has similar performance characteristics to the chrome box but the wiring is different. The final choice is the Chrysler gold box. It's an 8,000 rpm-plus unit. It should only be used on race cars both drag and oval track! It is a single ballast system.

For racing, there are two MSD systems that are recommended. The first is the MSD 6A, which is available with a rev control (MSD 6T). The second design is the MSD 7AL-2. It is a must for the "professional" racer who uses high rpm, high-output engines in his series. It is similar to the Chrysler gold box in performance. Many pros mount two boxes side-by-side. A switch is used to change from one to the other. If budget is of concern, you might consider the Chrysler chrome or gold boxes for this application. Keep in mind that both MSD race ignitions still use the standard electronic distributors described in an earlier section. The distributor doesn't need to be changed or modified in order to use these special race ignition systems.

SPARK PLUGS

The spark plug is really at the center of any ignition system. If it doesn't do its job, nothing happens and the engine or cylinder won't produce power. Despite this, many people don't give the plug much thought. But several areas, such as *heat range* and *plug gap* should be considered. They should also be chosen carefully.

Heat Range—The production A engines have used heat ranges from N9Y to N14Y (Champion) in the past. For racing purposes, the N9Y is the only one recommended. For racing and high rpm, high outputs, a colder plug should be used. The N9Y is the coldest of the production plugs and therefore makes a good choice for dual-purpose applications. The next coldest plug from the N9Y is the N66Y. For Super Stock racing use the N65Y, while an all-out, fully modified

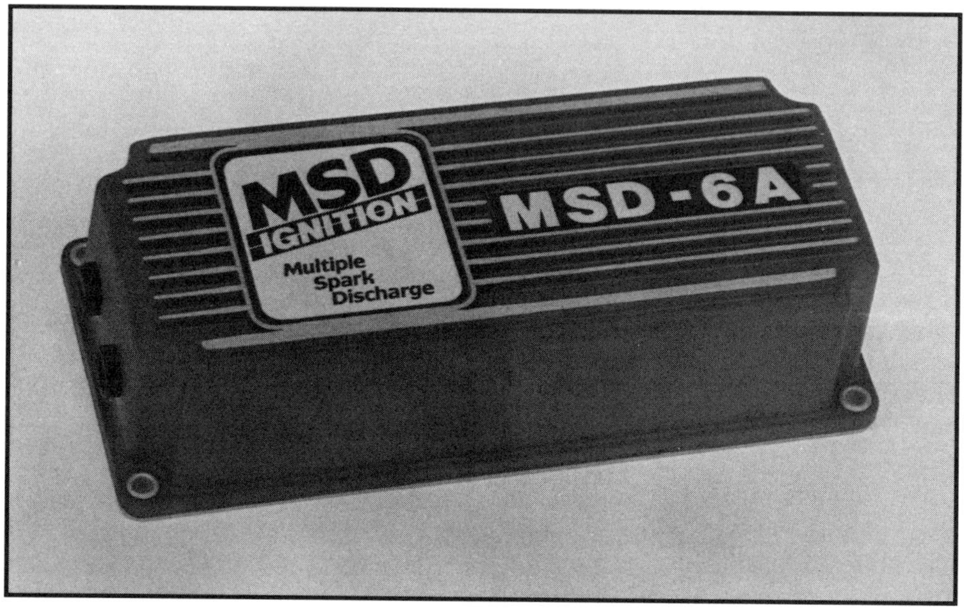

MSD 6A is a popular ignition for street machines, street rods and bracket racers. (Photo courtesy MSD)

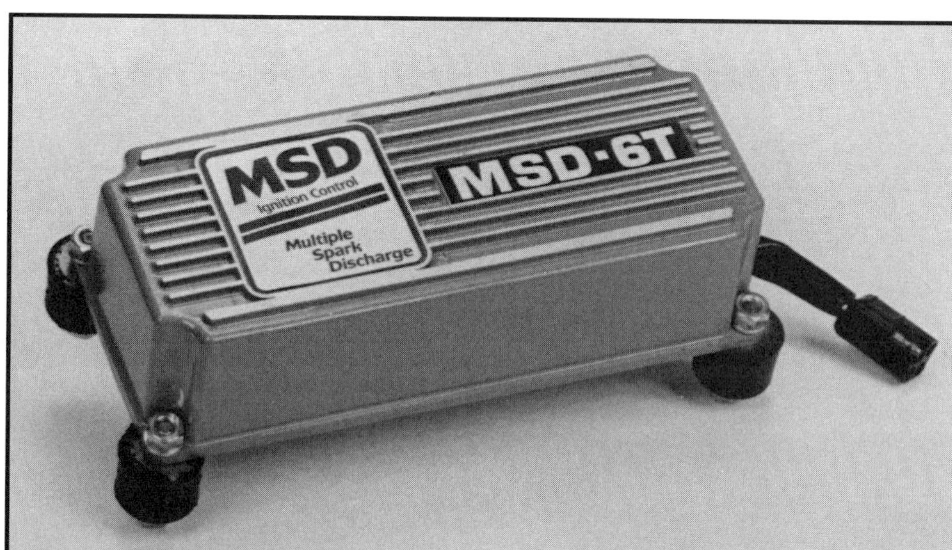

The MSD 6T is the choice of many oval track and road racers. It has the same basic circuitry of the 6A, but the 6T is reinforced with bracing and padding to withstand vibration. The most important feature is that it can accept MSD's soft touch rev control. (Photo courtesy MSD)

engine should run either the N63Y or N60Y plug. These designations are for Champion spark plugs only.

Plug Gap—The plug gap will be between 0.035 in. to 0.060 in. The wider gapped plugs between 0.055 in. to 0.060 in. are usually faster when used with the race ignitions (i.e. higher output), but the 0.035-in. to 0.038-in. gaps are more durable. You can experiment in your car with your driving style and chassis set-ups. Find out how much faster you go and compare that to the added cost and decreased durability. As a general rule, stick with smaller gapped plugs for bracket racing and wider gapped plugs for class racing.

Plug Wires—The plug wire has a fairly simple task. It must deliver the energy from the ignition system to the plug. They must be kept away from each other so that they don't *crossfire*. They must also be kept away from metal objects like valve covers, or they might short-out. Aside from the Chrysler production wires, there are two types of performance wires to consider; *suppression* and *solid-core*. Until recently all race engines used solid-core as a rule. The suppression wire was used in the street/dual-purpose engines. The solid-core wires were only used in racing. But within the last few years, many racers have started using on-board computer systems, which sometimes control the ignition system in addition to monitoring other aspects of the car. If the computer controls the ignition system, then suppression wires must be used because the solid-core wires drive the computer crazy.

The *size* of the wire is the next performance consideration. There are 7mm and 8mm wires. The number refers to the outside diameter of the wire. In truth, the wire itself stays the same and the amount of insulation gets larger. The 7mm wires are used in production. The 8mm wire should be used in all racing applications. This becomes important if a high-energy ignition system is used. At that point, the extra insulation is needed to keep the energy in the wire and deliver it to the plug.

CHRYSLER PARTS LIST

IGNITION PARTS

Electronic Distributor—Vac. advance	P3690430
Electronic Distributor—Mech. tach	P4120701
Electronic Ignition Kit (with vacuum advance dist.)	P3690426
Race Electronic Ignition Kit (with mech. tach dist.)	P4286510
Spark Plug Wires (Suppression)	P4529032
Spark Plug Wires (Metal Core)	P4829033 / P4120716
Ignition Wire Separator Kit	P4007667
Champion Spark Plugs	
N60Y	P4286562
N63Y	P4286563
N7YC (replaces N65Y)	P4286564
Chrome Distributor Clamp	P4349278

IGNITION COMPONENTS

Many people make the mistake of focusing on the components I've already covered: the distributor, coil, control box and sometimes the plugs. But there are many other smaller factors that should be considered. They all serve as links in a chain that comprise the ignition system.

Ballast Resistor—A ballast resistor is a basic part of all modern ignitions. They come in different sizes and designs. Basically, they are a resistance and are measured or graded in ohms. The .25 ohm resistor doesn't really look any different than a .50 ohm resistor or a .75 ohm resistor for that matter. The early production electronic ignitions used a dual-ballast resistor. This style of resistor is dictated by the control unit and the wiring. A dual-ballast looks like two regular ballast resistors side-by-side, except that it is a one-piece unit. There is a top and bottom to the dual-resistor which is usually denoted by a notch. Ballast resistors are not interchangeable. They

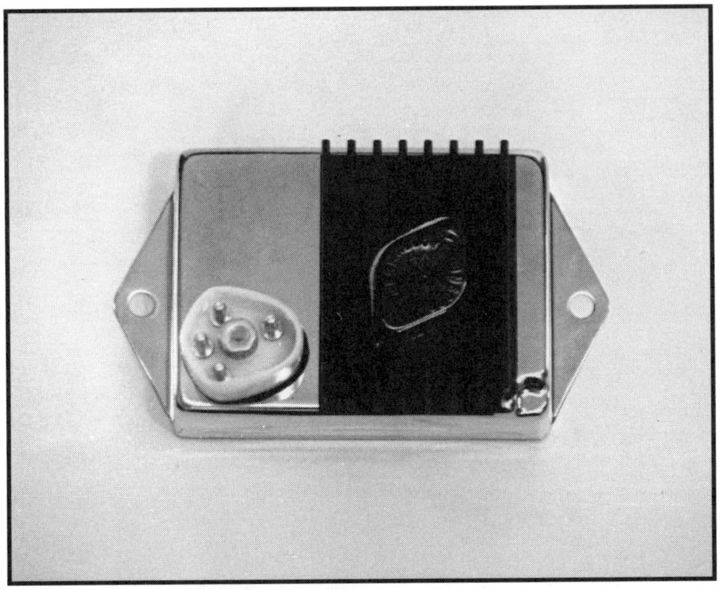

The Chrysler Gold Box. Note the special shape of the connector that allows the plug to go on one way only. Also, the race ignition boxes for the past 10 years use 4 pins only in the plug.

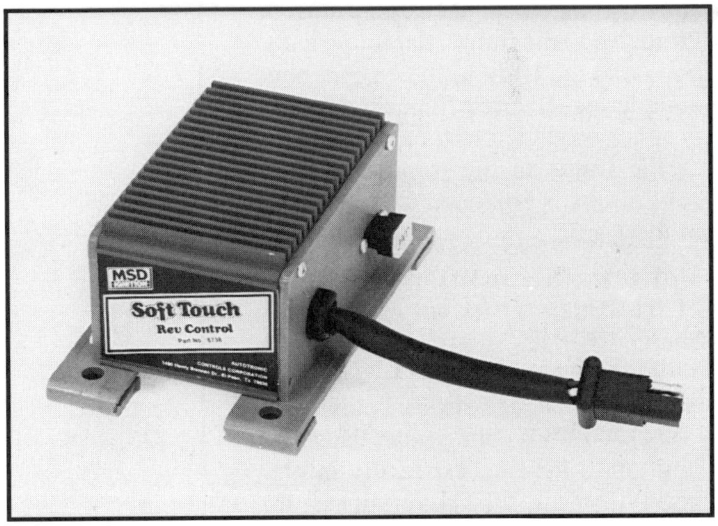

MSD's Rev Control unit doesn't just cut-off spark to avoid overreving. Instead, it uses computer circuitry to drop one cylinder at a time, then fire it on the next cycle. The result is smoother rev-limiting action that holds the engine at the pre-selected RPM limit without backfires, roughness or engine damage. (Photo courtesy MSD)

are dictated by the control box, wiring and by the coil itself.

Ignition Switch—In most performance vehicles there is no reason to replace the ignition switch. However, in some race cars with custom wiring, the production ignition switch is often replaced by a *toggle* switch. This isn't done because a toggle switch is better, but because it is more convenient. This change can cause problems. A high-amp rated switch is necessary, and most all the production switches are rated that way. But most toggle switches are not. The ones readily available are 14-16 amp rated. An ignition switch must be rated 18-20 amps. If the lower-rated toggle switch is used, it will cause more resistance in the ignition circuit, which will cut down on the voltage available to the ignition switch. The lower voltage available can cause the ignition to perform poorly and develop a miss.

Ground Straps—Production cars have several *ground straps*. On race cars they are often omitted, forgotten or removed because they are in the way. This can cause problems, because electronic ignitions must have the control box grounded. Painted surfaces act as poor grounds. Rubber also acts as insulation and will keep objects from grounding. For the ground to be effective, it must complete the "circle." By that I mean the engine must be grounded to the frame, the frame to the body and the body to the engine, thus completing the circle. Then the battery is either grounded to the engine block or the frame in the trunk area. The control box is grounded to the body or to the frame.

Battery—The most important aspect of the battery is its voltage. A fully charged battery should have close to 14 volts. There shouldn't be a problem unless the voltage drops below 12.0 volts. A lower voltage can cause the ignition to misfire, especially in high gear. It isn't just the battery itself. It's how much of the battery voltage is being delivered to the ignition system. It's not uncommon to have a 13-volt battery deliver 11.0 to 11.5 volts to the ignition system. You will have the same problem even though it isn't due directly to the battery. A .5 volt loss in wiring is acceptable but 1 to 2 volts is not. If you find a high voltage loss, you should find the cause and replace that part. Always measure both the battery's voltage and its output.

SPECIAL CONSIDERATIONS

There are always a few items that don't quite fit in with the general discussion. I will close-out the ignition discussion by covering the "specials."

Lean-Burn Conversion—The *lean-burn* ignition system is the official title for the Chrysler computer ignition that I have referred to throughout this Chapter. As mentioned, I don't recommend it for any performance applications. As such, the lean-burn must be removed and replaced with a standard electronic system, so

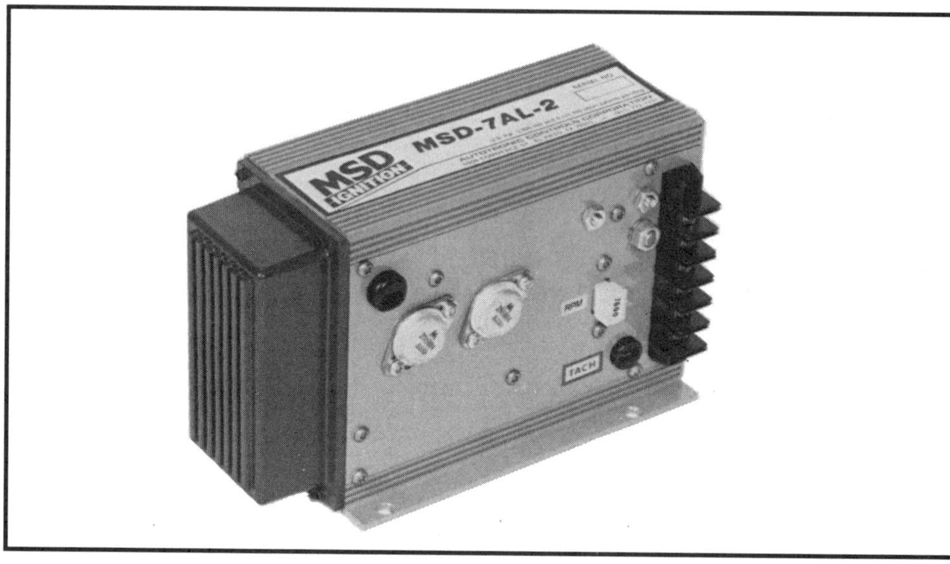

The 7AL-2 is the kingpin of MSD. Specifically designed for high rpm, high compression, all-out racing, it is popular in Pro Stock drag racing. The rev control is built in and can be adjusted by plugging in various RPM modules. (Photo courtesy MSD)

Accel's Laser II control box delivers ignition performance in high rpm ranges. (Photo courtesy Accel)

the cars and engines can be used in racing. This is not always covered in the instructions.

The lean-burn ignition system was introduced in the mid-70's on production engine vehicles. For off-road racing and high-performance applications, this ignition system is too complicated, has no adjustments and can't be utilized with high-output, race ignition systems which are based on an ECU system, as mentioned earlier. Mopar Performance carries a complete kit that is designed to convert point-type ignitions. The same kit can be used to convert lean-burn ignitions for racing except that the instructions do not refer to the lean-burn specifically. The problem is, that everything that is part of the original lean-burn has to go—distributor, computer, ballast resistor, etc. (except for the coil). The wires left at the ballast resistor location do not represent ignition #1 and ignition #2 (start and run positions on the ignition switch). This can be solved by finding ignition #1 and #2 wires from the ignition switch at the bulkhead connector (engine compartment side). A service manual for the car is helpful to locate wires and identify color codes.

Once the location of the ignition #1 and #2 wires are determined in the bulkhead connector, new wires (same size as original, but not included in the kit) can be run up to the new ballast resistor (supplied with the kit) and wired in by the instructions included. If you don't have the instructions, refer to earlier sections in this Chapter. The remaining "car" wiring (not related directly to the ignition system) is reinstalled in its original location as shown by the wiring diagram in the service manual.

Magnetos—Magnetos are commonly used on fuel engines or blown fuel engines such as the Chrysler Hemis used in Top Fuel and Funny Car racing. They have been much less popular or successful in gasoline-type racing. A magneto is a completely different ignition system from any of the above-mentioned systems. A properly set-up and tuned magneto will fire a very high output race engine. However, magnetos have more spark scatter than race electronic distributors and have a large amount of spark retard with speed. The spark retard can be as high as 1 deg. per thousand rpm. Another problem with magnetos is that they will "go away" in a short period of time and slow the car down, but they don't miss like other systems to give the racer a clue that his ignition may not be doing the job. This problem can be almost impossible to find. Therefore, I don't recommend them for gasoline engines but they are okay for engines burning nitromethane.

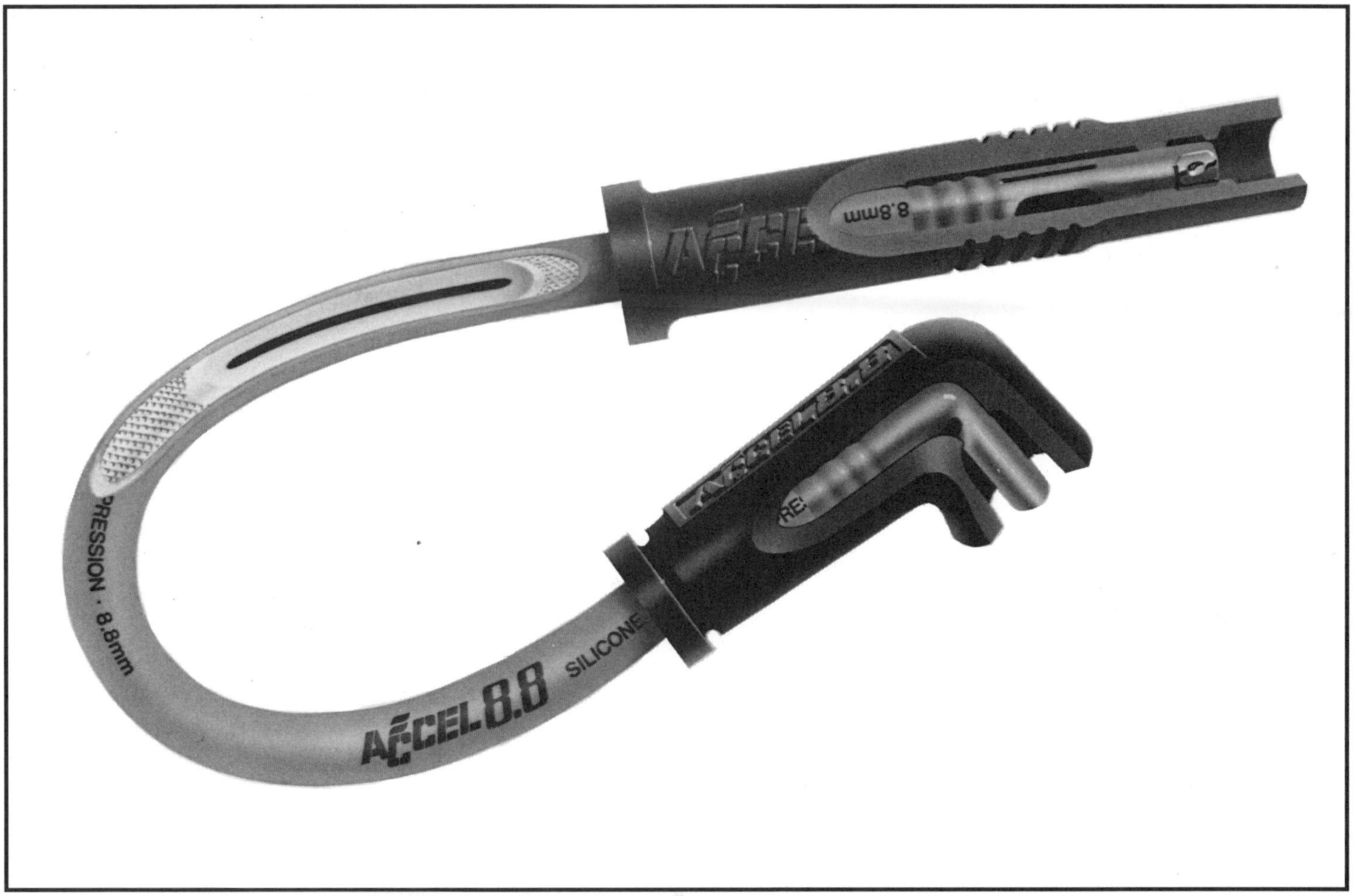

Accel's 8.8 wire is designed to ensure that spark energy required for drag, off-road, road and NASCAR racing actually reaches the plug. The thick silicone jacket can protect up to 500-deg. F. The 3-layer jacket surrounds the solid core of stainless steel and copper. (Photo courtesy Accel)

11 INDUCTION SYSTEMS

It may be a toss-up as to whether the camshaft or the induction system is changed more often. The camshaft is harder to change because it is located inside the short-block, but it is still a very popular modification. However, the induction system is the hands-down winner when it comes to "tinkering." And, while the heads may be where the most power can be gained, it's in the induction system where the most power can be lost! Many people make the mistakes of over-carbureting and tuning improperly.

Chances are the average racer doesn't "like" carburetors, and certainly doesn't like to take them apart. They don't mind bolting them on and off, but most people don't feel comfortable getting inside one. If a racer thinks he has a problem with one, he's likely to send it off to a carburetor tuner. These special tuners are very expensive, and may not solve the problem. It's hard to solve carburetor problems by remote control. It's much easier to be on-the-spot where you can listen to it and see it run and make changes immediately. The goal of this chapter is to keep you from making the big mistakes in hardware and to help keep you from sending everything to an expensive carburetor tuner.

The camshaft and the induction system are closely related. The camshaft can create problems for the carburetor, causing it to be readjusted. Solving these problems is not easy. The induction system can be a problem all by itself. Before attempting to solve potential problems, you should be familiar with all components of the induction system. There's the intake manifold, carburetor, fuel pumps, regulators and hood scoop. Most problems are related to the carburetor, which is often improperly adjusted.

Recommendations—Usually, I make recommendations at the end of the chapter, but this time I'm going to make them at the beginning. It is too confusing to keep speaking in the general terms of a carburetor. There are really only two—a Carter or a Holley. Since this is a hot rod book , I'm assuming that you are only interested in performance. Therefore, two-barrel carbs are out. For performance, the best choice for the average racer is a single four-barrel carburetor. Why? Because it's the easiest for the average guy to work on and it's the easiest to understand.

History—The performance A engine history goes back to the original 273 four-barrel package. It had a Carter AFB on it in stock trim. Since then, the A engines have used Carter AVS's; Carter TQ's (Thermo-Quad) and an occasional Holley. The early 273 engines used a single-plane intake manifold, but all the A engines went to two-level intakes in the later '60s for better torque and driveability. However, the single-plane manifold was reintroduced by the aftermarket as a performance manifold in the early '70s. The '68 340 engine sported the first Carter AVS four-barrel. The Holley six-barrel package was used on the 340 in 1970 on the special Trans-Am cars.

This is a typical single-plane 4-barrel race intake manifold. This particular one is a Holley for Carter TQ or other spread-bore carburetors.

It's a unique package, and fits in a class by itself. The Thermo-Quad was introduced on the 340 in 1971. The next year ('72) Carter went to a solid fuel design, which is the better choice for racing (the original was an air-bleed design). The 318 got its first production four-barrel in 1978.

Carter & Holley—As indicated by history, the majority of the production performance packages were based on one version or another of the Carter four-barrel. However, the most popular aftermarket carburetor is the Holley four-barrel. These two carbs don't adjust or modify the same way, nor do they use the same parts.

HOLLEY

The Holley four-barrel comes in two basic types—a vacuum secondary version and a double-pumper version. The double-pumper has mechanical secondaries and accelerator pumps in both the primary and secondary venturis. The vacuum secondary carbs have the accelerator pump only on the primaries and the secondary opening is controlled by vacuum and a spring located in the secondary diaphragm.

Vacuum Secondary—Vacuum secondary Holleys have the diaphragm spring, primary jets, power valve, secondary plate, primary accelerator pump size, pump nozzle size and pump cam—all can be adjusted or modified to gain more power. Ideally, you want to use the primary jets, the secondary plate and the power valve to get the best maximum power number. To get the car to launch harder, you want to use the accelerator pump size, nozzle opening, pump cam and diaphragm spring. If you have an idle problem, you have idle screw adjustments in the side of the primary metering plate. If this isn't enough adjustment, you may have to increase the primary jet size to get the idle where you want it.

Double-Pumper—The double-pumper has jets in the primary and secondary, power valves on the primary and secondary, accelerator pumps, nozzles and cams on both the primary and secondary sides of the carb. The jets and power valves are used to obtain maximum power. The accelerator pump system is used to improve the launch. There are idle adjustments on the side of both the primary and secondary metering blocks.

At this point if you want more information on the general design and adjustment of Holley carburetors, I suggest that you refer to the HPBooks' *Holley Carburetors* by Mike Urich & Bill Fisher, for more details.

CARTER

The Carter four-barrel carburetors have been popular on the A engine family since the early 60's. First there was the AFB, then in '68 on the 340's the AVS and then in '71 on the 340 the Thermo-Quad (TQ). The '71 TQ is an oddball to a certain extent. It's the only air-bleed version in the TQ family. From '72 on they have all been solid fuel versions because of EPA requirements. The parts and modifications are much easier on the '72 and newer versions. In stock form there's no real difference in performance. However in modified/racing form, it's hard to get the parts to obtain the proper air/fuel ratio on the air-bleed version. It's much easier to switch to the '72 and

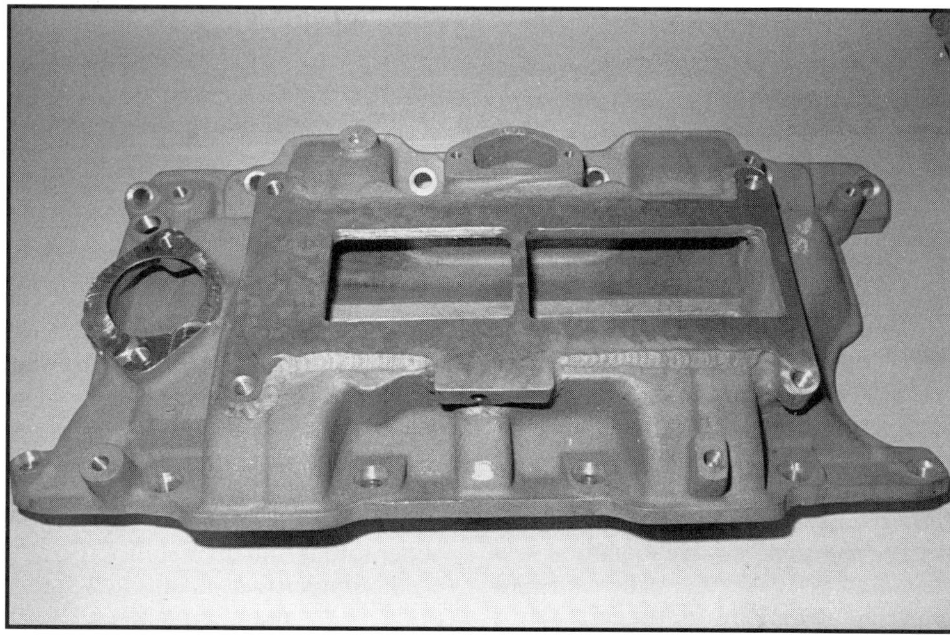

Dick Landy Industries offers this modified dual-quad. A cast insert is heli-arc welded in to accept an underhood B&M Supercharger. (Photo by Michael Lutfy)

This manifold is the production 340 Trans Am 6BBL intake. Front is to the right.

MODEL 4500

Holley Typical View 42–1

NOTE: General view is useful for visualizing relationship of various parts in carburetor. Specific details will vary with Part Numbers because each carburetor is made to fit a particular application. Current production 4500s don't have chokes and air horns don't protude as in this drawing.

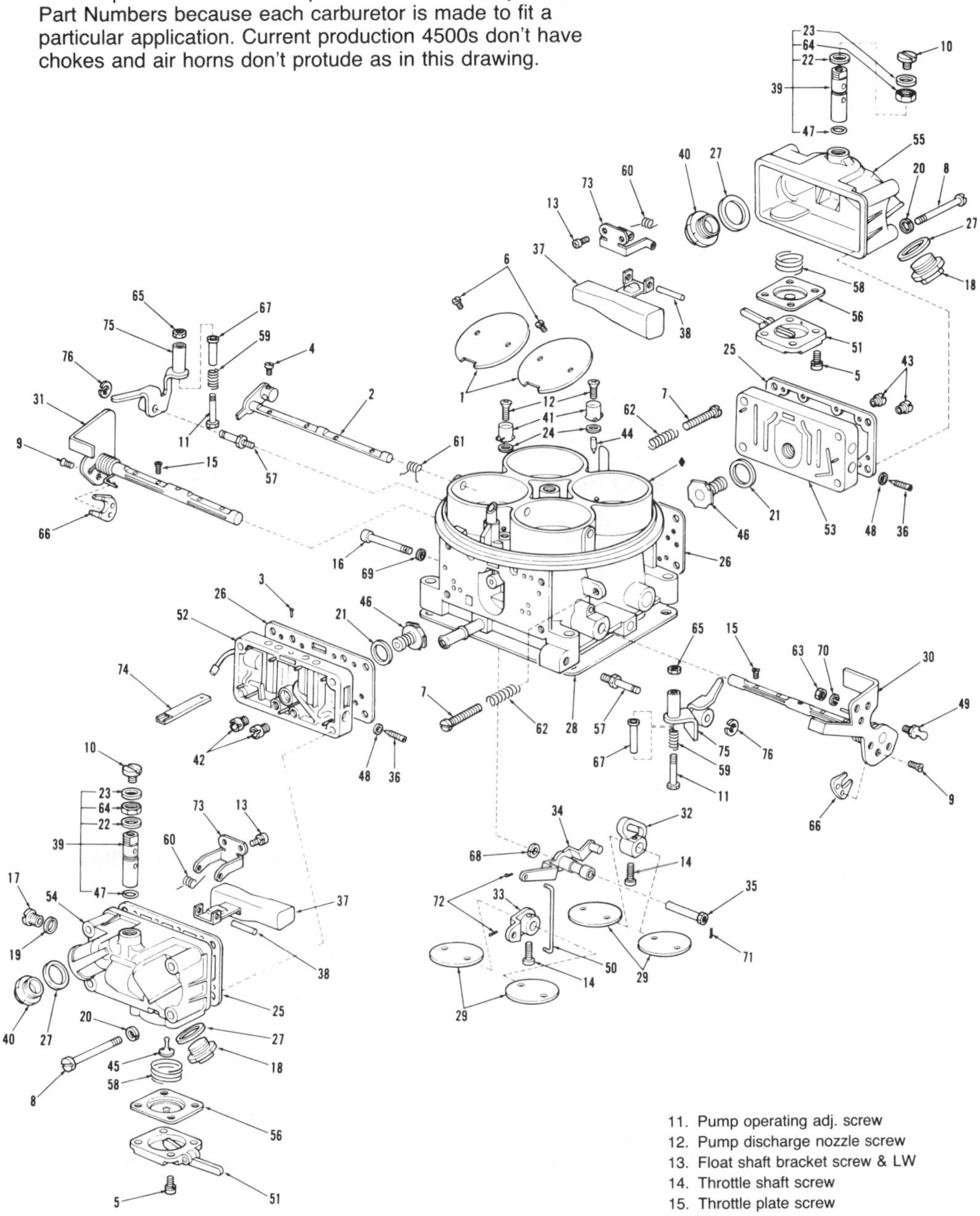

1. Choke plate
2. Choke shaft assy.
3. Fuel bowl vent baffle drive screw
4. Choke shaft swivel screw
5. Fuel pump cover screw & LW
6. Choke plate screw & LW
7. Throttle stop screw
8. Fuel bowl screw
9. Pump cam lock screw
10. Fuel valve seat lock screw
11. Pump operating adj. screw
12. Pump discharge nozzle screw
13. Float shaft bracket screw & LW
14. Throttle shaft screw
15. Throttle plate screw
16. Pivot screw
17. Fuel level check plug
18. Fuel inlet plug
19. Fuel level check plug gasket
20. Fuel bowl screw gasket
21. Power valve gasket
22. Fuel valve seat adj. nut gasket
23. Fuel valve seat lock screw gasket
24. Pump discharge nozzle gasket
25. Fuel bowl gasket
26. Metering body gasket—pri. & sec.
27. Fuel inlet fitting & plug gasket
28. Flange gasket
29. Throttle plate
30. Throttle shaft assy.—primary
31. Throttle shaft assy.—secondary
32. Primary throttle lever (internal)
33. Secondary throttle lever & bushing assy.
34. Intermediate throttle lever assy (comp.)
35. Threaded guide bushing
36. Idle adjusting needle
37. Float & hinge assy.
38. Float shaft
39. Fuel valve & seat assy.
40. Fuel inlet fitting
41. Pump discharge nozzle
42. Main jet—primary
43. Main jet—secondary
44. Pump discharge needle valve
45. Pump check valve
46. Power valve assy.
47. Fuel valve seat O-ring seal
48. Idle adjusting needle seal
49. Throttle lever ball
50. Connecting link
51. Fuel pump cover assy.
52. Metering body & plugs assy.—primary
53. Metering body & plugs assy.—secondary
54. Fuel bowl—primary
55. Fuel bowl—secondary
56. Pump diaphragm assy.
57. Pump lever stud
58. Diaphragm return spring
59. Pump operating adj. screw spring
60. Float spring
61. Choke spring
62. Throttle stop screw spring
63. Throttle lever ball nut
64. Fuel valve seat adj. nut
65. Pump operating adj. nut
66. Pump cam
67. Pump operating lever screw sleeve
68. Pivot screw LW
69. Pivot screw washer
70. Throttle lever ball LW
71. Pivot screw cotter pin
72. Connecting link cotter pin
73. Float shaft retaining bracket
74. Fuel bowl vent baffle
75. Pump operating lever & guide assy.
76. Pump operating lever retainer

newer carb and use the parts that are available for this carb.

AFB, AVS & TQ—The AFB was only used on the 273. It uses jets (primary and secondary), metering rods and accelerator pump. The AVS was introduced in 1968, and was the same as the AFB except that it had an adjustable air door over the secondaries. It had a locked screw (double screw if you prefer) on the air door shaft that turned a spring that adjusted the tension on the air door itself. This allowed the airflow through the carburetor to be adjusted to the engine's actual requirements at that time. It didn't affect wide-open throttle power. It helped get the car to launch better under different conditions. As such, it was much more helpful to cars equipped with automatic transmissions.

The Thermo-Quad was similar to the AVS in general hardware but had a thermo-plastic body and giant secondaries. Up to this point most four-barrel carburetors, both Carter and Holley, had secondary throttle bore diameters that were only slightly larger (about 1/4-in.) than the primaries. But on the Thermo-Quad the secondaries were almost *twice* as large in diameter as the primaries. The actual increase in the secondary bore diameter was over 1-in. larger than the primary. The TQ uses an air door similar to the AVS. When it opens it's controlled by the adjustable spring. Additionally, the amount of opening can be controlled. This can change the air/fuel ratio of the carburetor for maximum power. This is done externally to the carb and requires no disassembly of the carb itself. This can be a very helpful feature at the race track for performance tuning.

If you want more information on adjustment and modification of Carter carburetors, about the best place to look is the Chrysler book on *Engines,* which has an extensive section on Carter TQ modification and adjustment. It is available from Mopar Performance.

CARB SELECTION

There are so many different sizes of four-barrel carburetors that you need to be somewhat more specific than just "four-barrel." One carburetor is *not* best for all applications. To this end, I have come up with the following chart. It lists what size carburetor to use with which engine, for performance/racing applications.

Engine Size	Carb cfm
273	600
318	650
340	750
360	750

These numbers are for a typical package with an automatic transmission. If you have a four-speed or five-speed manual transmission, you could move up 50 to 100 cfm on each engine in drag racing. The 50 would be for the 273 and the 100 for the other three engines. Additionally, if you select a Holley four-barrel carb and you have an automatic transmission, you must use the vacuum secondary design. The double-pumpers are used on the manual transmission packages. The double-pumpers with their mechanical secondaries are hard to launch with the automatic transmissions. The vacuum secondary carbs will work on the manual transmissions, however. The double-pumpers are almost impossible to work out with an automatic because you can't delay the secondaries.

The spread-bore, TQ-style carburetors, are exceptions to this general "four-barrel" rule. The six-barrel package doesn't fit this rule either. Both will be covered in a little more detail shortly.

Slot and discharge nozzle for intermediate-idle system (arrows) (Photo courtesy Mike Urich).

THERMO-QUAD

The Carter Thermo-Quad, commonly called the "TQ," was introduced in 1971 on the 340 A engine. It's a unique four-barrel because it's much different than the previous four-barrel carbs such as the Carter AFB/AVS and vacuum-secondary/double-pumper Holleys. There are two other carburetors that are similar to the TQ: the Rochester QuadraJet and the Holley Spread-Bore. Since the TQ was installed on most Chrysler high-performance A engines, I'll concentrate on that specific carb.

Holes (arrows) feed fuel from bowl for intermediate-idle system. Whistle type vent is standard on 4500s (Photo courtesy Mike Urich).

Accelerator pump uses umbrella-type inlet valve to ensure quick pump shot when throttle is actuated (Photo courtesy Mike Urich).

TQ DESIGN

What makes the TQ unique from other four-barrels is the size of the secondaries. They are enormous at 2.25 in. On the other hand, the primaries are quite small at 1.38 to 1.50 in. To give you a comparison, a typical 750 cfm carb might have secondaries that are only 1.75 in. Although a TQ flows around 800 cfm, its biggest advantage isn't its ability to make power, but its versatility. Typically, an 800 cfm carb would drive around (driveability) on the primaries at slightly under 400 cfm. In the TQ, the primaries hardly flow 200 cfm. This means that you can drive around town with the fuel economy of a two-barrel, yet have the power of an 800 cfm hot rod on demand! This makes the Carter TQ an ideal carburetor for dual-purpose street machines. Furthermore, the small primaries allow the TQ to work on small displacements. However, if small displacements are going to be used, be sure that you use the model with 1.38 primaries. This allows one carburetor to work for almost all A engines. It also works well with either a manual or automatic transmission. The obvious drawback is they aren't made anymore.

Special Features—The TQ has primary and secondary jets, metering rods and an accelerator pump like other carburetors. However, the secondaries are so large that they don't fit on a standard intake manifold. Most of the newer manifolds have big secondary holes so that the TQ's can bolt on directly with no adapter.

If the manifold you have is for standard four-barrel carbs, then you will need a TQ adapter, which is readily available at your local speed shop.

The second feature of the TQ is the *air door*. The tension is adjustable by the double screw set-up on the left of

Three members of the Holley 4500 family: 0-7320, 0-8082 and 0-8896 (left to right). Airflow capacity is 1150, 1050 and 1050 cfm respectively. Two others 0-9377 and 0-9375 offer 1050 and 1150 cfm and feature annular-discharge boosters. There is also a 750 cfm version not shown (Photo courtesy Mike Urich).

Underside of 4500 shows rugged enclosed linkage connecting primary and secondary throttles. Accelerator pumps are actuated by levers operating on cams at back of throttle lever and on the secondary shaft (arrows). (Photo courtesy Mike Urich)

Velocity stacks add 3-in. rams to Model 4500. Center keeper holds stacks in place and locks securing capscrew (Photo courtesy Mike Urich).

the carb, which is the same as on the AVS model. The unique part is the opening adjustment. By increasing the opening, you can change the air/fuel ratio that the engine is operating at. In other words, you can richen up the carburetor without taking it apart. You can lean it out the same way.

Adjustment—To adjust the TQ, you need to know how to change the adjustment and how to measure it. The air door opening should be measured from the rear tip of the air door back to the flat of the thermoplastic housing with the air door fully open, i.e. as close to vertical as it goes until it hits the stop. Usually the numbers are in the 0.750 in. to 0.900 in. area. This method differs from that used in the service manual, which measures from the top front of the air door (at full open) to the center of the carb. I have found that the inside measurement is more accurate. Once you know how to measure using a snap-gauge and dial-vernier calipers, the actual adjustment is quite easy. To adjust the door, all you need is a pair of pliers. On the top front of the air door, in the open position, there is a slot in the air door that forms a tab. The tab strikes a stop located over the metering rod which limits the travel of the air door. By using the pliers and bending the tab, the opening can be changed. The metal is quite stiff, so the pliers are needed. Opening the air door wider (bigger number), leans the engine down; closing the air door down (smaller number), richens the engine.

INTAKE MANIFOLD

Proper selection of the intake manifold is not as complicated as the carburetor. Obviously, the type of carburetor has to be chosen before you can pick the manifold. For example, you can't pick a 340 6-barrel intake manifold and then put a 4-barrel on top. So at this point, you should have selected the carburetor. You should also know what type of racing you'll be running, and what type of cylinder heads you have. The W-2 heads require special intake manifolds.

SELECTION

For my example, I'm going to use a single four-barrel. For street/strip applications always use a dual-plane intake manifold. This could be either the stock cast-iron production manifold or an aluminum version. Edelbrock offers the Performer, and at one time manufactured the LD340—both aluminum dual-plane.

In a race car, you have to look at the rest of the engine to a certain extent. If you have a hydraulic cam, stock torque converter and/or mild axle ratio (2.76 to 3.55), I would suggest sticking with the dual-plane manifolds. With me-

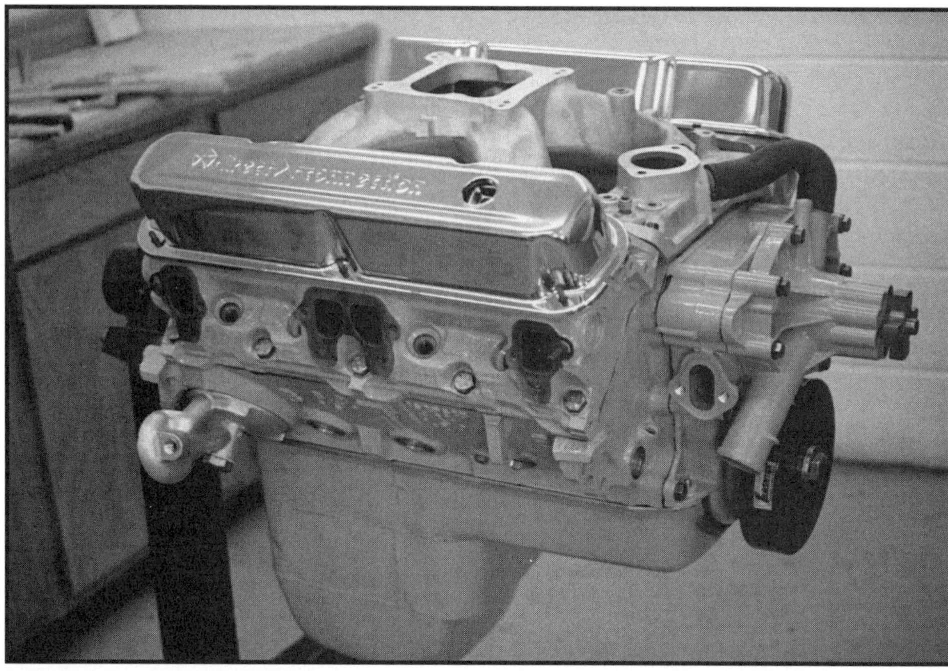

Holley single-plane installed on the engine. Note that the A engine manifolds have a water passage at the front.

An experimental 6BBL package adapted to fit the top of a tunnel ram.

chanical cams, race torque converter, and/or 3.91 and lower axle ratios, the single-plane is going to be the best. It will make the most horsepower. However it can't be just any single-plane. Most are not up to the power level of the stock cast-iron, dual-plane manifold, obviously, a 4BBL version for the 340/360's built from '68-'89.

Single-Plane—The two horsepower choices for single-planes are the Holley Strip Dominator and the Offenhauser Port-O-Sonic. The Port-O-Sonic needs some distribution help before it can equal the Holley, but once done it's a real good intake. If the four-barrel carb is a TQ or Holley Spread-Bore, then you'll need a TQ-style adapter to allow you to bolt these giant secondary carbs onto these manifolds. The race manifold would still be a Holley or a Port-O-Sonic. For the street the '71-'86 cast-iron TQ production manifolds are best, but the

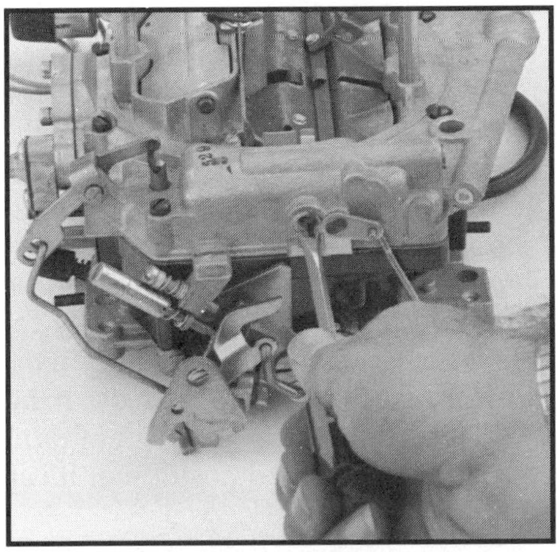

There are many adjustments on the Thermo-Quad carburetor. Here, I am adjusting the spring tension on the air door. Less tension, the faster it opens. Usually 1-1/2 to 2 turns is a good starting place in a race car.

Be sure that you get the most pump shot available by having the linkage in the closest hole to the pivot.

One of the hardest things to understand about setting up a race Thermo-Quad is to wire the secondary cancellation linkage shut (down) so the secondaries always open with the throttle. It is not wired yet is this photo. The pencil is holding the linkage down against the secondary throttle shaft which is the position it should be wired in.

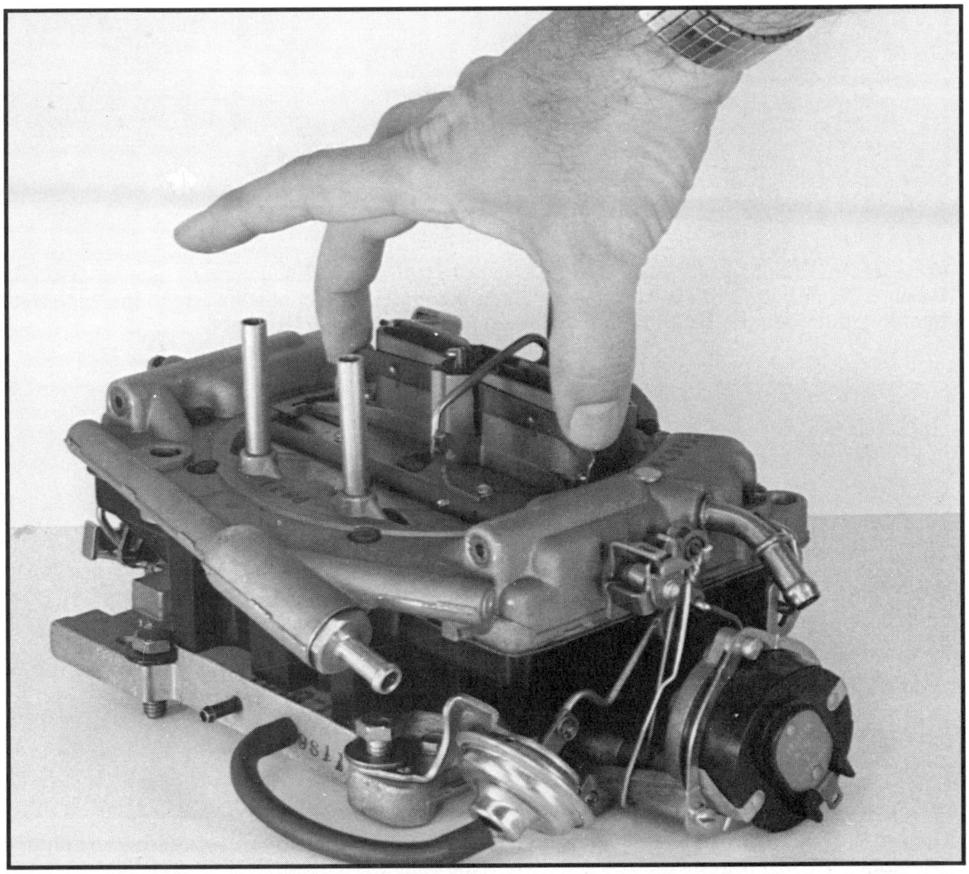

The metering rods on the TQ are on the end of the center-pivoted bar. It must be free to move up and down easily.

I recommend that the air door setting measurement be taken from the tip of the air door bump in the open position to the rear of the carburetor which is a flat section of the black thermoplastic. This setting usually runs in the 0.750 in. to 0.900 in. area.

The air door opening is adjusted by bending the tab on the left side of the air door. It is bent using a pair of pliers. Note that the fixed stop is attached to the carb housing and is not to be bent.

LD340/Performer can be used with a TQ adapter if lighter weight is desired.

Special Manifolds—That pretty much covers the four-barrel manifolds. There is only one A engine six-barrel manifold and that's the 1970 aluminum production unit. It was made for Chrysler by Edelbrock. For eight-barrel manifolds, the obvious choice is the *tunnel ram*. There are no longer any "low-rise" eight-barrels available. This is a race-only package. It will make more horsepower than any of the other induction systems. It generally works better with a manual transmission than an automatic because it's harder to launch with the automatic. With the big "cc" accelerator-pumps and with some tuning, the launch problems can be cured for the most part. With a 340 or 360 engine and an automatic, limit a tunnel ram package to twin 650 cfm carbs, maximum!

Spacer—The final choice in manifolds is the spacer that goes between the manifold itself and the carburetor. These are only used in single four-barrel applications. They should only be used on race cars with single-plane intake manifolds. You can install them on dual-plane intakes, but you are better off to switch manifolds. Spacers should be looked at as fine-tuning tools. Evaluate these spacers only on race cars that are consistent and well-developed. As a general rule the four-hole spacer helps torque while the one-hole spacer helps top end. You may have to rejet the carburetor before any gain from the spacer is found, whether it is a one-hole or a four-hole.

SIX-BARREL

The 340 six-barrel package was introduced in 1970 on Trans-Am cars. It was an outgrowth of the very popular 440-6 that was introduced in 1969. The 340 carburetors are basically the same as the 440's except that one of the end carburetors in the 340 package has a fuel distribution tab on the nozzle. This package was rated at 290 hp by the factory—15 more than the four-barrel. It was the highest rated output on any A engine in production. There is only one intake manifold and one basic set of carburetors. Edelbrock still has the intake manifold available new and Chrysler and Holley still have brand new carbs.

The six-barrel package is similar in many performance characteristics to the TQ. It drives around on the small center carb and races on the two big end carbs. This allows the engine to deliver the fuel economy and driveability of a two-barrel and yet perform at maximum power better than the biggest four barrels (production-style). In more technical terms, the center carb flows 250 cfm while the whole package flows approximately 1,000 cfm. The advantages are obvious. The 1,000 cfm allows for more power than the big four-barrels. The 250 cfm allows for the economy, driveability and versatility. It's almost the ideal package for the dual-purpose performance machine and a very high output race package at the same time.

Features—There are several unique features of the 340 six-pack. There have been other six-barrel packages produced at Chevy, Ford and Pontiac, but unlike the others, Chrysler carburetors have an adjustable idle circuit in the end carbs and they open by vacuum. The idle circuit is adjusted by two screws behind lead plugs located in the base-plate of the carb. Once the lead plugs are popped out, the small adjustment screws are easily turned, although they should be turned very gently. The rear carb as installed on the manifold, can't be adjusted easily for idle because the center carb gets in the way. It can be adjusted off the manifold by estimating the number of turns from the fully seated position or it can

Be sure the upper tip of the TQ air door blade doesn't hit the choke linkage in the wide open position. If it does hit, bend the choke linkage back enough so it clears.

The 340 6BBL linkage can be difficult to adjust. The front and rear rods are threaded and should be adjusted to hold the end carbs closed when the center is closed. To tight an adjustment and the center carb won't be able to go to idle. Too loose an adjustment and the ends won't close all the way.

be adjusted in the front position (front carb or rear) and then swapped with front carb back to its normal (rear) position. Having an idle circuit in the end carbs allows for many advantages in racing, since everything doesn't have to be done on the center carb.

Having the end carbs open by vacuum can be a plus and a minus at the same time. In reality, it is a plus because it performs like a vacuum-secondary four barrel, i.e. it works well with automatic transmissions and launches hard at the strip while it maintains good driveability. These are some pretty big "pluses." As with any vacuum-secondary system, if you want to change *when* the end carbs open, you change the spring inside the diaphragm. There are two springs to change in the six-barrel. The problem with all this is, most people don't understand a vacuum-secondary system, which is compounded by the additional carburetors.

Tuning—The root of the six-barrel carburetor problem for most people is complexity. The six-barrel system consists of three individual carbs. To most racers, this is three times as complex and confusing as a four-barrel setup because the four-barrel is only one carb. Most racers that can adjust a single four-barrel carb themselves can handle the vacuum-secondary situation on their four-barrel. But three carbs are too complex to figure out, right? The trick to handling this problem successfully is to look at the six-barrel package as a single six-barrel carburetor, not as three two-barrel carbs. This means that you have one carburetor with four secondaries instead of two but only two primaries, just like a four-barrel. Whatever you do to the front secondaries, you do to the rear secondaries. Now it is a lot easier to master the carb package if you need to make changes. Be careful of the throttle linkage and adjust it by the Chrysler service manual or Mopar Performance engine book. It's easy to make an error in linkage adjustment that can cause poor performance that is not really the carburetor's fault.

RACING MANIFOLDS

Race intake manifolds come down to the single-plane four-barrel and the eight-barrel tunnel ram. The characteristics of the manifold you would like to change are the *plenum volume*, the *runner length* and the runner volume. These characteristics change as a function of engine displacement, rod ratio, rpm range and camshaft specifications. Small changes in plenum size and runner volume can be accomplished by porting, but these changes result in very small gains, if any. Small changes in runner length can be accomplished by milling the manifold face. Again, these small changes, even though in the right direction, will be impossible to find. Also, the manifold wouldn't seal well to the cylinder head because this "easy" approach has made the manifold flange too thin. It will flex too much or may even crack. That is a bad trade-off. Also, keep in mind that as you shorten the runner you decrease its volume. The short runner tunes to a higher rpm. The smaller runner volume tunes to a lower rpm. However, you can't really change these aspects anyway—at least not easily.

Plenums—Plenums are easier to modify since they are on top of the manifold. Their volumes on four-barrel intakes can be increased by adding one-hole carburetor spacers. In tunnel rams, there are also big and small plenums, generally available from the manufacturer. Plenum volumes can be decreased by adding plenum stuffers inside the existing plenum or by sectioning a tunnel ram plenum. In two-piece manifolds, spacers

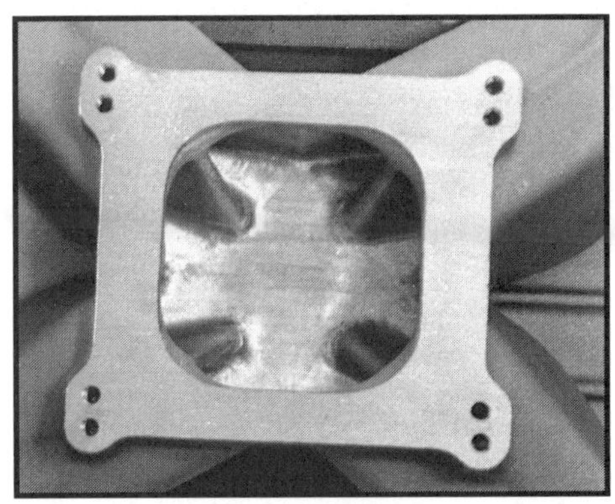

Plenum and runner entrances are polished and radiused for maximum flow.

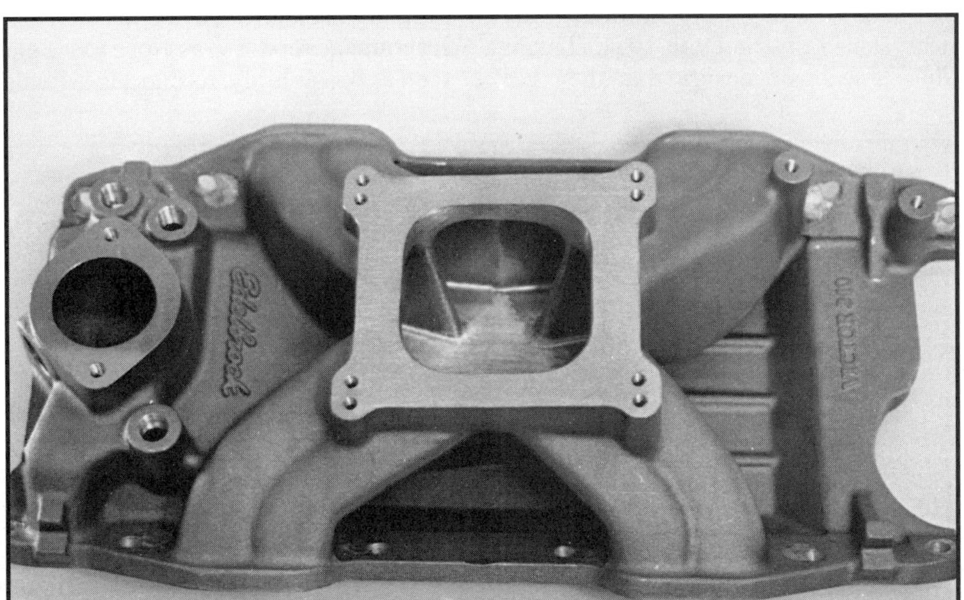

This is an Edelbrock Victor 340, single-plane race manifold.

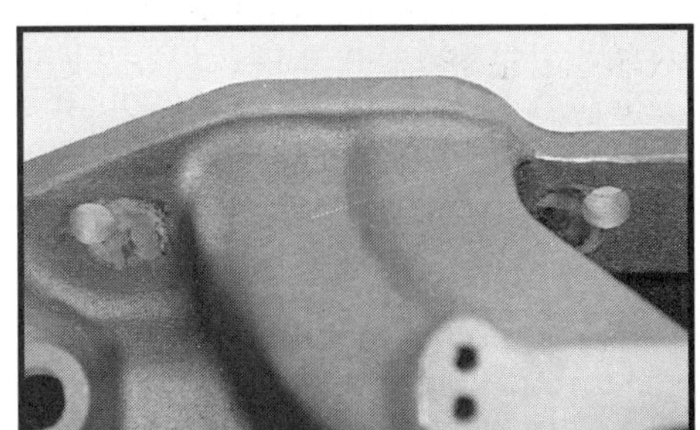

Intake manifold holes are welded up (std. position) and redrilled (W-2, W-5 position). This is a common method to convert standard manifolds to use with W-2 and W-5 heads.

can be made to go between the top and the bottom to increase the plenum size. With a given carburetor, the larger plenums tend to make more power, but may not launch as hard. In drag strip numbers the speed would improve but the E.T. gets slower or doesn't improve as much. The large plenum requires more accelerator pump shooters than the smaller plenum size. To fully evaluate a plenum size change, expect to change the jets to get back to the proper air/fuel ratio.

Port-Matching—With any race intake manifold, you want to match the ports of the intake manifold to the ports of the head. There are times when this isn't reasonable. For example, using the 360 intake on the 318 heads. The 360 intake ports are larger than the ports in the 318 head (stock, small-port version). In this case the manifold is better in performance and the effect of the "mismatched" ports is not large enough to be considered. Port-matching is particularly important on the top of the runner, or roof, and on either side of the center divider.

On the tunnel ram, each individual runner entry can be radiused and smoothed out. This will help flow. Caution must be used on the four-barrel intake because radiusing the runners will change the fuel distribution and flow characteristics of the individual runners. So unless you can get on a dyno that can do fuel distribution work, it is best to leave the four-barrel runner entrances alone.

Alignment—The final item to check on the intake manifold before installation is the alignment of the carburetor on the manifold itself. This alignment is always assumed to be correct if the carb bolts on; therefore it is rarely checked. On a cast-iron intake you may be able to get away with this assumption. However, any aluminum intake should be checked for throttle-bore alignment very carefully. The easiest way to do this is to bolt the carb(s) onto the manifold. By holding the throttles open by hand and looking down the throttle bores themselves, you can check the alignment. If you can see an edge of the manifold's bore, then it doesn't align properly for racing. Sometimes, loosening the bolts and pushing the carb in the proper direction will shift it into alignment. Then you have to remember this position when the final installation of the carburetor is done. Reboring the hole in the manifold is the best solution to misalignment unless it's way off. A slight misalignment can be ground-out by hand before the manifold is installed.

HOOD SCOOPS

Chrysler may not have invented the hood scoop but they certainly did develop it to the level that it is today. Virtually all the Chrysler high-performance cars from 1963 on had hood scoops, including the 1970 340 Challenger T/A and AAR Cuda. Hood scoops really are not part of the engine package but they do affect the carburetor and the overall output of the engine. A typical bracket car could pick up as much as a tenth-of-a-second and 1.5 mph in the quarter-mile by adding a hood scoop. This translates to 10 to 15 hp. However, the cold air and extra entrance velocity can get the carburetor confused. The general rule of thumb is that the carb (single 4-barrel) needs to be richened up 3 to 4 steps (jet sizes) for use with fresh air.

The biggest problem relating to hood scoops usually occurs with the big, tunnel-ram type scoops. These are called Bauman boundary-layer bleed-off scoops. They were introduced by Chrysler in Pro Stock in 1971. This is the same basic design that they use today although the current version is more rounded than the original. These scoops are fine for the tunnel ram crowd but can create real problems to the single four-barrel racers. Basically, the carburetor doesn't want to be at the bottom of a well. The entrance to the carb should be depressed approximately .50 in. to .75 in. below the bottom floor of the snorkel of the scoop. The four-barrel manifold is so short that these Pro Stock-style scoops have to be brought down much closer to the hood surface. If this can't be done, then a different-style scoop should be used. The same .50-.75 depression distance has to be applied to the two carbs in the tunnel ram but the tunnel ram manifold itself is so tall that it is a lot easier to do.

FUEL DELIVERY

For the carburetor to function properly in a race car, it must have a full load of fuel all the time. This requires a good fuel delivery system and on-track performance is something that cannot be checked on a dyno. A fuel delivery system consists of everything between the carburetor itself and the gas tank. Like the hood scoop, it is really part of the car, not the engine. But if it malfunctions, then the carburetor or ignition system are usually blamed first.

Fuel Pump—The first step is to add an electric fuel pump. For a one-pump system, an electric Carter 4594 is best. It should be mounted as close to the gas tank as possible and its entrance should

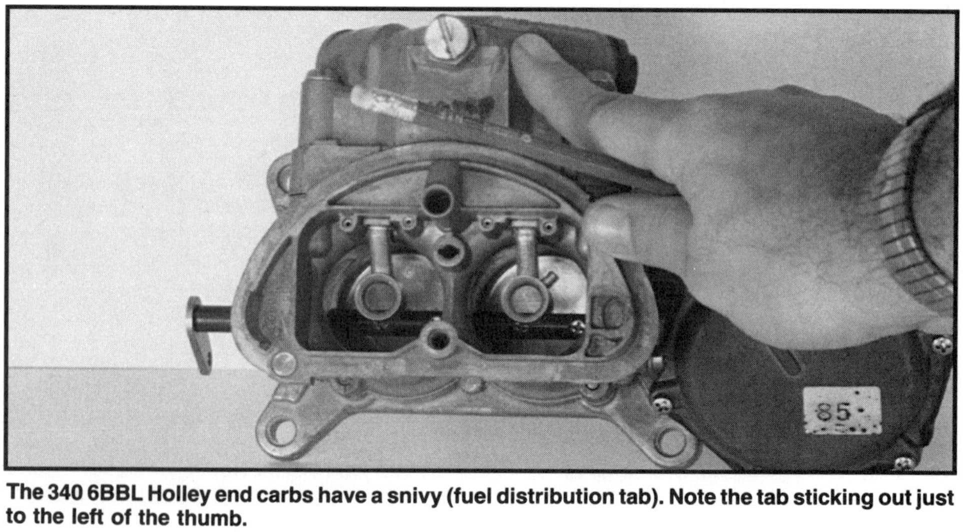

The 340 6BBL Holley end carbs have a snivy (fuel distribution tab). Note the tab sticking out just to the left of the thumb.

not be higher than the outlet from the tank. It also should be kept away from the exhaust system, protected and shielded by the frame rail as much as possible. On this system, the stock mechanical pump and the stock fuel line are used. On any serious race car you should have a fuel pressure gauge. It should read the pressure as close to the carburetor as possible.

The next consideration or step-up in the fuel delivery system is to increase the capacity of the mechanical pump. Carter has a high-volume mechanical pump for the A engine. They all use the same pump and they all fit in the same place. At this point the package should be able to run into the mid-11's.

Flow Rate—From here the fuel delivery gets a little more complicated. Keep in mind that the fuel delivery system doesn't make you go fast by itself, but it sure can keep you *from* going fast. There's a correlation between the car's performance (speed and E.T.) and the engine's horsepower. There is a similar correlation between the engine's horsepower and the induction system's cfm or airflow. The more airflow (cfm) then the more fuel that's required to maintain the proper air/fuel ratio. What all this comes down to is, that as the car gets faster, you have to increase the fuel delivery rate, i.e. pump more gas or maintain a higher pressure. Increasing this *flow rate* is done by adding another electric fuel pump. For most applications, two Carters in parallel are your best choice. Parallel means two "ins" and two "outs" joined by a tee on both ends. You still need to maintain the mechanical pump.

Pro System—With a high cfm induction system like a 4500 Holley, or two Holleys on a tunnel ram manifold, you need to step-up the fuel delivery system again. This "Pro" system should use two Holley electric fuel pumps in parallel with a 1/2-in. fuel line. In this system the mechanical pump is removed. You should have a regulator for each carburetor. The regulators should be adjusted to give 7-8 psi at the carburetors. For the most part, a 6BBL can use one regulator—3 aren't needed.

Fuel Check—The fuel check on the fuel system is an in-the-shop performance test. The point of the test is to see if one of the pumps has failed or fallen off in its ability to deliver fuel. To perform this test, you disconnect the fuel line at a convenient place in the engine compartment. Holding the open fuel line into a measured container, turn on the electric fuel pump(s) and time how long it takes to pump a given quantity of fuel. As a rough guide, two electric pumps should pump one quart in 10-15 seconds. If it takes over 20 seconds, then disconnect one pump (pull wire) and test again. Then switch pumps by pulling first wire and connecting up the previously pulled second wire and test again. Replace the bad pump if found. Also replace fuel filters frequently.

CHRYSLER PARTS LIST

INTAKE MANIFOLD	
Alum. Dual-Plane 1-4BBL	P4286531
Alum. Single-Plane 1-4BBL	P4007664
1-4BBL W-2 Single-Plane (water headed)	P4120160
1-4BBL W-2 Single Plane (no heat—Race Only)	P4007659
2-4BBL Tunnel Ram—W-2 Heads	P4120580
Intake Gasket & Head Block-Off Package	P4286569
INTAKE MANIFOLD GASKET SETS	
Std. Heads 0.060 in.	P4007570
Race Intake Manifold Gasket (Std. heads)	P4452101
W-2 Heads 0.060 in.	P4007571
W-2 Heads 0.030 in.	P4120210
W-2 Heads 0.045 in.	P4120211

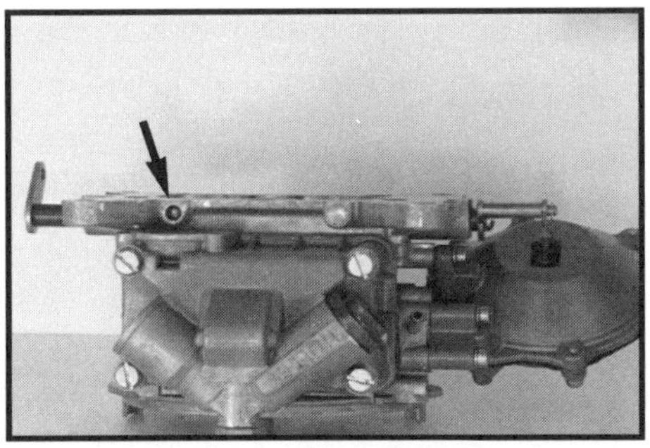

One of the things that make the Chrysler 6BBL package unique is the idle circuit in the end carbs. The idle adjustment is in the baseplate. The photo shows the idle screw exposed to the right of the center of the carb. To the left the lead plug is still in place covering the adjusting screw.

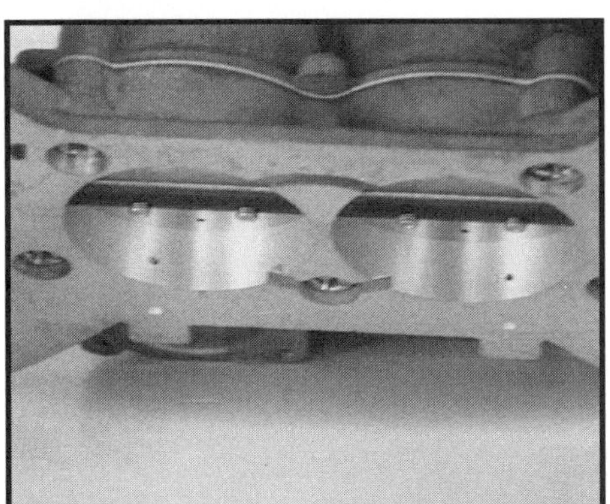

The idle adjustment screw in the end carbs comes out as a needle into each throttle bore just below the throttle plates. A lot of needle showing is a lean setting; no needle showing is a rich setting.

A properly installed hood scoop can add as much as a 1/10 of a second and 1.5 mph to a bracket car, as long as the carb has been adjusted to compensate for the cold air and extra entrance velocity. This is Paul Rossi's 360-powered Dodge Rampage. (Photo by Jim Kelso)

12 EXHAUST SYSTEMS

The typical stock exhaust system on all A engines consists of two cast-iron manifolds, exhaust pipe, muffler and tailpipe. In most cases, these engines used a single exhaust. Only the high-performance models and some police engines used dual-exhaust. Of all the A engines, only the 1966 273 Dodge "D" Dart came from the factory with tubed headers. In general, the cast-iron exhaust manifolds installed on high-performance 340 and 360 engines from 1968 through 1974 are the best selections for performance.

These cast-iron manifolds were well-designed by Chrysler for performance and maximum output. Only the 426 Hemi and several early 60's RB exhaust manifolds could flow as freely and efficiently.

The exhaust system consists of the parts that bolt to the head and conduct the exhaust gases to the rear of the car. The exhaust port in the cylinder head is *not* part of the exhaust system. The special raised W-2 port considerations will be discussed shortly.

The biggest advantage cast-iron manifolds have over headers is that they are quieter. They are also more durable in most applications. Indirectly, they offer better driveability, especially in colder climates. This is caused by the heat cross-over built into the production intake manifold. The heat-control valve is located in the right side exhaust manifold. For racing purposes, the heat cross-over is blocked at the intake manifold. This allows the intake to run cooler during competition. If the cross-over is blocked, be sure to remove the heat-control valve or wire it open.

In the late '70s, most manufacturers introduced an air pump as part of the emissions package. Obviously they are not used in racing. However, many of these late 70's heads and almost all A engine heads built after 1980 have a small hole in the bottom of the exhaust port that is used with air pumps. It is best to plug these holes (one per port) before using them in racing applications. Some exhaust gaskets cover these holes but this is too risky. Be sure these plugs are flush with the flange once installed.

Production engines use heat shields around the spark plugs. Be sure to use them in all applications if possible, but especially with the cast-iron manifolds, which tend to run much hotter.

TUBING HEADERS

Steel-tubing headers generally make more power than cast-iron production units. They also deliver a lot of performance for little money. Headers are probably second (next to a good ignition system) in terms of cost-effective horsepower.

There are two basic categories of headers—street and race. Race obviously goes all-out for maximum power or performance. Street headers are usually smaller in diameter, cost less and are easier to install.

STREET HEADERS

Street headers usually run through mufflers rather than an open collector used on racing headers. Generally, street headers feature the smallest primary tube design. The primary length is also longer. The small primary diameter is good for torque, which is advantageous for the street. The smaller size also makes it easier to install in the engine compartment. This style of header is not usually equal in length from one primary to another. The engine will average the various lengths and perform to this average.

"H" Pipe—Street headers should be used with an "H" pipe. This pipe reduces exhaust noise. The "H" pipe

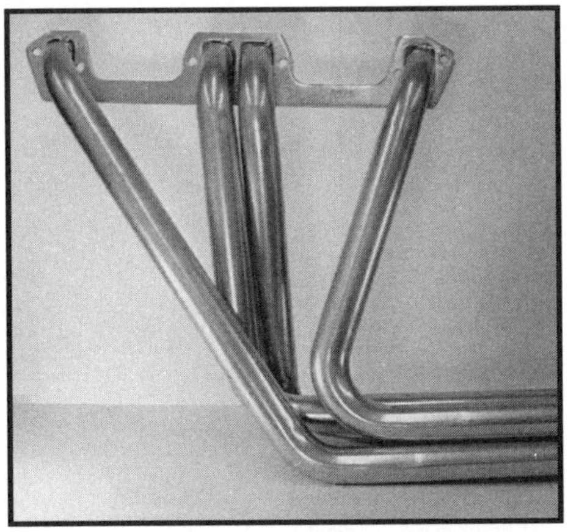

Smaller tube headers take up less space and make a much neater package. These headers are specially coated which adds a bit of shine.

serves as a connector between the two sides. It should be 2 in. in diameter or slightly smaller than the main exhaust pipe size. It should be located 12 to 18 in. behind (rearward) the beginning of the header collector. These "H" pipes were introduced in the early '60's on the 426 Max Wedge Mopars. They were used on early 340's, but most A engine packages used single exhaust or a standard dual system. Because of the driveshaft and transmission, the best location for the "H" pipe is rearward of the transmission mount but not farther back than the front U-joint. Compare this to the 12 to 18-in. dimension listed earlier on your specific headers for the best overall location.

The specific dimensions of street headers are much less critical than race headers because of the mufflers used for the street. The collectors should also be kept small. This means 2-1/2 to 3 in. in diameter maximum! Longer collectors, smaller primary diameters and longer primary length all help street headers perform better.

RACE HEADERS

For some time now, race headers have followed a "4-into-1" design, and most are adjustable. This means that the primary length can be changed by inserting four short tubes between the end of the main primaries and the beginning of the transition. The extensions are usually 3 and 6 in. lengths. The adjustment allows the racer to fine-tune the headers for specific racing conditions.

The real secret of the racing exhaust system is that it is part of a team and all the parts of the team must work together. For example, putting Pro Stock headers on a 318 stocker will result in a slower car. Why? Because not all the parts are working together. Pro Stock headers are designed for maximum rpm (9500 and more) while the stocker is only going to go 5500.

Selection—Many factors need to be considered before choosing the best header to suit a particular race application. These factors include overall car weight, transmission type, axle ratio, intake manifold design, camshaft and cylinder head configuration. This is obviously very complicated. The table *Performance Packages* on page 151 lists some very good information. These specific recommendations for headers hide a lot of data in the general description. For example, a 340/360 Super Stock package is known to have a roller cam. It must use stock cylinder heads, but with a Holley or Offy single-plane aluminum intake manifold. The carb will be a Carter TQ. The car weight will be within 200 pounds of 3300 pounds. The tire size and axle ratio are also known.

Size—The size of the race header dictates the trade-off between power and torque. The main dimension relating to size is the diameter of the primary tube. The smaller the diameter, the more torque; the larger diameter yields more power but less torque. In a drag racing vehicle, more torque usually gives lower E.T.s while more power gives higher MPHs, but higher E.T.s as well. The recommendations shown in the *Performance Packages* table on page 151 are the best compromise for that specific application.

Collector—The collector on the racing header can also be adjusted to change the engine performance characteristics. Since it is relatively easy to do, most racers adjust them while at the track. It is considered a fine-tuning technique. It gives the last few hundredths in E.T., which may result in the win. The starting line capabilities (co-efficient of friction) are not the same at all drag strips. The collector can be used to adjust the car to the track

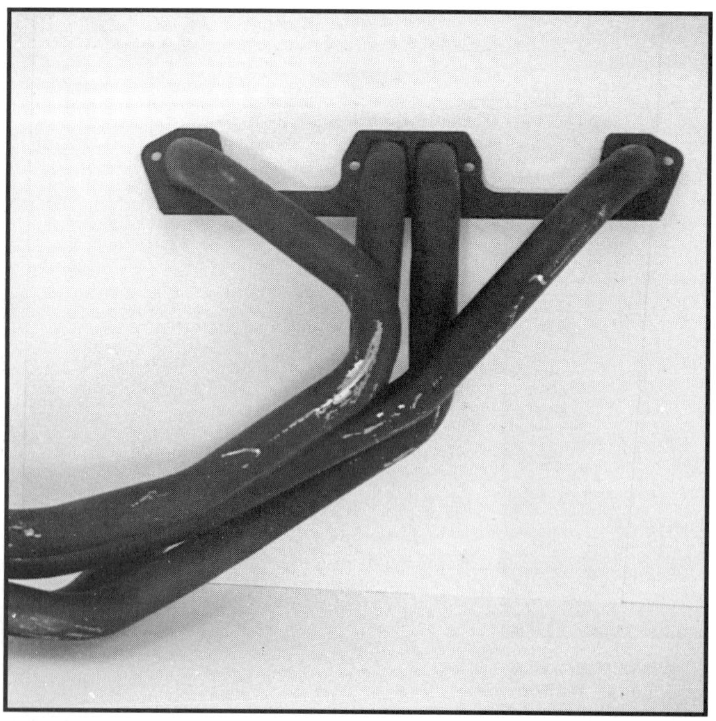

Right side headers are much easier to make into a neat package even though the tubes may get larger.

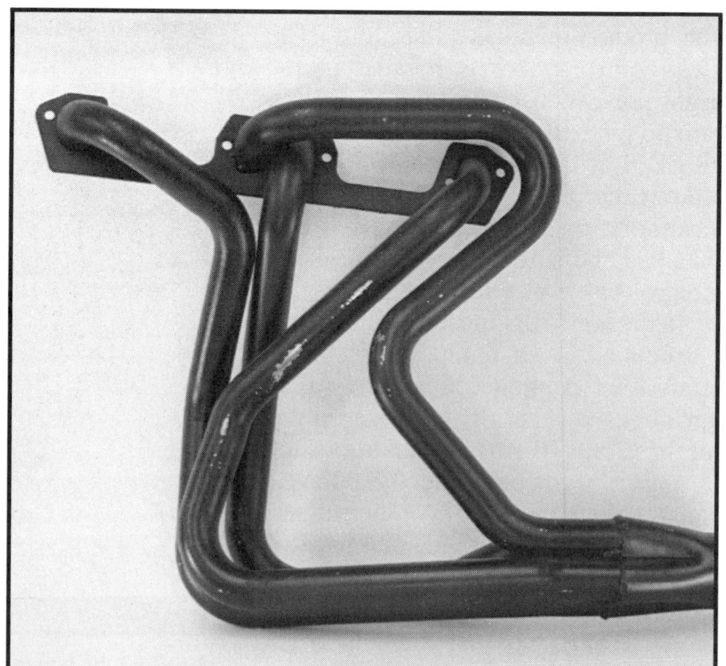

The left side of the bigger tube headers that match the right side in the previous photo shows the added complexity of getting all the tubes into the chassis.

for improved times. Collectors adjust similar to the primaries—longer makes more torque, shorter makes more power. But don't forget that it's torque that usually produces a lower elapsed time.

Types—All race headers fall into two basic categories—*under chassis* and *fenderwell*. Under chassis headers have all four tubes under the chassis. Fenderwell headers run one tube on each side (sometimes two or more) through the fenderwell. This is done strictly for room. Small engine compartments do not have enough room for four large-diameter primary tubes.

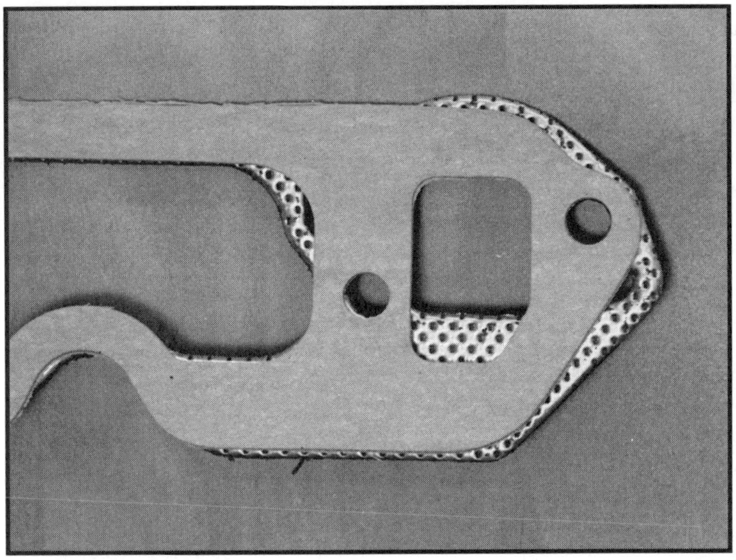

The production exhaust gasket is on top. The shiny one on the bottom is the W-2 gasket. Note how much the bottom of the W-2 exhaust port has been raised.

The header collector is on the right. The transition is on the left. The bottom 3-4 in. is actually part of the collector length (tuning-wise).

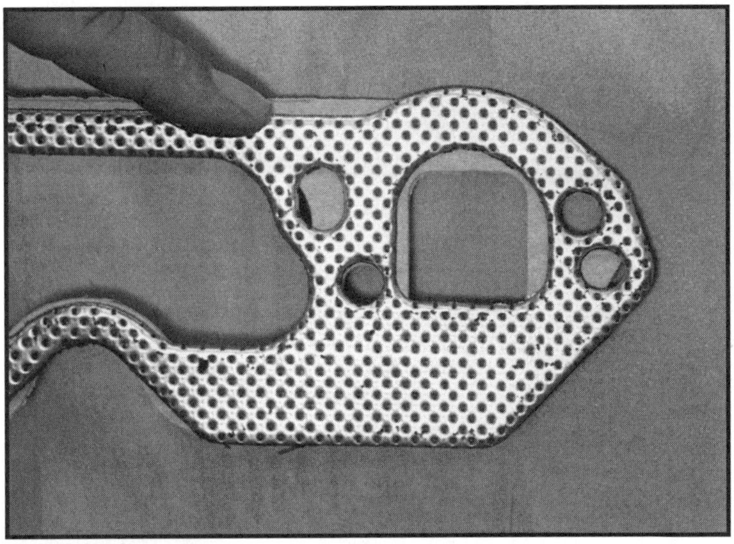

With the W-2 exhaust gasket on top, note how much taller and wider the W-2 port is in comparison.

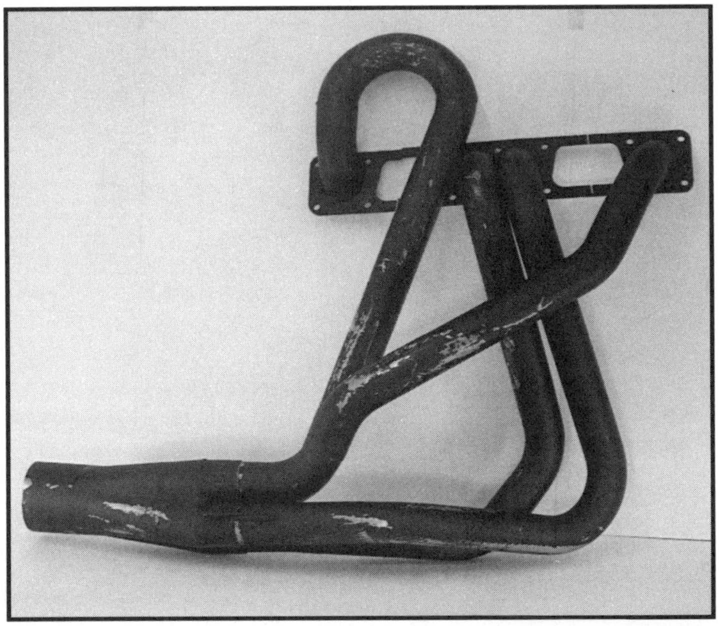

These are larger tube W-2 headers for the right side. They get more complicated and come in more pieces.

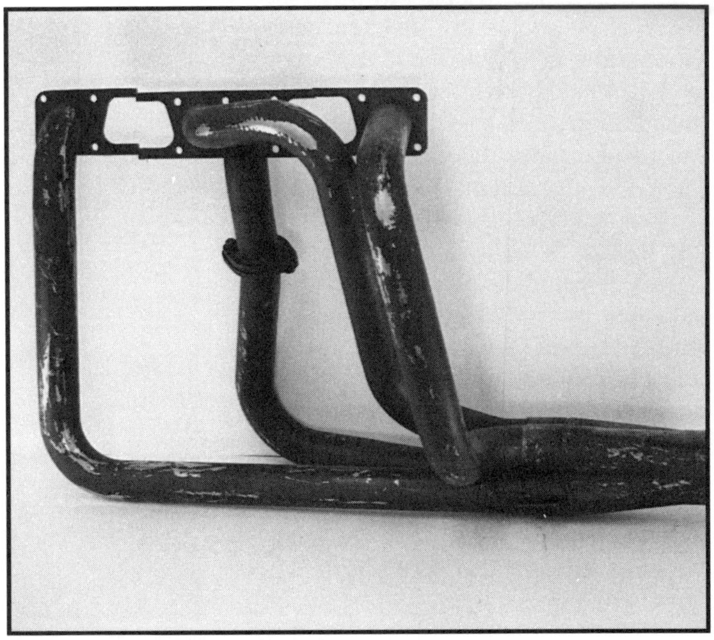

Left side W-2 headers now come in many sections.

There is no real advantage of one over the other, but most racers don't have a choice. Fenderwell headers obviously require a hole cut in the fender shield. They also require that the turning radius (steering) be cut down or restricted so the large-diameter, skinny drag racing tire will not hit the exhaust tube at full lock.

MUFFLERS

Much has been written about mufflers over the years, mostly on how they affect street packages. But many race tracks today require racing mufflers to run. To meet this demand, a new generation of efficient race mufflers has been developed. These new mufflers are superior in performance over all the old standbys.

There is a popular belief among racers relating to mufflers that louder equals more horsepower. This type of "mental horsepower" is largely false. The reverse is actually true. The loudest muffler generally will *not* make the most power.

INSTALLATION

There is no such thing as a V-8 header in a full-bodied car that is easy to install. Some are just easier to install than others. Usually the right (passenger) side isn't too difficult. But the left side can be very tricky. If you have not installed headers before, invite a friend with experience to help. In this case, there is no substitute for experience. Sometimes unbolting the motor mount and jacking the engine up makes the installation much easier. If this is helpful, it is usually recommended in the included instructions.

Headers typically use bolts to attach the header to the cylinder head. Cast-iron exhaust manifolds are sometimes attached by studs. Often, the headers can use the stud but in some locations it must be replaced by a bolt. Studs left in the head can make header installation more difficult.

Once the headers have been installed, be sure that the plug wires do not touch or rest directly on the header tubes or valve covers. Use plastic wire looms to support them.

High-temperature header paint is readily available at all speed shops. It makes the headers look nicer and keeps them from rusting. Chrome headers tend to change color quickly. Stainless-steel headers are expensive.

Carburetion— High-performance headers will usually lean-out the engine. If no other change is made, this extra leanness can be beneficial or harmful. On the older engines that run too rich to begin with, the leaner mixture makes the engine go from too rich to just right and results in more power. Since 1972, production cars have gradually been tuned to run much leaner. Some are controlled with ECMs (electronic control modules). Headers can confuse these systems to the point where they over-compensate by dumping more fuel than necessary to enrichen the air/fuel ratio. Or they may not compensate at all, in which case the air/fuel ratio will be too lean. If that is the case, the jet sizes can be changed on the primaries to enrichen or lean the carburetor.

Maintenance—Many production A engines do not use a gasket between the exhaust manifold and the cylinder head. For racing and other high-performance applications, exhaust gaskets are mandatory. There are two different ones—standard and W-2.

When installing the exhaust manifold, whether it be headers or a cast-

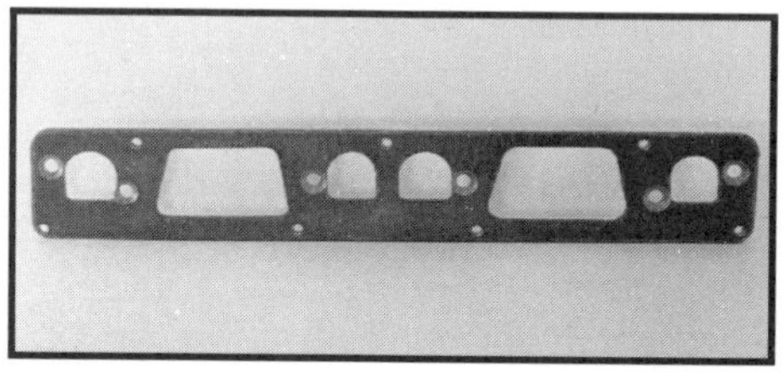

This is the W-2 header adapter flange. It adapts the standard exhaust flange to the wider spaced special flange. Special W-2 headers use this wider spacing. If you want to use large primary tube diameters, then you have to use the wide spacing.

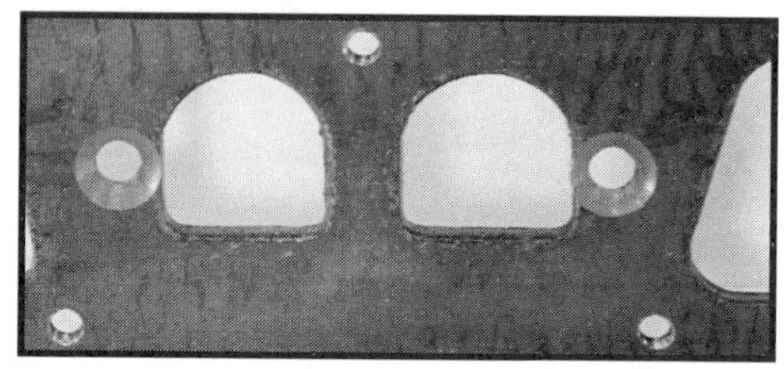

This is a close up of the adapter flange. The holes (actually chamfered) on the port-opening centerline attach the flange to the head. The smaller holes at the top and bottom of the flange (threaded) are used to attach the header to the flange.

Left side headers installed on engine. These headers were made for in-car use. Installation looks pretty easy on a dyno.

iron design, use sealer on the bolt threads. Some exhaust bolt holes are machined into the waterjacket and will leak if sealer isn't used.

After the engine has been run and warmed up, the exhaust manifold bolts should be retightened. After the first retorque, they should only need to be checked occasionally.

W-2 Tips—The W-2 high-performance cylinder head has a unique exhaust port. It's raised and somewhat D-shaped. Most of the race W-2 heads have *two* exhaust manifold attaching patterns machined into each flange. One pattern is for stock and the other is for the W-2. The exhaust manifold gasket set onto the cylinder head will point out the holes that should be used. The ones that are not going to be used, should be plugged. This is very important and should be done before attempting to install the headers.

The special W-2 pattern has an adapter plate. These plates bolt to the cylinder head and the header then bolts to the adapter. As mentioned earlier, these adapter plates are not mandatory with W-2 heads, but are commonly used. Adapter plates are required with the extra large primary tube diameters—commonly 2 in. O.D.'s and up.

Adapter plates should be matched to the cylinder head's exhaust port before installation. Loctite all the countersunk screws at installation.

SPECIAL STYLES

Up to this point, all headers discussed have been 4-into-1 designs. The Tri-Y headers are another style used a lot in the early '60's, but they are not very common any longer. The Tri-Y made good torque in its day, but is not as trick as the 4-into-1, nor do they look as good.

Another special type header design is the 180 deg. header. These headers sound funny, but they can produce power. They are usually only used in NASCAR racing on high-speed ovals. They are difficult to construct because one pipe from each cylinder bank must be run over to the opposite side. This makes the headers quite long, which isn't desirable in drag racing.

Several years ago, the evacu-sump system became very popular in drag racing. This system connected the valve cover breather to the collector. This created a vacuum inside the engine. By itself, this engine vacuum didn't increase horsepower output. But it didn't hurt anything as long as it didn't suck-in the valve cover gaskets. It did however, keep the inside of the engine very clean. In the last couple of years, the vacuum-sump has dropped in use due mainly to the cost and complexity of its design.

CHRYSLER PARTS LIST

EXHAUST HEADER GASKET SETS	
Std. Heads	P4120099
W-2 Heads	P4120100
RACE HEADER GASKETS	
Std. Heads	P4452107

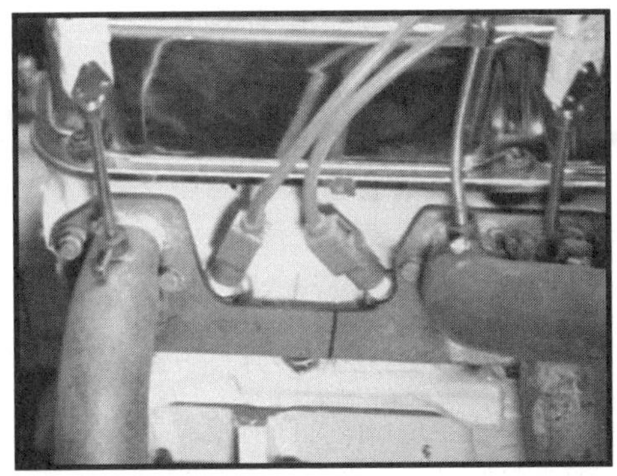

Most A engine spark plugs have adequate boot clearance to the headers. The tubes sticking out of the header pipes next to the head are temperature probes used in fuel-distribution work on dynos.

Right side headers installed on the engine. Note the #8 pipe.

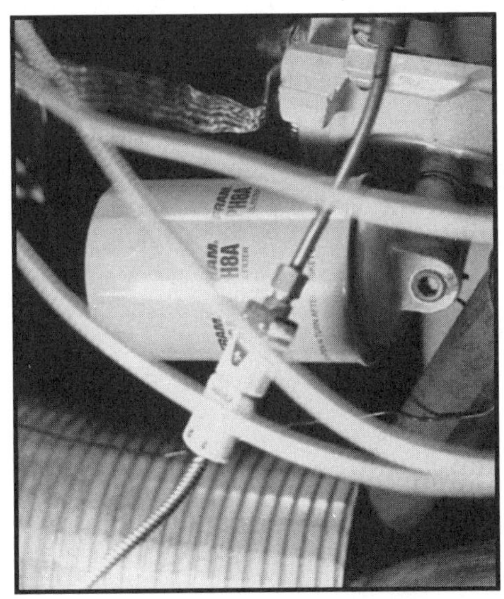

Close-up view of #8 pipe as it passes the oil filter area. Obviously on this style header the right-angle oil filter adaptor is required.

PERFORMANCE PACKAGES

I will divide each engine displacement into six categories. In each category (I call them "Stages" which is stolen from the 62-64 B engines) I will list the engine hardware that's being used except for the camshaft itself. I will keep it fairly general. At the end of the parts list, I will suggest the general camshaft selection. Then I will list the special applications and any revision to the cam still using the same group of basic parts. For the most part, I will list the 273 and 318 together. I will also cover the 340 and 360 together.

DISPLACEMENT: 273 & 318

STAGE I

CYLINDER HEADS:	Stock 273 or 318
VALVES:	Stock 1.74 Intake and 1.50 Exhaust
COMPRESSION RATIO:	8.0-to-1; gas—unleaded
EXHAUST SYSTEM:	Stock Manifolds, dual exhaust
CARBURETION:	Big 2BBL or Small 4BBL
INTAKE MANIFOLD:	Stock cast iron
IGNITION:	Stock electronic
CAMSHAFT:	240 to 250 advertised duration 18-24 deg. overlap—Hydraulic 0.400- to 0.420-in. lift at valve
SPECIAL APPLICATIONS:	None. Too mild. Use above cams for all.
EST. BRACKET E.T.	16 seconds

STAGE II

CYLINDER HEADS:	Stock 273 or 318
VALVES:	Stock
COMPRESSION RATIO:	8.5 to 1; gas—unleaded
EXHAUST SYSTEM:	Stock Manifolds, dual exhaust
CARB:	Small 4BBL
INTAKE MANIFOLD:	Stock cast iron
IGNITION:	Stock electronic
CAMSHAFT:	255 to 260; duration 30-35 deg. overlap—Hydraulic 0.420 to 0.435-in. lift at valve
EST. BRACKET E.T.	15 seconds
For more torque at lower rpm:	Use Stage I cam
Truck and Vans:	Use Stage I cam
Street Machine and Street Rod:	Use above

STAGE III

CYLINDER HEADS:	Stock 273 or 318
VALVES:	Stock 1.74 and 1.50 Backcut
COMPRESSION RATIO:	9.0 to 1
EXHAUST SYSTEM:	340 Manifolds, dual exhaust
CARBURETOR:	Carter TQ or Holley Spread-Bore
INTAKE MANIFOLD:	Stock 318-340-360; cast iron TQ
IGNITION:	High-performance electronic ignition with vac. advance
CAMSHAFT:	264 to 272; duration 40-50 deg. overlap—Hydraulic 0.435-in. to 0.455-in. lift
EST. BRACKET E.T.	14 seconds
For more torque at lower rpm:	Use Stage II cam
Truck and Vans:	Use Stage II cam
Street Machine and Street Rod:	Use above

STAGE IV

CYLINDER HEADS:	360
VALVES:	1.88 and 1.60 with bracket valve job
COMPRESSION RATIO	9.5 to 1; gas—unleaded premium
EXHAUST SYSTEM:	Street Headers
CARBURETOR:	Carter TQ or Holley Spread-Bore
INTAKE MANIFOLD:	Holley single-plane Dominator
IGNITION:	High-performance electronic with race coil
CAMSHAFT:	280 to 285 duration 60-70 deg. overlap—Hydraulic 0.470- to 0.485-in. lift
EST. BRACKET E.T.	13 seconds
For more torque at lower rpm:	Use Stage III cam
Truck and Vans:	Use Stage III cam
Street Machine:	Use above
Street Rod:	Use Stage III cam
Bracket Racer:	Automatic transmission—use above Manual transmission—use 285-292 cam
Supercharger:	Use 285-292 cam

STAGE V

CYLINDER HEADS:	Ported 360
VALVES:	1.88 to 1.60 with bracket valve job
COMPRESSION RATIO:	9.5 to 1; gas—unleaded premium
EXHAUST SYSTEM:	Race Headers
CARBURETOR:	750 cfm Holley
INTAKE MANIFOLD:	Holley single plane
IGNITION:	High-performance electronic with race coil
CAMSHAFT:	280 to 290; duration 55-70 deg. overlap—mechanical 0.500 to 0.530-in. lift
EST. BRACKET E.T.:	12 seconds
For more torque at lower rpm:	Use wider lash
Street Machine:	Use above
Street Rod:	Use Stage IV Hyd. cam
Bracket Racer:	Automatic transmission—use above Manual transmission—use 290-300 mech. cam

STAGE VI (318 only)

CYLINDER HEADS:	W-2 Econo
VALVES:	2.02 and 1.60
COMPRESSION RATIO:	11.0 to 1; gas—racing
EXHAUST SYSTEM:	Race headers
CARBURETOR	750 Holley
INTAKE MANIFOLD:	Holley single plane
IGNITION:	Race electronic
CAMSHAFT:	300 to 315 duration 95-105 deg. overlap 0.575 to 0.600-in. lift
EST. BRACKET E.T.:	11 seconds
For more torque at lower rpm:	Use 290-300 cam
Street Machine:	Use 290-300 cam
Street Rod:	Use Stage V cam
Bracket Racer:	Automatic transmission—Use 290-300 cam Manual transmission—Use above

DISPLACEMENT: 340 & 360

STAGE I

CYLINDER HEADS:	Stock 340 or 360
VALVES:	1.88 and 1.60
COMPRESSION RATIO:	8.5 to 1; gas—unleaded
EXHAUST SYSTEM:	Stock manifolds, dual exhaust
CARB:	Stock 4BBL
INTAKE MANIFOLD:	Stock cast iron 4BBL
IGNITION:	Stock electronic
CAMSHAFT:	255 to 265 duration 30-40 deg. overlap—hydraulic 0.430 to 0.445-in. lift
EST. BRACKET E.T.:	15 seconds
For more torque at lower rpm:	Use 245-255 cam

STAGE II

CYLINDER HEADS:	Stock 340 or 360
VALVES:	1.88 and 1.60 Backcut or 2.02 and 1.60
COMPRESSION RATIO:	9.0 to 1; gas—unleaded premium
EXHAUST SYSTEM:	Stock manifolds, dual exhaust
CARBURETOR	Carter TQ or Holley Spread-Bore
INTAKE MANIFOLD:	Stock TQ cast iron
IGNITION:	High performance electronic
CAMSHAFT:	270 to 280 duration 45-60 deg. overlap—hydraulic 0.450 to 0.475-in. lift
EST. BRACKET E.T.:	14 seconds
For more torque at lower rpm:	Use Stage I cam
Street Machine:	Use above
Street Rod:	Use Stage I cam
Trucks & Vans:	Use Stage I cam
Bracket Racer:	Automatic Transmission: Use above Manual Transmission: Use 280-290 cam
Supercharger:	Use 280-295 cam

STAGE III

CYLINDER HEADS:	Stock 340 or 360
VALVES:	2.02 and 1.60 w/Bracket Valve Job
COMPRESSION RATIO:	9.5 to 1; gas—unleaded premium
EXHAUST SYSTEM:	Street headers
CARBURETOR:	Carter TQ or Holley Spread-Bore
INTAKE MANIFOLD:	Holley single-plane Dominator
IGNITION:	High performance electronic w/race coil
CAMSHAFT:	280 to 295 duration 60-80 deg. overlap—hydraulic 0.480 to 0.510-in. lift
EST. BRACKET E.T.:	13 seconds
For more torque at lower rpm:	Use Stage II cam
Street Machine:	Use 280-285 cam
Street Rod:	Use Stage II cam
Trucks & Vans:	Use Stage II cam
Bracket Racer:	Automatic transmission: use 280-285 cam Manual transmission: use 280-295 cam
Supercharger:	Use 285-295 cam

This set up with Stage II CAM

PERFORMANCE PACKAGES (CONT.)

STAGE IV

CYLINDER HEADS:	Ported 340 and 360
VALVES:	2.02 and 1.60
COMPRESSION RATIO:	9.5 to 1; gas—unleaded premium
EXHAUST SYSTEM:	Street headers
CARBURETOR:	Carter TQ or 750/850 cfm Holley
INTAKE MANIFOLD:	Holley Dominator
IGNITION:	Race electronic
CAMSHAFT:	280 to 290 duration Mechanical 55-65 degrees overlap 0.520-in. to 0.540-in. lift
EST. BRACKET E.T.	12 seconds
For more torque at lower rpm:	Use wider lash
Street Machine:	Use above
Street Rod:	Use Stage III Hydraulic cam.
Bracket Racer:	Automatic transmission—use above Manual transmission—use 290-300 Mechanical camshaft.

STAGE V

CYLINDER HEADS:	Ported 340 or 360
VALVES:	2.08 and 1.60
COMPRESSION RATIO:	10.0 to 1; gas—50% racing fuel
EXHAUST SYSTEM:	Race headers
CARBURETOR:	Carter TQ or 750/850 cfm Holley
INTAKE MANIFOLD:	Holley Dominator
IGNITION:	Race electronic
CAMSHAFT:	295 to 315 duration Mechanical 75-105 deg. overlap 0.550-in. to 0.600-in. lift
EST. BRACKET E.T.	11 seconds
For more torque at lower rpm:	Use Stage IV cam
Street Machine:	Use 295-300 cam
Street Rod:	Use Stage IV cam
Bracket Racer:	Automatic transmission—use 295-300 cam Manual transmission—use 310-315 cam

STAGE VI

CYLINDER HEADS:	W-2 or W-5
VALVES:	2.02 and 1.60, Bracket valve job optional
COMPRESSION RATIO:	11.0-to-1; gas—racing gas
EXHAUST SYSTEM:	Race headers
CARBURETOR:	Carter TQ or 750/850 cfm Holley
INTAKE MANIFOLD:	Holley Dominator
IGNITION:	Race electronic
CAMSHAFT:	310 to 315 duration Mechanical 100-105 deg. overlap 0.580-in. to 0.600-in. lift
EST. BRACKET E.T.	10 seconds
For more torque at lower rpm:	use 295-300 cam

CARBURETOR MODIFICATIONS

Every carburetor, manifold, engine and hardware package requires a different set-up—jets, accelerator pumps, etc. The actual set-up of the carburetor is quite specific to the package. Although there are an almost infinite number of different packages, I will pick the most popular ones or the packages that are most likely to be raced. This information is based on years of testing by Chrysler on the dyno and at the track.

340-360 BRACKET RACER

CARBURETOR:	Carter TQ 4-Barrel	
INTAKE MANIFOLD:	Edelbrock LD 340	
HEADERS:	Yes	
CAMSHAFT:	0.550-in. lift or larger—Mechanical	
HEADS:	340-360 Stock	
MANIFOLD MODIFICATIONS:	Blocked Head Crossover Milled Center Divider to 3/8 in.	
CARB MODIFICATIONS:		
Primary Jet	0.095 (left)	0.095 (right)
Secondary Jet	0.149 (left)	0.149 (right)
Metering Rods	1962 (.045 x .067)	

'72-'73 340 SUPER STOCK

CARBURETOR:	Carter TQ 4-Barrel	
INTAKE MANIFOLD:	Offenhauser Port-O-Sonic	
HEADERS:	Yes	
CAMSHAFT:	0.600-in. lift and larger. Roller	
HEADS:	Std. 340—1.88 & 1.60 Valves	
HOOD SCOOP:	None	
MANIFOLD MODIFICATIONS:	4 Distribution Sticks Added to Bottom of Plenum	
CARB MODIFICATIONS:		
Primary Jet	0.092 (left)	0.092 (right)
Secondary Jet	0.154 (left)	0.154 (right)
Accelerator Pump Shooters	0.040 in.	
Air Door	0.850-0.900	

'74-'76 360 SUPER STOCK

CARBURETOR:	Carter TQ 4-Barrel	
INTAKE MANIFOLD:	Offenhauser Port-O-Sonic	
HEADERS:	Yes	
CAMSHAFT:	0.600-in. lift and larger Roller	
HEADS:	Std. 340/360	
HOOD SCOOP:	None	
MANIFOLD MODIFICATIONS:	4 Distribution Sticks Added to Bottom of Plenum	
CARB MODIFICATIONS:		
Primary Jet	0.095 (left)	0.095 (right)
Secondary Jet	0.169 (left)	0.169 (right)
Accelerator Pump Shooter	0.040 in.	
Air Doors—0.850 to 0.900 in.		
Drill center hole in each secondary nozzle bar to 0.101 in.		

CARBURETOR MODIFICATIONS (CONT.)

340-360 MODIFIED OR BRACKET

CARBURETOR:	Holley List 4779-750 cfm	
INTAKE MANIFOLD:	Offenhauser Port-O-Sonic or Holley Strip Dominator	
HEADERS:	Yes	
CAMSHAFT:	Any mechanical over 0.550-in. lift	
HEADS:	340-360 Stock with 2.02 & 1.60 Valves	
MANIFOLD MODIFICATIONS:	Offy—4 Distribution Sticks Added Holley—None	
CARB MODIFICATIONS:		
Primary Jet	#75 (left)	#75 (right)
Secondary Jet	#76 (left)	#76 (right)

340-360 TUNNEL RAM

CARBURETOR:	Two Holley 660's List 4224	
INTAKE MANIFOLD:	Edelbrock TR5 with High Plenum	
HEADERS:	Yes	
CAMSHAFT:	Any roller over 0.600-in. lift	
HEADS:	Ported Stock 340/360	
MANIFOLD MODIFICATIONS:	None Front & Rear Carburetors	
CARB MODIFICATIONS:		
Primary Jet	#76 (left)	#76 (right)
Secondary Jet	0.076 in. (left)	0.076 in. (right)
Accelerator Pump Shooters	0.025	
Cut off choke towers		

340-360 PRO STOCK/MODIFIED

CARBURETOR:	Two Holley 4500 List 4664—1050 cfm	
INTAKE MANIFOLD:	Edelbrock TR5 Base with Short Plenum	
HEADERS:	Yes	
CAMSHAFT:	Any roller over 0.600-in. lift	
HEADS:	Fully ported 340-360 Production	
MANIFOLD MODIFICATIONS:	None Front & Rear Carburetor	
CARB MODIFICATIONS (both front and rear):		
Primary Jet	#80 (left)	#80 (right)
Secondary Jet	#80 (left)	#80 (right)
Accelerator Pump Shooter	0.035	

GENERAL INFORMATION

CYLINDER BLOCKS

CASTING NUMBER	ENGINE	YEARS
2465330	273	64-66
2536130	273	65
2806130	273	65-69
2536030	318	67-75
4006730	318	76-79
4104230	318	80-83
2780930	340	68-73
3577130	340 T/A	70
3418496	360	71-74
3870230	360	75
4006830	360	76-80

CRANKSHAFTS

YEAR	ENGINE	SERVICE PN	FORGED OR CAST	STROKE	MAIN JOURNAL DIAMETER	ROD JOURNAL DIAMETER
64-66	273 Auto	2463551	Forged	3.31	2.50	2.125
64-66	273 Man.	2268119	Forged	3.31	2.50	2.125
67	273 & 318	2268119	Forged	3.31	2.50	2.125
68-71	273/ 318	2843699	Cast	3.31	2.50	2.125
72-74	318	3698267	Cast	3.31	2.50	2.125
74-86	318	3751831	Cast	3.31	2.50	2.125
80-86	318 4BBL	4227880	Cast	3.31	2.50	2.125
68-71	340	2843868	Forged	3.31	2.50	2.125
72-73	340	3751162	Cast	3.31	2.50	2.125
71-73	360	3418995	Cast	3.58	2.81	2.125
74-80	360	375160	Cast	3.58	2.81	2.125

FLEX PLATES

YEAR	ENGINE	TORQUE CONVERTER	FLEXPLATE
64-69	273	10 3/4 11 3/4	2400885 2204768
67-77	318	10 3/4 11 3/4	2400885 2204768
68-73	340	10 3/4	2400885
71-76	360	11 3/4	2204768
73-77	360	10 3/4	2400885
78-86	318-360	All	4058429

FLYWHEELS

YEAR	ENGINE	FLYWHEEL	CLUTCH SIZE	BALANCE
64-67	273	2463256	9.5 in.	Neutral
67-69	273	2463256	9.5 in.	Neutral
67-86	318	2264597	10.5 in.	Neutral
68-72	340	2264597	10.5 in.	Neutral
72-73	340	3681639	10.5 in.	External
71-74	360	3410916	10.5 in.	External

PUSHRODS

YEAR	ENGINE	PUSHROD	LENGTH	DIAMETER	STYLE
64-67	273	2465343	7.344	.312	Mech.
68-69	273	2899561	7.500	.28	Hyd.
67	318	2532152	7.500	.28	Hyd.
68-74	318	2899561	7.500	.28	Hyd.
75-77	318	3780148	7.500	.28	Hyd.
78-86	318	4095291	7.515	.28	Hyd.
85-86	318 Roller	4323205	6.766	.28	Hyd.
68-73	340	2899567	7.534	.28	Hyd.
71-77	360	2899567	7.534	.28	Hyd.
78-80	360	4095291	7.515	.28	Hyd.
70	340 6BBL	3577121	7.250	.375	Mech.

GENERAL INFORMATION (CONT.)

INTAKE MANIFOLDS

YEAR	ENGINE	SERVICE PN	CASTING NUMBER	MATERIAL
64-65	273 2BBL	2463253	2463252	Cast Iron
65	273 4BBL	2465727	2465726	Cast Iron
66-69	273 2BBL	2536560	2536559	Cast Iron
66-67	273 4BBL	2536564	2536563	Cast Iron
67-69	318 2BBL	2468959	2468960	Cast Iron
70-71	318 2BBL	2951184	2951185	Cast Iron
72	318 2BBL Std 318 2BBL CA	3614032 3614033	2951185 2951185	Cast Iron Cast Iron
73	318 2BBL	3698432	3698431	Cast Iron
74	318 2BBL	3769991	3698431	Cast Iron
75-78	318 2BBL	3870021	3830941	Cast Iron
78	318 4BBL	3870023	3830945	Cast Iron
79	318 2BBL 318 4BBL	4100329 4100338	4100331 4100340	Cast Iron Cast Iron
80	318 2BBL	4201135	4100331 & 4201120	Cast Iron
	318 4BBL	4179457	4100340 & 4173915	Cast Iron
81-86	318 2BBL	4179456 & 4201135	4201120 & 4104506	Cast Iron
81-84	318 4BBL	4179457 & 4179967	4100340 & 4173915	Cast Iron
81-83	318 EFI	4201399	4201397	Cast Iron
85-86	318 4BBL	4323354	4323352	Cast Iron
68-69	340	2531914	2531915	Cast Iron
70	340 4BBL 340 6BBL	3462847 3418681	3462848 3418682	Cast Iron Aluminum
71	340 4BBL	3512099	3512100	Cast Iron
72	340 Std 340 CA	3671462 3671878	3512100 3512100	Cast Iron Cast Iron
73	340	3698656	3671918	Cast Iron
71	360 2BBL	3418434	3418435	Cast Iron
72	360 2BBL Std 360 2BBL CA	3614039 3614040	3418435 3418435	Cast Iron Cast Iron
73	360 2BBL 360 2BBL Air Pump	3671548 3698438	3698437 3698437	Cast Iron Cast Iron
74	360 2BBL	3769993	3698437	Cast Iron
74	360 4BBL	3698435	3698434	Cast Iron
75-78	360 2BBL	3870022	3830943	Cast Iron
75-78	360 4BBL	3870023	3830945	Cast Iron
79	360 2BBL 360 4BBL	4100332 4100338	4100334 4100340	Cast Iron Cast Iron
80	360 2BBL	4201135	4100334 & 4201120	Cast Iron Cast Iron
	360 4BBL	4179967	4100340 & 4173915	Cast Iron Cast Iron

GENERAL INFORMATION (CONT.)

CONNECTING RODS

YEAR	ENGINE	SERVICE PN	FORGING NUMBER	PIN TYPE	CENTER-TO-CENTER LENGTH
64-71	273-318	2406785	1618699	Floating	6.123 in.
72-73	318	2899061 & 2899495	*2899062 & 2899486	Floating	6.123 in.
74-75	318	2899061	2899063	Pressed	6.123 in.
76-86	318	3751015	3418645 & 3751016	Pressed	6.123 in.
68-73	340	2899495	2899496	Floating	6.123 in.
71-74	360 2BBL	2899061	*2899062	Pressed	6.123 in.
74-75	360 H.P.	3751015	3418645	Pressed	6.123 in.
76-80	360	3751015	3418645	Pressed	6.123 in.

*3418645 and 3870404 also used.

PISTONS

YEAR	ENGINE	SERVICE PN	APPROX. WEIGHT	PIN HEIGHT
64-67	273 2BBL	2495072	530 g.	1.79 in.
65-67	273 4BBL	2585267-8	569 g.	1.79 in.
68-69	273	3004502	547 g.	1.82 in.
67	318	2808405	592 g.	1.75 in.
68-69	318	3004509	592 g.	1.77 in.
70-72	318	3420222	592 g.	1.74 in.
73-84	318	3685751	594 g.	1.74 in.
85-86	318	4342772	594 g.	1.76 in.
68-71	340	3004516-7	720 g.	1.84 in.
72-73	340	3621352	720 g.	1.74 in.
71-72	360	3549366	584 g.	1.59 in.
73-75	360 2BBL	3685737	584 g.	1.59 in.
74-75	360 H.P.	3780071	584 g.	1.59 in.
76-80	360	4026451	584 g.	1.59 in.

GENERAL INFORMATION (CONT.)

ROCKER SHAFTS		
YEAR	ENGINE	ROCKER SHAFT
64-65	273	2463321
66-67	273	2536526
68-69	273	2532149
67-86	318	2532149
68-73	340	2532149
70	340 6BBL	3577086
71-80	360	2532149

ROCKER ARMS			
YEAR	ENGINE	ROCKER ARM	STYLE
64-66	273	2465332	Adjustable
67	273	2806988	Adjustable
68-69	273	3004716-7	Stamped
67	318	2536145-6	Stamped
68-73	318	3004716-7	Stamped
74-78	318	3418424-5	Stamped
68-69	340	3004716-7	Stamped
70-73	340	3418424-5	Stamped
70	340 6BBL	3577076-7-8	Adjustable
71-78	360	3418424-5	Stamped
79-84	318-360	4095708-9	Stamped
85-86	318	4100774-5	Stamped

VIBRATION DAMPERS			
YEAR	ENGINE	PN	CRANK TYPE
64-65	273	1947235	Forged
66-69	273	2658458	Forged
67	318 Auto	2658459	Forged
68-69	318	2658458	Cast
70-71	318	2951673	Cast
72-74	318	3614369	Cast
75-77	318 Lean Burn	3830481	Cast
76-77	318	3614369	Cast
78	318	3830481	Cast
79-80	318	4095633	Cast
81-86	318	4173443	Cast
68-69	340	2658458	Forged
70-71	340	2951673	Forged
72	340 Forged	3614369	Forged
72-73	340 Cast	3751160	Cast
71	360	3512972	Cast
72-73	360	3614370	Cast
74	360	3751808	Cast
75	360	3830480	Cast
76	360	3751808	Cast
77	360	3830480	Cast
78	360	4071743	Cast
79	360	4095632	Cast
80-81	360	4173447	Cast

GENERAL INFORMATION (CONT.)

CYLINDER HEADS

YEAR	ENGINE	SERVICE PN	CASTING NUMBER	COMBUSTION CHAMBER VOLUME (cc's)
64-65	273	2532080	2465315	57.0-64.5
66	273	2536176	2536178	57.0-64.5
67	273/318	2806213	2658920	57.0-64.5
68-71	273/318	2843674	2843675	60.0-67.5
72	318	3671637	2843675	61.0-67.5
73-74	318	3698615	2843675	61.0-67.5
75	318	3769950	3769973	63.0-69.0
76	318	3769950- 4027591 4006335- 4041001	3769973	63.0-69.0
77-79	318	4027591- 4041001	4027163- 4027593	64.5-71.5
80	318	4100405	4027163- 4027593	64.5-71.5
81-83	318 Std.	4100405	4027163- 4027593	64.5-71.5
81-83	318 H.P.	4100408-9	4027596- 4071051	66.0-72.5
85-87	318	4323423	4323302	56.0-65.0
85-87	318 4BBL	4323346	4323475	69.0-77.0
68-71	340	2531902	2531894	63.0-73.5
70	340 6BBL	3577053	3418915	63.0-73.5
72	340	3671639	3418915	65.0-73.0
73	340	3698617	3671587	65.0-73.0
71	360	3418432	3418915	65.0-73.5
72	360	3671639	3418915	65.0-73.5
73-74	360	3698617	3671587	65.0-73.0
73	360 Air Pump	3671873	3671587	65.0-73.0
75	360	3769952	3769974	65.0-73.0
76	360	3769952- 4006337	3671587- 3769974	65.0-73.0
77-79	360	4041003- 4027594 4071049- 4071047	4027596- 4071051	66.0-72.5
80	360	4100408-9 4071051	4027596-	66.0-72.5

GENERAL INFORMATION (CONT.)

VALVES

YEAR	ENGINE	INTAKE PN	INTAKE DIAMETER	EXHAUST PN	EXHAUST DIAMETER
64-74	273/318	2465335	1.75	2465339	1.50
75-78	318	2465335	1.75	3830705	1.50
79-86	318	2465335	1.75	4100387	1.50
78-86	318 4BBL	3683801	1.88	4100632	1.60
68-71	340	2863520	2.02	2863521	1.60
72-73	340	3683801	1.88	2863521	1.60
71	360 Sngl. Groove	3418688	1.88	2863521	1.60
71	360 Dbl. Groove	3549971	1.88	2863521	1.60
72-73	360	3577948	1.88	2863521	1.60
74	360 Std.	3683801	1.88	2863521	1.60
74	360 H.P.	3683801	1.88	3751673	1.60
75	360 Std.	3683801	1.88	3830710	1.60
75	360 H.P.	3683801	1.88	3751673	1.60
76-78	360 Std.	3683801	1.88	3751673	1.60
76-78	360 H.P.	3683801	1.88	4071008	1.60
79-80	360 Std.	3683801	1.88	4100377	1.60
79-80	360 H.P.	3683801	1.88	4100632	1.60

CAMSHAFTS

YEAR	ENGINE	CAMSHAFT	TYPE	VALVE LIFT INTAKE/EXHAUST	DURATION INTAKE/EXHAUST
64-67	273 2BBL	2463360	Mech.	.395/.405	240/240
65-67	273 4BBL	2465679	Mech.	.415/.425	248/248
68-69	273	2899144	Hyd.	.373/.400	240/248
67	318	2668362	Hyd.	.390/.390	244/244
68-70	318	2899144	Hyd.	.373/.400	240/248
71-79	318	3512635	Hyd.	.373/.400	240/248
80-83	318	4179768	Hyd.	.373/.400	240/248
80-86	318 4BBL	4227879	Hyd.	.373/.400	240/248
85-86	318 Roller	4323239	Hyd.	.391/.391	240/240
68	340 Manual	2899205	Hyd.	.444/.453	276/284
68-73	340	2899206	Hyd.	.430/.444	268/276
70	340 6BBL	2899206	Hyd.	.430/.444	268/276
71-74	360 2BBL	3512589	Hyd.	.410/.412	252/256
74-75	360 4BBL H.P.	2899206	Hyd.	.430/.444	268/276
75-78	360 2BBL	3830025	Hyd.	.410/.410	252/252
76-80	360 4BBL	4041998	Hyd.	.430/.444	268/276
76-78	360 H.P.	4042000	Hyd.	.410/.410	252/252
79	360	4042000	Hyd.	.410/.410	252/252
80	360	5214658	Hyd.	.410/.410	252/252

INDEX

HP AUTOMOTIVE BOOKS

HANDBOOK SERIES

Auto Electrical Handbook
Baja Bugs & Buggies
Brake Handbook
Metal Fabricator's Handbook
Mustang Restoration Handbook
Off-Roader's Handbook
Paint & Body Handbook
Sheet Metal Handbook
Small Trucks
Street Rodder's Handbook
Turbochargers
Turbo Hydra-matic 350
Welder's Handbook

CARBURETORS

Holley 4150
Holley Carburetors & Manifolds
Rochester Carburetors
Weber Carburetors

PERFORMANCE SERIES

How To Hot Rod Big-Block Chevys
How To Hot Rod Small-Block Chevys
How To Hot Rod VW Engines
How To Make Your Car Handle

REBUILD SERIES

How To Rebuild Big-Block Chevys
How To Rebuild Big-Block Fords
How To Rebuild Your Ford V-8—351C-351M-400-429-429-460
How To Rebuild Small-Block Chevys
How To Rebuild Small-Block Fords
How To Rebuild Small-Block Mopars
How To Rebuild Air-Cooled VW Engines

SPECIAL INTEREST

Fast Fords

Books are available from your local bookstore,
auto store or order direct from publisher,
Price Stern Sloan, Inc., 360 N. La Cienega Blvd., Los Angeles, CA 90048.
Call Toll Free: 800-421-0892 (Inside California) 800-227-8801.